A Physicist's
Guide to
Mathematica®

A Physicist's Guide to *Mathematica*®

Patrick T. Tam

Physics Department
Humboldt State University
Arcata, California

ACADEMIC PRESS

San Diego   London   Boston
New York   Sydney   Tokyo   Toronto

ACADEMIC PRESS
525 B Street, Suite 1900, San Diego, CA 92101-4495, USA
1300 Boylston Street, Chestnut Hill, MA 02167, USA
http://www.apnet.com

Academic Press Limited
24–28 Oval Road, London NW1 7DX, UK
http://www.hbuk.co.uk/ap/

Library of Congress Cataloging-in-Publication Data

Tam, Patrick.
   A physicist's guide to Mathematica / Patrick Tam.
      p.   cm.
   Includes bibliographical references and index.
   ISBN 0-12-683190-4
   1. Physics—Data processing.   2. Mathematica (Computer file)   I. Title.
QC20.7.E4T36   1997
530'.0285'53—dc20   97-9900
                                                                         CIP

Printed in the United States of America
97 98 99 00 01 IC 9 8 7 6 5 4 3 2 1

To
P. T. N. H. Jiyu-Kennett, Shunryu Suzuki
He Tin and May Yin Tam
Sandra, Teresa
Harriette, Frances

Contents

PART II: Physics with *Mathematica*

4   Mechanics   349

Appendices

Preface

Traditionally, the upper-division theoretical physics courses teach the formalisms of the theories, the analytical technique of problem-solving, and the physical interpretation of the mathematical solutions. Problems of historical significance, pedagogical value, or if possible, recent research interest are chosen as examples. The analytical methods consist mainly of working with models, making approximations, and considering special or limiting cases. The student must master the analytical skills, because they can be used to solve many problems in physics and, even in cases where solutions cannot be found, can be used to extract a great deal of information about the problems. As the computer has become readily available, these courses should also emphasize computational skills, since they are necessary for solving many important, real, or "fun" problems in physics. The student ought to use the computer to complement and reinforce the analytical skills with the computational skills in problem-solving and, whenever possible, use the computer to visualize the results and observe the effects of varying the parameters of the problem in order to develop a greater intuitive understanding of the underlying physics.

The pendulum in classical mechanics serves as an example to elucidate these ideas. The plane pendulum is used as a model. It consists of a particle under the action of gravity and constrained to move in a vertical circle by a massless rigid rod. For small angular deviations, the equation of motion can be linearized and solved easily. For finite angular oscillations, the motion is nonlinear. Yet it can still be studied analytically in terms of the energy integral and the phase diagram. The period of motion is expressed in terms of an elliptic integral. The integral can be expanded in a power series, and for small angular oscillations the expansion converges rapidly. However, numerical methods and computer programming are necessary for determining the motion of a damped, driven pendulum. The student can use the computer to explore and simulate the motion of the pendulum with different sets of values for the parameters in order to gain a deeper intuitive understanding of the chaotic dynamics of the pendulum.

Normally, physics juniors and seniors have taken a course in a low-level language such as FORTRAN or Pascal and possibly also a course in numerical analysis. Nevertheless, attempts to introduce numerical methods and computer programming into the upper-division theoretical physics courses have been largely unsuccessful. Mastering the symbols and syntactic rules of these low-level languages is straightforward; but programming with them requires too many lines of complicated and convoluted code in order to solve interesting problems. Consequently, rather than enhancing the student's problem-solving skills and physical intuition, it merely adds a frustrating and ultimately nonproductive burden to the student already struggling in a crowded curriculum.

Mathematica, a system developed recently for doing mathematics by computer, promises to empower the student to solve a wide range of problems including those that are important, real, or "fun," and to provide an environment for the student to develop intuition and a deeper understanding of physics. In addition to numerical calculations, *Mathematica* performs symbolic as well as graphical calculations and animates two- and three-dimensional graphics. The numerical capabilities broaden the problem-solving skills of the student; the symbolic capabilities relieve the student from the tedium and errors of "busy" or long-winded derivations; the graphical capabilities and the capabilities for "instant replay" with various parameter values for the problem enable the student to deepen his or her intuitive understanding of physics. These astounding interactive capabilities are sufficiently powerful for handling most problems and are surprisingly easy to learn and use. For complex and demanding problems, *Mathematica* also features a high-level programming language that can make use of more than a thousand built-in functions and that embraces many programming styles such as functional, rule-based, and procedural programming. Furthermore, to provide an integrated technical computing environment, the Macintosh and Windows versions for *Mathematica* support documents called "notebooks." A notebook is a "live textbook." It is a file containing ordinary text, *Mathematica* input and output, and graphics. *Mathematica*, together with the user-friendly Macintosh and Windows interfaces, is likely to revolutionize not only *how* but also *what* we teach in the upper-division theoretical physics courses.

Purpose

The primary purpose of this book is to teach upper-division and graduate physics students as well as professional physicists how to master *Mathematica*, using examples and approaches that are motivating to them. This book does not replace Stephen Wolfram's **Mathematica: A System for Doing Mathematics by Computer** [Wol91] for *Mathematica* version 2 or **The Mathematica Book** [Wol96] for version 3. The encyclopedic nature of these excellent references is formidable, indeed overwhelming, for novices. My guidebook prepares the reader for easy access to Wolfram's indispensable references. My book also shows that *Mathematica* can be a powerful and wonderful tool for learning, teaching, and doing physics.

Uses

This book can serve as the text for an upper-division course on *Mathematica* for physics majors. Augmented with chemistry examples, it can also be the text for a course on *Mathematica* for chemistry majors. (For the last several years, a colleague in the chemistry department and I have team-taught a *Mathematica* course for both chemistry and physics majors.) Part I, "*Mathematica* with Physics," provides sufficient material for a two-unit, one-semester course. A three-unit, one-semester course can cover Part I, sample Part II, "Physics with *Mathematica*," require a polished *Mathematica* notebook from each student reporting a project, and include supplementary material on introductory numerical analysis discussed in many texts (see [KM90], [DeV94], [Gar94], and [Pat94]). Exposure to numerical analysis allows the student to appreciate the limitations (i.e., the accuracy and stability) of numerical algorithms and understand the differences between numerical and symbolic functions, for example, between **NSolve** and **Solve**, **NIntegrate** and **Integrate**, as well as **NDSolve** and **DSolve**. Experience suggests that a three-hour-per-week laboratory is essential to the success of both the two- and three-unit courses. For the degree requirement, either course is an appropriate addition to, if not replacement for, the existing course in a low-level language such as C, Pascal, or FORTRAN.

If a course on *Mathematica* is not an option, a workshop merits consideration. A two-day workshop can cover Chapter 1, "The First Encounter," and Chapter 2, "Interactive Use of *Mathematica*," and a one-week workshop can also include Chapter 3, "Programming in *Mathematica*." Of course, further digestion of the material may be necessary after one of these accelerated workshops.

For students who are *Mathematica* neophytes, this book can also be a supplemental text for upper-division theoretical physics courses on mechanics, electricity and magnetism, and quantum physics. For *Mathematica* to enrich rather than encroach upon the curriculum, it must be introduced and integrated into these courses gradually and patiently throughout the junior and senior years, beginning with the interactive capabilities. While the interactive capabilities of *Mathematica* are quite impressive, in order to realize its full power the student must grasp its structure and master it as a programming language. Be forewarned that learning these advanced features as part of the regular courses, while possible, is difficult. A dedicated *Mathematica* course is usually a more gentle, efficient, and effective way to learn this computer algebra system.

Finally, the book can be used as a self-paced tutorial for advanced physics students and professional physicists who would like to learn *Mathematica* on their own. While the sections in Part I should be studied consecutively, those in Part II, each focusing on a particular physics problem, are independent of each other and can be read in any order. The reader may find the solutions to exercises and problems in Appendices D and E helpful.

Organization

Part I gives a practical, physics-oriented, and self-contained introduction to *Mathematica*. Chapter 1 shows the beginner how to get started with *Mathematica* and discusses the notebook front end. Chapter 2 introduces the numerical, symbolic, and graphical capabilities of *Mathematica*. Although these features of *Mathematica* are dazzling, *Mathematica*'s real power rests on its programming capabilities. While Chapter 2 considers many elements of *Mathematica*'s programming language, Chapter 3 treats in depth five key programming elements: expressions, patterns, functions, procedures, and graphics. It also examines three programming styles: procedural, functional, and rule-based. It shows how a proper choice of algorithm and style for a problem results in a correct, clear, efficient, and elegant program. This chapter concludes with a discussion of writing packages. Examples and practice problems, many from physics, are included in Chapters 2 and 3.

Part II considers the application of *Mathematica* to physics. Chapters 4 through 6 illustrate the solution with *Mathematica* of physics problems in mechanics, electricity and magnetism, and quantum physics. Each chapter presents several examples of varying difficulty and sophistication within a subject area. Each example contains three sections: The Problem, Physics of the Problem, and Solution with *Mathematica*. Experience has taught that the Physics of the Problem section is essential because the mesmerizing power of *Mathematica* can distract the student from the central focus, which is, of course, physics. Additional problems are included as exercises in each chapter.

Appendix A relates the latest news on *Mathematica* version 3.0 before this book goes to press. Appendix B tabulates many of *Mathematica*'s operator input forms together with the corresponding full forms and examples. Appendix C provides information about the books, journals, conferences, and electronic archives and forums on *Mathematica*. Appendices D and E give solutions to selected exercises and problems.

Suggestions

The reader should study this book at a computer with a *Mathematica* notebook opened, key in the commands, and try out the examples on the computer. Although all of the code in this book is included on an accompanying diskette, directly keying in the code greatly enhances the learning process. The reader should also try to work out as many as possible of the exercises at the end of the sections and the practice problems at the end of the chapters. The more challenging ones are marked with an asterisk, and those requiring considerable effort are marked with two asterisks.

Prerequisites

The prerequisites for this book are calculus through elementary differential equations, introductory linear algebra, and calculus-based physics with modern physics. Some of the physics in Chapters 5 and 6 may be accessible only to seniors. Basic Macintosh or Windows skills are assumed.

Computer Systems

This book, compatible with *Mathematica* versions 3.0 and 2.2, is to be used with Macintosh and Microsoft-Windows-based IBM-compatible computers. While the front end or the user interface is optimized for each kind of computer system, the kernel, which is the computational engine of *Mathematica*, is the same across all platforms. As over 95% of this book is about the kernel, the book can also be used, with the omission of the obviously Macintosh- or Windows-specific comments, for all computer systems supporting *Mathematica*, such as NeXT computers and UNIX workstations.

Acknowledgments

I wish to express my deepest gratitude to Mervin Hanson (Humboldt State University), who is my partner, friend, mentor, and benefactor. Saying that I wrote this book with him is not an exaggeration. Bill Titus (Carleton College), to whom this book owes its title, deserves my heartfelt gratitude. His involvement, guidance, support, and inspiration in the writing of this book is beyond the obligation of a colleague and a friend. I am indebted to Zenaida Uy (Millersville University), whose great enthusiasm and considerable labor for my project invigorated me when I was weary and feeling low, and to her students for testing my manuscript in their *Mathematica* class. I am most grateful to Jim Feagin (California State University, Fullerton) for his careful reading of my manuscript, for being my friend and stern master, and for sharing his amazing insight into physics and *Mathematica*. Special recognition is due to my students who put up with the numerous errors in my innumerable editions of the manuscript, submitted to being the subjects of my experiments, and gave me their valuable feedback. I am eternally grateful to my wife, Sandra, of more than 30 exciting years for her labor of love in editing and proofreading the evolving manuscript and for keeping faith in me during those dark nights of writer's blues. I am thankful to my friend, David Cowsky, for revealing to me some of the subtleties of the English language.

I would like to acknowledge and thank the following reviewers for their constructive criticisms, invaluable suggestions, and much needed encouragement:

Anthony Behof, DePaul University
Wolfgang Christian, Davidson College
Robert Dickau, Wolfram Research, Inc.
Richard Gaylord, University of Illinois
Jerry Keiper, Wolfram Research, Inc.
Peter Loly, University of Manitoba
David Withoff, Wolfram Research, Inc.
Amy Young, Wolfram Research, Inc.

To Nancy Blachman (Variable Symbols, Inc., and Stanford University) and Vithal Patel (Humboldt State University), I am grateful for their interest, advice, and friendship. Special appreciation is due to my colleagues who covered my classes while I was away on many *Mathematica*-related trips. I am most appreciative of my department chair, Richard Stepp, for his support, and my dean, James Smith, for cheering me onto my *Mathematica* ventures. For their assistance, guidance, and patience in the production and marketing of this book, I would like to thank Abby Heim, Kenneth Metzner, and Zvi Ruder at Academic Press, Inc., and Joanna Hatzopoulos and her associates at Publication Services. I am much indebted to Prem Chawla, Chief Operating Officer of Wolfram Research, Inc., for granting me permission to include *Mathematica* usage statements and help messages in this book.

A special commendation to my daughter, Teresa, is in order for her patience with the sparse social calendar of our family during the development of this book. Finally, I am grateful to my physicians, David O'Brien and John Biteman, for improving and maintaining my health.

Patrick T. Tam

How to Use the Disk

The accompanying disk contains all of the *Mathematica* input in this book. The Version2 and Version3 folders (or directories) include files for *Mathematica* versions 2.2 and 3.0, respectively. Each file corresponds to a chapter with all of the *Mathematica* input statements organized by section and subsection exactly as they appear in the book.

The 3.5-inch, 1.4 MB high-density disk, which is in DOS format, can be used by both Macintosh and Microsoft Windows-based IBM-compatible computers.

- Macintosh
    - System Software Version 7.*x* Before Version 7.5

        Use the Apple File Exchange program that comes on a system software disk to translate the files in the Version2 or Version3 folder to Macintosh format. Convert the new files to notebooks by opening and saving them with *Mathematica*, that is, with the **Open** and **Save** commands in the File menu of *Mathematica*.
    - System Software Version 7.5 or Later

        Copy the Version2 or Version3 folder to a hard disk that has a copy of *Mathematica*. Open a notebook by double-clicking its icon.

- Windows
    - Version 3.1 Operating System

        Copy the Version2 folder to a hard disk. To generate notebooks from the files, open and save them with *Mathematica*, that is, with its **Open** and **Save** commands in the File menu.
    - 95 Operating System

        Copy the Version2 or Version3 folder to a hard disk that has a copy of *Mathematica*. To open a notebook, double-click its icon.

Updates

This book has been developed with the most recent releases of *Mathematica* version 2.2 and prereleases of version 3.0. For its latest updates reflecting the changes in the current releases of *Mathematica* version 3, visit APNet of Academic Press at http://www.apnet.com/, look for the book update information link, and check out the author's Web page.

PART 1

Mathematica with Physics

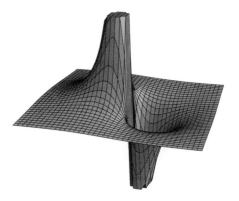

CHAPTER 1

The First Encounter

Mathematica consists of two parts: the kernel and the front end. The kernel conducts the computations, and the front end provides the interface between the user and the kernel. Whereas the kernel remains the same, the front end is optimized for each kind of computer system.

This chapter shows the neophyte how to get started with *Mathematica*. Macintosh-specific comments are enclosed in boxes marked with the ⌘ icon; Windows-specific comments are enclosed in boxes marked with the ☐ icon. Version 3.0- and 2.2-specific comments are labeled with the ✲ **3.0** and ✲ **2.2** icons, respectively. Platform-specific subsections are entitled "Macintosh Specifics" or "Microsoft Windows Specifics"; version-specific subsections are entitled "*Mathematica* Version 3.0 Specifics" or "*Mathematica* Version 2.2 Specifics." The appearance of *Mathematica* input and output may vary from version to version and platform to platform.

1.1 The First Ten Minutes

Let us begin at the very beginning. To open a new *Mathematica* document, called a "notebook," double-click the *Mathematica* icon. A *Mathematica* window appears. We are now ready to do several computations.

✲ **3.0**   To reproduce *Mathematica* output resembling that in this book, choose **OutputForm** in the Default Output FormatType submenu of the Cell menu. (Section 2.5, "Some New Features in *Mathematica* Version 3.0," will discuss **InputForm, OutputForm, StandardForm,** and **TraditionalForm.**)

This book has been developed with prereleased versions of *Mathematica* 3.0. For the latest updates just before it goes to press, see Appendix A, "The Last Ten Minutes."

Type exactly as shown:

```
In[1] := 2 + 3
```

Without moving the text insertion point, evaluate the input. To evaluate an input, press Enter for the Macintosh or Shift+Enter for Windows. (For further discussion of evaluating input, see Section 1.6.4, "Sending Input to the Kernel.") If *Mathematica* beeps, use the mouse to pull down the Help menu and select **Why the Beep?** to see what is happening. Otherwise, *Mathematica* performs the calculation and returns the result below the input. It appears that *Mathematica* takes a long time to add two integers. That really is not so. It takes time for the kernel to be loaded at the initial evaluation request. Computations for subsequent requests are much faster.

Type exactly as shown:

```
In[2] := 100!
```

Without moving the text insertion point, evaluate the input. *Mathematica* computes 100 factorial and returns the output

```
93326215443944152681699238856266700490715968264381621468\
9296389521759999322991560894146397615651828625369792082\
7223758251185210916864000000000000000000000000
```

Type and evaluate the input

```
In[3] := Expand[(x + y)^30]
```

Mathematica expands $(x + y)^{30}$. The output is

$$
x^{30} + 30 \, x^{29} \, y + 435 \, x^{28} \, y^2 + 4060 \, x^{27} \, y^3 +
$$
$$
27405 \, x^{26} \, y^4 + 142506 \, x^{25} \, y^5 + 593775 \, x^{24} \, y^6 +
$$
$$
2035800 \, x^{23} \, y^7 + 5852925 \, x^{22} \, y^8 + 14307150 \, x^{21} \, y^9 +
$$
$$
30045015 \, x^{20} \, y^{10} + 54627300 \, x^{19} \, y^{11} +
$$
$$
86493225 \, x^{18} \, y^{12} + 119759850 \, x^{17} \, y^{13} +
$$
$$
145422675 \, x^{16} \, y^{14} + 155117520 \, x^{15} \, y^{15} +
$$
$$
145422675 \, x^{14} \, y^{16} + 119759850 \, x^{13} \, y^{17} +
$$
$$
86493225 \, x^{12} \, y^{18} + 54627300 \, x^{11} \, y^{19} +
$$
$$
30045015 \, x^{10} \, y^{20} + 14307150 \, x^9 \, y^{21} + 5852925 \, x^8 \, y^{22} +
$$
$$
2035800 \, x^7 \, y^{23} + 593775 \, x^6 \, y^{24} + 142506 \, x^5 \, y^{25} +
$$
$$
27405 \, x^4 \, y^{26} + 4060 \, x^3 \, y^{27} + 435 \, x^2 \, y^{28} + 30 \, x \, y^{29} + y^{30}
$$

Mathematica distinguishes between uppercase and lowercase letters for input. For example, `Expand` and `expand` are different.

Enter and evaluate

In[4] := `Integrate[1/(Sin[x]^2 Cos[x]^2), x]`

Mathematica performs the integration and returns the result

`-(Cos[2 x] Csc[x] Sec[x])`

In the preceding input, be sure to leave a space between the number "2" and the letter "C". Again, *Mathematica* distinguishes between uppercase and lowercase letters for input. It also insists that parentheses, curly brackets, and square brackets are different.

Enter and evaluate

In[5] := `Plot3D[Sin[x y], {x, -Pi, Pi}, {y, -Pi, Pi}, PlotPoints -> 25];`

Mathematica displays the following graphic:

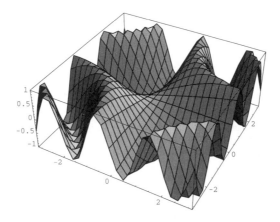

> ⬛　The meter at the lower left-hand corner of the notebook window indicates the amount of memory available to the front end. To see the meter showing the amount of memory available to the kernel, choose *Mathematica* **Kernel** or **MathKernel** in the Applications menu as well as **Show Meter** or **Show Memory Usage** in the File menu, move the meter to a convenient location, and click the notebook window.

> ⬚　In Windows version 3.1, the amount of free memory available to the combined front end and kernel is indicated in the status bar. For the location of the memory indicator in Windows 95, see Appendix A.

Before the available memory becomes low, save the notebook and quit *Mathematica;* otherwise, *Mathematica* crashes and all the work is lost! To quit *Mathematica,* select **Quit** for the Macintosh or **Exit** for Windows in the File menu. If the computer has limited memory, we may have to quit and restart often.

1.2   A Touch of Physics

1.2.1   Numerical Calculations

EXAMPLE 1.2.1   Find the eigenvalues and eigenvectors of the Pauli matrix

$$\sigma_x = \begin{pmatrix} 0 & 1 \\ 1 & 0 \end{pmatrix}$$

```
In[6] := pauliMatrix = {{0, 1}, {1, 0}};
```

```
In[7] := Eigensystem[pauliMatrix]
```

```
Out[7] = {{-1, 1}, {{-1, 1}, {1, 1}}}
```

The eigenvalues of the Pauli matrix σ_x are -1 and 1, and the corresponding eigenvectors
are $(-1, 1)$ and $(1, 1)$.   ∎

1.2.2   Symbolic Calculations

EXAMPLE 1.2.2   Consider an object moving with constant acceleration a in one dimension. The initial displacement and velocity are x_0 and v_0, respectively. Determine the
displacement x as a function of time t.

```
In[8] := DSolve[{x''[t] == a, x'[0] == vo, x[0] == xo}, x[t], t]
```

$$Out[8] = \left\{\left\{x[t] \rightarrow \frac{a\ t^2}{2} + t vo + xo\right\}\right\}$$

In the input, the symbol `''` in the second derivative `x''[t]` consists of two single quotation
marks.   ∎

1.2.3   Graphics

EXAMPLE 1.2.3   For an acoustic membrane clamped at radius a, the $n = 2$ normal mode
of vibration is

$$z(r, \theta, t) = J_2(\omega r/v) \sin(2\theta) \cos(\omega t)$$

where z, (r, θ), and t denote the membrane displacement, polar coordinates, and time,
respectively. J_2 is the Bessel function of order 2; v, the acoustic speed; ω, the frequency.
The boundary condition at $r = a$ requires that $J_2(\omega a/v) = 0$, which is satisfied if $\omega a/v$
$= 5.13562$. Let $\omega = 1$ and $v = 1$. Display graphically the vibration at time $t = \pi$.

```
In[9] := Needs["Graphics`ParametricPlot3D`"]
```

We loaded the package `Graphics`ParametricPlot3D`` with the command `Needs[`
`"Graphics`ParametricPlot3D`"]` so that we could use the function `CylindricalPlot3D`.
The backquote `` ` `` rather than the single quotation mark `'` is used in the command. For an
exposition of packages, see Section 1.5, "Packages."

```
In[10] := CylindricalPlot3D[BesselJ[2, r] Sin[2 theta] Cos[Pi],
                  {r, 0, 5.13562}, {theta, 0, 2Pi},
                  PlotPoints -> 25,
                  BoxRatios -> {1, 1, 0.4},
                  ViewPoint -> {2.340, -1.795, 1.659}];
```

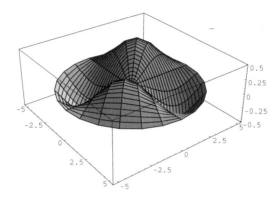

(For a brief discussion of the vibration of the circular acoustic planar membrane, refer to [Cra91].) ∎

1.3   On-Line Help

We can access information about any kernel object by typing a question mark followed by the name of the object and then pressing Enter for the Macintosh or Shift+Enter for Windows. The question mark must be the first character of the input line. To obtain information on Pi, for example, enter

```
In[11] := ?Pi
```

```
Pi is the constant pi, with numerical value approximately equal to
    3.14159.
```

For the **DSolve** function used in Example 1.2.2, enter

```
In[12] := ?DSolve
```

```
DSolve[eqn, y, x] solves a differential equation for the function y,
    with independent variable x. DSolve[{eqn1, eqn2, ... }, {y1, y2,
    ... }, x] solves a list of differential equations. DSolve[eqn, y,
    {x1, x2, ... }] solves a partial differential equation.
```

※ **3.0**   Section A.10, "Listing of Major Built-in *Mathematica* Objects," of Appendix in Wolfram's **The *Mathematica* Book** [Wol96] provides a printed alternative to the on-line help.

❋ **2.2**   Section A.8, "Listing of Built-in *Mathematica* Objects," of Appendix in Wolfram's ***Mathematica*: A System for Doing Mathematics by Computer,** often referred to as the "*Mathematica* book" [Wol91], and Blachman's ***Mathematica* Quick Reference** [Bla92] provide printed alternatives to the on-line help.

To get more information about an object, use `??` instead of `?`:

```
In[13] := ??DSolve
```

```
DSolve[eqn, y, x] solves a differential equation for the function y,
    with independent variable x. DSolve[{eqn1, eqn2, ...  },  {y1, y2,
    ...  }, x] solves a list of differential equations. DSolve[eqn, y,
    {x1, x2, ... }] solves a partial differential equation.
Attributes[DSolve] = {Protected, ReadProtected}

Options[DSolve] = DSolveConstants -> C
```

When used in conjunction with `?`, the symbol `*` is a wild card that matches any sequence of alphanumeric characters:

```
In[14] := ?ND*
```

```
NDSolve[eqns, y, {x, xmin, xmax}] finds a numerical solution to the
    ordinary differential equations eqns for the function y with the
    independent variable x in the range xmin to xmax. NDSolve[eqns, y,
    {x, xmin, xmax}, {t, tmin, tmax}] finds a numerical solution to
    the partial differential equations eqns. NDSolve[eqns, {y1, y2,
    ... }, {x, xmin, xmax}] finds numerical solutions for the
    functions yi.
```

In this example, the symbol `*` matches the characters "`Solve`". If the specification matches more than one name, *Mathematica* returns a list of the names:

```
In[15] := ?*Plot*
```

ContourPlot	MovieParametricPlot	PlotJoined
CylindricalPlot3D	MoviePlot	PlotLabel
DensityPlot	MoviePlot3D	PlotPoints
ListContourPlot	ParametricPlot	PlotRange
ListDensityPlot	ParametricPlot3D	PlotRegion
ListPlot	Plot	PlotStyle
ListPlot3D	Plot3D	PointParametricPlot3D
MovieContourPlot	Plot3Matrix	SphericalPlot3D
MovieDensityPlot	PlotDivision	

Here the names all include the characters "`Plot`".

We can use `?` to ask for information about most operator input forms. To obtain information on `->`, for example, enter

In[16] `:= ?->`

`lhs -> rhs represents a rule that transforms lhs to rhs.`

1.4 Error Messages

When *Mathematica* finds an input questionable, it prints one or more error messages each consisting of a *symbol*, a message *tag*, and a brief message, in the form *symbol*`::`*tag*`:` *message text*. For instance, entering *y* instead of *t* in an equation of Example 1.2.2 triggers error messages:

In[17] `:= DSolve[{x''[y] == a, x'[0] == vo, x[0] == xo}, x[t], t]`

`DSolve::nvld:`
   `The description of the equations appears to be ambiguous or`
   `   invalid.`

`DSolve::deqx:`
   `Supplied equations are not differential equations of the given`
   `   functions.`

Out[17] `= DSolve[{x''[y] == a, x'[0] == vo, x[0] == xo}, x[t], t]`

Mathematica **Warning Messages** contains a list of all error messages for built-in *Mathematica* objects along with explanations of the messages [Wit91].

1.5 Packages

Mathematica versions 2.2 and 3.0 have, respectively, about 1000 and 1500 built-in functions. Yet we often need a function that is not already built into *Mathematica*. In that case, we can define the function in the notebook or use one contained in a package which is a file consisting of functional definitions written in the *Mathematica* language. Many standard packages come with *Mathematica*. (To obtain information about these packages, consult **Guide to Standard *Mathematica* Packages** for version 2.2 [AAB93] or ***Mathematica* 3.0 Standard Add-on Packages** for version 3.0 [WR96].)

To use a function in a package, we must first load the package. The command for loading a package is

`<<`*context*`

or equivalently, `Get[`"*context*`"]`. The context name, *context*`, of a package is system-independent. (For a discussion of *Mathematica* contexts, see Section 3.7.1.)

✵ **3.0 &** ✵ **2.2**   The context name, *context*`, of a package is *FolderName*` *FileName*`
where *FileName* is the name, excluding the suffix ".m", of the file and *FolderName* is

the name of the folder containing the file. For example, the context
`Calculus`DiracDelta`` refers to the package `DiracDelta.m` in the Calculus folder. For
version 3.0, the package folders are in the StandardPackages folder of the AddOns folder;
for version 2.2, they are in the Packages folder.

☐ ✿ **2.2**   The context name, *context`*, of a package for the corresponding
DirectoryName\ FileName is given in the appendix of **Guide to Standard *Mathematica*
Packages**. *FileName* is the name of the file and *DirectoryName* is the name of the directory
containing the file. In the file naming conventions of pre-95 versions of Windows, directory names and file names (excluding the suffix " .m") have at most eight characters. For
example, the appendix indicates that the context `Calculus`DiracDelta`` corresponds to
CALCULUS\ DIRACDEL.M and refers to the package `DiracDel.m` in the Calculus directory. The package directories are in the Packages directory.

The backquote ` rather than the single quotation mark ' is used in context names. The
backquote, or grave accent character, is called a "context mark" in *Mathematica*.

Another command for loading a package is

Needs ["*context`*"]

In Example 1.2.3, the package `Graphics`ParametricPlot3D`` was loaded with the command
Needs ["Graphics`ParametricPlot3D`"] so that the function `CylindricalPlot3D` could be
used.

The "`<<`" command requires fewer key strokes to enter than the **Needs** command does. On
the other hand, **Needs** has the advantage over "`<<`" in that it reads in a package only if the
package is not already loaded whereas "`<<`" reads in a package even if it has been loaded.
This book will use **Needs** to read in a *Mathematica* package. (For further comparison of the
commands "`<<`" and **Needs**, see Problems 5 and 6 in Section 3.7.5.)

After loading a package, we can obtain information about the functions defined in the
package with the ? and ?? operators. For example, we can access information about the
function `CylindricalPlot3D` defined in the package `Graphics`ParametricPlot3D`` that
was loaded in Example 1.2.3:

```
In[18] := ?CylindricalPlot3D

CylindricalPlot3D[z, {r, rmin, rmax}, {phi, phimin, phimax},
    (options)] plots z as a function of r and phi.
    CylindricalPlot3D[{z, style},  ...] uses style to render each
    surface patch.
```

1.5.1 *Mathematica Version 3.0 Specifics*

There is an important point to remember. If we intend to use the variables and functions defined in a package, we must not invoke their names before loading the package.

For example, let us first enter

```
In[19] := Xx

Out[19] = Xx

In[20] := Div

Out[20] = Div
```

and define

```
In[21] := Grad[f_] := f^3
```

Then, load the package `Calculus`VectorAnalysis``:

```
In[22] := Needs["Calculus`VectorAnalysis`"]
```

```
Grad::shdw: Symbol Grad appears in multiple contexts
    {Calculus`VectorAnalysis`, Global`}; definitions in context
    Calculus`VectorAnalysis`
     may shadow or be shadowed by other definitions.

Div::shdw: Symbol Div appears in multiple contexts
    {Calculus`VectorAnalysis`, Global`}; definitions in context
    Calculus`VectorAnalysis`
     may shadow or be shadowed by other definitions.

Xx::shdw: Symbol Xx appears in multiple contexts
    {Calculus`VectorAnalysis`, Global`}; definitions in context
    Calculus`VectorAnalysis`
     may shadow or be shadowed by other definitions.
```

Mathematica warns that the definitions for `Grad`, `Div`, and `Xx` in the package may shadow or be shadowed by other definitions. For this package, `CoordinateSystem` gives the name of the default coordinate system and `Coordinates[]` shows the default names of the coordinate variables in the default coordinate system:

```
In[23] := CoordinateSystem

Out[23] = Cartesian

In[24] := Coordinates[]

Out[24] = {Calculus`VectorAnalysis`Xx, Yy, Zz}
```

To illustrate the idea of shadowing, let us evaluate

```
In[25] := Div[{Xx^2, Yy^2, Zz^2}]

Out[25] = Div[{Xx^2, Yy^2, Zz^2}]
```

In[26] := `Grad[Xx^2 Yy^2 Zz^2]`

Out[26] = $Xx^6 Yy^6 Zz^6$

Mathematica returns `Div[{Xx^2, Yy^2, Zz^2}]` unevaluated because it cannot access a definition for `Div` and gives a result for `Grad[Xx^2 Yy^2 Zz^2]` in accordance with our earlier definition of `Grad`. In other words, the definitions for `Div` and `Grad` in the package are shadowed or ignored. To use the definitions of `Div` and `Grad` in the package, we must first execute the command `Remove[name1, name2, ...]`:

In[27] := `Remove[Div, Grad, Xx]`

Let us evaluate again

In[28] := `Div[{Xx^2, Yy^2, Zz^2}]`

Out[28] = `2 Xx + 2 Yy + 2 Zz`

In[29] := `Grad[Xx^2 Yy^2 Zz^2]`

Out[29] = $\{2\ Xx\ Yy^2\ Zz^2,\ 2\ Xx^2\ Yy\ Zz^2,\ 2\ Xx^2\ Yy^2\ Zz\}$

Mathematica returns the correct results with the definitions of `Div` and `Grad` contained in the package `Calculus`VectorAnalysis``, which is consistent with vector analysis in mathematics and physics.

1.5.2   *Mathematica Version 2.2 Specifics*

There is an important point to remember. If we intend to use the variables and functions defined in a package, we must not invoke their names before loading the package. For example, let us first enter

`Y`

Y

`Div`

Div

and define

`Grad[f_] := f^3`

Then, load the package `Calculus`VectorAnalysis``:

`Needs["Calculus`VectorAnalysis`"]`

```
Grad::shdw:
     Warning: Symbol Grad appears in multiple contexts
        {Calculus`VectorAnalysis`, Global`}; definitions in
          context Calculus`VectorAnalysis`
        may shadow or be shadowed by other definitions.

Div::shdw:
     Warning: Symbol Div appears in multiple contexts
        {Calculus`VectorAnalysis`, Global`}; definitions in
          context Calculus`VectorAnalysis`
        may shadow or be shadowed by other definitions.

x::shdw:
     Warning: Symbol x appears in multiple contexts
        {Calculus`VectorAnalysis`, Global`}; definitions in
          context Calculus`VectorAnalysis`
        may shadow or be shadowed by other definitions.

y::shdw:
     Warning: Symbol y appears in multiple contexts
        {Calculus`VectorAnalysis`, Global`}; definitions in
          context Calculus`VectorAnalysis`
        may shadow or be shadowed by other definitions.

r::shdw:
     Warning: Symbol r appears in multiple contexts
        {Calculus`VectorAnalysis`, Global`}; definitions in
          context Calculus`VectorAnalysis`
        may shadow or be shadowed by other definitions.

theta::shdw:
     Warning: Symbol theta appears in multiple contexts
        {Calculus`VectorAnalysis`, Global`}; definitions in
          context Calculus`VectorAnalysis`
        may shadow or be shadowed by other definitions.
```

Mathematica warns that the definitions for **Div** and **Grad** in the package may shadow or be shadowed by other definitions. The same warnings are given for the symbols **x**, **y**, **r**, and **theta**. To illustrate the idea of shadowing, let us evaluate

Div[{x^2, y^2, z^2}]

$\mathrm{Div}[\{x^2,\ y^2,\ z^2\}]$

Grad[x^2 y^2 z^2]

$x^6\ y^6\ z^6$

Mathematica returns `Div[{x^2, y^2, z^2}]` unevaluated because it cannot access a definition for `Div` and gives a result for `Grad[x^2 y^2 z^2]` in accordance with our earlier definition of `Grad`. In other words, the definitions for `Div` and `Grad` in the package are shadowed or ignored. To use the definitions of `Div` and `Grad` in the package, we must first execute the command `Remove[name1, name2, ...]`:

```
Remove[Div, Grad, x, y]
```

Let us evaluate again

```
Div[{x^2, y^2, z^2}]
```
```
2 x + 2 y + 2 z
```
```
Grad[x^2 y^2 z^2]
```
$$\{2\ x\ y^2\ z^2,\ 2\ x^2\ y\ z^2,\ 2\ x^2\ y^2\ z\}$$

Mathematica returns the correct results with the definitions of `Div` and `Grad` contained in the package `Calculus`VectorAnalysis``, which is consistent with vector analysis in mathematics and physics.

1.6   *Notebook Interfaces*

This section briefly discusses notebook interfaces, or notebook front ends. Presently, notebook interfaces are available for the Macintosh, Microsoft Windows, NeXT, and X Window System. Though notebook front ends share many standard features, a front end is customized for each kind of computer system. Here we focus on the Macintosh and Microsoft Windows front ends. To learn more about these interfaces, see

> ❈ **2.2**   *Mathematica:* **A System for Doing Mathematics by Computer** [Wol91], *Mathematica* **User's Guide for the Macintosh** [DG94], or *Mathematica* **User's Guide for Microsoft Windows** [WW94].

> ❈ **3.0**   The *Mathematica* **Book** [Wol96], **Getting Started with** *Mathematica* **on Macintosh Systems** [BG96], or **Getting Started with** *Mathematica* **3.0 under Microsoft Windows** [BWG96].

1.6.1   *Notebooks*

For notebook interfaces, *Mathematica* documents are called notebooks. A *Mathematica* notebook contains ordinary text, *Mathematica* input and output, and graphics. Animated graphics can be produced from a sequence of graphics cells, and existing or modified input can be sent to the kernel for actual computations.

The basic unit of organization in a notebook is a cell. The bracket to the right of a cell marks its extent. For a new notebook, a new cell is created when we start typing. A new cell in a new notebook is shown here:

```
╔═════════════════ Lorenz Equations ═══════════════╗
║  NDSolve[{x'[t]  ==  -3 {x[t]  -  y[t]},          ║ ⇧
║         y'[t]  ==  -x[t] z[t] +                   ║ ▓
║                26.5 x[t]  -  y[t],                ║
║         z'[t]  ==  x[t] y[t]  -  z[t],            ║
║         x[0]  ==  z[0]  ==  0,                    ║
║         y[0]  ==  1},                             ║
║         {x, y, z}, {t, 0, 20},                   ║
║         MaxSteps -> 3000]                         ║
║                                                   ║
║                                                   ║
║                                                   ║ ⇩
╟───────────────┬─────────┬─────────────────────────╢
║ ▪ ' ' ' ' ' ' │ 100%  ▼ │ ⇦▒                    ⇨▢║
╚═══════════════╧═════════╧═════════════════════════╝
```

To produce another cell above or below this one or between any two cells, click when the pointer turns into a horizontal I-beam above or below this cell or between two cells and then type.

1.6.2 Getting Help

Section 1.3 discussed getting help directly from the kernel; this section considers the help provided by the notebook front end.

1.6.2.1 *MATHEMATICA* VERSION 3.0 SPECIFICS

The Help Browser provides several types of help:

Built-in Functions	information and examples on the built-in functions
Add-ons	information and examples on functions defined in the standard packages
The *Mathematica* Book	on-line version of **The *Mathematica* Book** [Wol96]
Getting Started	on-line version of **Getting Started with *Mathematica* on Macintosh Systems** [BG96] or **Getting Started with *Mathematica* 3.0 under Microsoft Windows** [BWG96]
Other Resources	information about the menu commands and other *Mathematica* resources
Master Index	on-line search of all items in the Help Browser

To explore the Help Browser, choose **Help** in the Help menu.

1.6.2.2 *MATHEMATICA* VERSION 2.2 SPECIFICS

> ▲   To access information about a menu command, choose **Help Pointer** in the Help menu and then select the menu command with the question mark pointer. To obtain information on built-in functions or functions defined in the standard packages, choose **Open Function Browser** in the Help menu. Clicking the Help button in the browser opens a dialog box that explains the use of the Function Browser.

> ⬚   Windows Help offers Front End Help and Kernel Help. Front End Help gives information on dialog boxes, key combinations, and command menus. Kernel Help provides information on built-in functions. To get help, choose **Contents** in the Help menu and click the appropriate underlined topics.

1.6.3   Preparing Input

Input lines can be edited with standard Macintosh or Windows techniques. *Mathematica* is very strict with input spelling. The **CompleteSelection** command can help.

Given the initial characters of the name of a kernel object, the **CompleteSelection** command returns the full name. If, for example, we type "**Fou**", highlight the characters, and choose **CompleteSelection** in the Input menu for version 3.0 or in the Prepare Input submenu of the Action menu for version 2.2, the full name appears as

`Fourier`

which is the name of the function that finds the discrete Fourier transform of a list of complex numbers. If there are several possible completions, a list of the names is displayed. For example, if we type "**Plot**" and choose **Complete Selection,** the following menu pops up:

(In *Mathematica* version 2.2 for Windows, the Complete Selection dialog box appears with a list of the possible completions.) Click a name to have it pasted in the cell.

A function's template specifies the number, type, and location of its arguments. The template of the **Do** function is

Do[*expr*, {*imax*}]

The first argument is an expression; the second argument, an iterator.

⌘ **3.0** & ✿ ⌘ **2.2**   The **Make Template** command returns a function's template. To obtain, for example, the template of the `DensityPlot` function, type the name of the function, leave the text insertion point at the end of the name, and choose **Make Template** in the Input menu for version 3.0 or in the Prepare Input submenu of the Action menu for version 2.2. The following is pasted in the cell:

DensityPlot[*f*, {*x*, *xmin*, *xmax*}, {*y*, *ymin*, *ymax*}]

We can now replace the arguments with our function and values. **Make Template** usually returns the simplest of several possible forms for the function. To access more information on the function, use the ? or ?? operator discussed in Section 1.3, "On-Line Help," or consult the Help Browser for version 3.0 or the Function Browser for version 2.2.

✿ ⌘ **2.2**   The Function Browser also provides function templates. To open the browser, choose **Open Function Browser** in the Help menu. Clicking the Help button in the browser opens a dialog box that describes the use of the browser.

1.6.4 Sending Input to the Kernel

To send *Mathematica* input to the kernel, put the text insertion point anywhere in the input cell or highlight the cell bracket, and press

---

✿ ⌘ **3.0**   Enter or Shift-Return.

✿ ⌘ **2.2**   Enter, Shift-Return, or Command-Return.

⊡ ⌘ **3.0**   Shift+Enter or Enter on the numeric keypad.

⊡ ⌘ **2.2**   Shift+Enter, Insert, or 5 key on the numeric keypad.

The kernel evaluates the input and returns the result in one or more cells below the input cell.

1.6.5   *Interrupting Calculations*

❋ **3.0**   To interrupt a calculation, choose **Abort Evaluation** in the Kernel menu. *Mathematica* aborts the current calculation, sometimes after some delay, and we may continue with further calculations. If the kernel does not respond to this command, we can interrupt the calculation by choosing **Local** or another appropriate item in the Quit Kernel submenu of the Kernel menu and clicking **Quit** in the dialog box to disconnect the kernel from the front end and terminate the current *Mathematica* session. The notebook, however, is unaffected. Sending another input to the kernel restarts the kernel and begins a new *Mathematica* session.

 ❋ 2.2   To interrupt a calculation, choose **Abort Calculation** in the Action menu or press Command-Period (-.). *Mathematica* aborts the current calculation, sometimes after some delay, and we may continue with further calculations. If the kernel does not respond to this command, we can interrupt the calculation by selecting **Quit/Disconnect Kernel** in the Action menu to disconnect the kernel from the front end and terminate the current *Mathematica* session. The notebook, however, is unaffected. For further computations, select *Mathematica* **Kernel** in the Applications menu and **Quit** in the File menu. Sending another input to the kernel restarts the kernel and begins a new *Mathematica* session.

⬚ ❋ **2.2**   To interrupt a calculation, select **Interrupt** in the Action menu or press Alt+Period (Alt+.). In the Interrupt dialog box, click **Abort**. *Mathematica* aborts the current calculation, sometimes after some delay, and we may continue with further calculations. If the kernel does not respond to this request, we can interrupt the calculation by saving the notebook and quitting *Mathematica*. For further computations, begin a new *Mathematica* session.

Interactive Use of *Mathematica*

This chapter covers the use of *Mathematica* as a super-calculator. It does what an electronic calculator can do, and it does a lot more. We enter input and *Mathematica* returns the output. As seen in Chapter 1, the *n*th is labeled `In[n]:=` and the corresponding output, `Out[n]`.

> ✻ **3.0**   To reproduce *Mathematica* output resembling that in this book, make sure **OutputForm** in the Default Output FormatType submenu of the Cell menu is checked at the beginning of each *Mathematica* session. (Section 2.5, "Some New Features in *Mathematica* Version 3.0," will discuss **InputForm, OutputForm, StandardForm,** and **TraditionalForm.**)
>
> This book has been developed with prereleased versions of *Mathematica* 3.0. For the latest updates just before it goes to press, see Appendix A, "The Last Ten Minutes."

2.1   Numerical Capabilities

2.1.1   Arithmetic Operations

Table 2.1.1 lists the arithmetic symbols in *Mathematica*. Here are some examples of their use:

```
In[1] := 2.1+3.72

Out[1] = 5.82


In[2] := 6.882/2

Out[2] = 3.441
```

Mathematica Operation	Symbol
Addition	+
Subtraction	–
Multiplication	*
Division	/
Exponentiation	^

Table 2.1.1   Arithmetic operations in _Mathematica_

```
In[3] := 2^3
```
```
Out[3] = 8
```

2.1.2   Spaces and Parentheses

A space can replace the symbol * for multiplication. For example, `2 5` is an alternative form of `2*5`:

```
In[4] := 2 5
```
```
Out[4] = 10
```

We can put spaces before and after the arithmetic symbols to make the input easier to read. For example, `(3+4)^2` can be entered as

```
In[5] := (3 + 4) ^ 2
```
```
Out[5] = 49
```

Parentheses are used for grouping. Though _Mathematica_ observes the standard mathematical rules for precedence of arithmetic operators, parentheses should be used generously to avoid ambiguity about the order of operations. For example, it is not obvious whether `2^3^4` means `(2^3)^4` or `2^(3^4)`. Parentheses are needed for clarity:

```
In[6] := (2^3)^4
```
```
Out[6] = 4096
```

```
In[7] := 2^(3^4)
```
```
Out[7] = 2417851639229258349412352
```

```
In[8] := 2^3^4
```
```
Out[8] = 2417851639229258349412352
```

As it turns out, `2^3^4` stands for `2^(3^4)`.

2.1.3   Common Mathematical Constants

Table 2.1.2 lists some built-in mathematical constants.

Mathematica Name	Constant
Pi	π
E	e
Degree	$\pi/180$
I	$\sqrt{-1}$
Infinity	∞
GoldenRatio	$(1 + \sqrt{5})/2$

Table 2.1.2   Some mathematical constants known to *Mathematica*

2.1.4   Some Mathematical Functions

Some common mathematical functions built into *Mathematica* are

Sqrt[x]	square root
Exp[x]	exponential
Log[x]	natural logarithm
Log[b, x]	logarithm to base b
Sin[x]	sine
Cos[x]	cosine
Tan[x]	tangent
ArcSin[x]	inverse sine
ArcCos[x]	inverse cosine
ArcTan[x]	inverse tangent
Sinh[x]	hyperbolic sine
Cosh[x]	hyperbolic cosine
Tanh[x]	hyperbolic tangent
ArcSinh[x]	inverse hyperbolic sine
ArcCosh[x]	inverse hyperbolic cosine
ArcTanh[x]	inverse hyperbolic tangent
Factorial[n] or n!	factorial
Round[x]	closest integer to x
Random[]	pseudorandom number between 0 and 1
Abs[x]	absolute value

2.1.5 *Cases and Brackets*

Mathematica is case sensitive. The names of built-in *Mathematica* objects all begin with capital letters. For example, `Sqrt` and `sqrt` are different, and the former is a built-in *Mathematica* object:

```
In[9] := Sqrt[5.0]

Out[9] = 2.23607

In[10] := sqrt[5.0]

Out[10] = sqrt[5.]
```

Since `sqrt` has not been defined, *Mathematica* returns the input unevaluated.

There are five different kinds of brackets in *Mathematica*:

(*term*)	parentheses for grouping
f[*expr*]	square brackets for functions
{*a*, *b*, *c*}	curly brackets for lists
v[[*i*]]	double square brackets for indexing list elements
(*comment*)	commenting brackets for comments to be ignored by the kernel

2.1.6 *Ways to Refer to Previous Results*

One way to refer to previous results is by their assigned names. We can assign values to variables with the operator "`=`":

variable = *value*

For example, we can assign a value to the variable `t`:

```
In[11] := t = 3 + 4

Out[11] = 7
```

The variable `t` now has the value 7:

```
In[12] := t + 2

Out[12] = 9
```

To avoid confusion, names of user-created variables should always start with lowercase letters, because built-in *Mathematica* objects have names beginning with capital letters. In the previous example, we chose the name `t` rather than `T` for the variable.

In each *Mathematica* session, the assignments of values to variables remain in effect until the values are removed from or new values are assigned to these variables. It is a good practice to remove the values as soon as they are no longer needed. We can use t =. or Clear[t] to remove the value assigned to t:

In[13] := t =.

Another way to reference previous results is by using one or more percent signs:

%	the previous result
%%	the second previous result
%%%	the third previous result
%*n*	the result on output line Out[*n*]

Let us illustrate their use:

In[14] := 2^3

Out[14] = 8

The symbol % refers to the previous result:

In[15] := % + 5

Out[15] = 13

The symbol %14 stands for the result on the fourteenth output line:

In[16] := %14^2

Out[16] = 64

2.1.7 *Standard Computations*

Mathematica can do standard computations just like an electronic calculator. For example, it can evaluate the expression

$$\sqrt{\frac{(1.1 \times 10^{-23})(6.8 \times 10^{-2})}{1.4 \times 10^{-24}}}$$

In[17] := Sqrt[(1.1 10^-23)(6.8*10^-2)/(1.4 10^-24)]

Out[17] = 0.730949

The function ScientificForm expresses the result in scientific notation:

In[18] := ScientificForm[%]

Out[18]//ScientificForm=
7.30949 $\times 10^{-1}$

2.1.8   Exact versus Approximate Values

Mathematica treats integers and rational numbers as exact. `Precision[x]` gives the number of digits of precision in the number x. For exact numbers, `Precision` returns ∞ or `Infinity`. For example, 5 and 345/678 are exact:

    In[19] := Precision[5]

    Out[19] = ∞

    In[20] := Precision[345/678]

    Out[20] = ∞

When we give *Mathematica* exact values as input, it tries to return exact results:

    In[21] := 123/456 + 456/789 + 2 Sqrt[5]
    Out[21] = 33887/39976 + 2 Sqrt[5]

Mathematica considers $\sqrt{5}$ to be exact:

    In[22] := Precision[Sqrt[5]]

    Out[22] = ∞

We can obtain an approximate numerical result with the function `N`:

    In[23] := N[%%]

    Out[23] = 5.31982

We can also use the function `N` in its postfix form:

    In[24] := %%%//N

    Out[24] = 5.31982

Mathematica regards an expression containing one or more approximate values as an approximate expression:

    In[25] := Sqrt[1.0 + 4]

    Out[25] = 2.23607

The default machine precision is 16:

    In[26] := Precision[123.0/456 + 456/789]

    Out[26] = 16

(The default machine precision is 19 in *Mathematica* version 2.2 for the Macintosh.)

The arguments of trigonometric functions must be in radians. The constant `Degree` is used to convert degrees to radians. Consider, for example, the evaluation of `Cos[20 Degree]`:

    In[27] := Cos[20 Degree]

    Out[27] = Cos[20 Degree]

Why does *Mathematica* return the expression unchanged? Built-in mathematical constants such as E, Pi, and Degree are exact, and *Mathematica* does not automatically convert them to approximate numbers. We can use the function N to obtain an approximate numerical result:

In[28] := **Cos[20 Degree]//N**

Out[28] = 0.939693

2.1.9 Arbitrary Precision

Mathematica is not limited by the machine precision of a computer. N*[expr, n]* evaluates *expr* numerically using numbers with *n* digits of precision. For example, we can determine the volume of a sphere of radius 2 m to 200 digits:

In[29] := **N[(4 Pi/3)(2 m)^3, 200]**

Out[29] = 33.510321638291127876934862754981364098103140260001\
          1128757066075651283375000386229318699038136982512\
          8205847624625614167793756900100916853954380507050\
          0035337046525280303042882067755845930287597812212\
          18592407 m^3

2.1.10 Special Functions

All the familiar special functions of mathematical physics are built into *Mathematica*. Here are some of them:

LegendreP[*n*, *x*]	Legendre polynomials $P_n(x)$
LegendreP[*n*, *m*, *x*]	associated Legendre polynomials $P_n^m(x)$
SphericalHarmonicY[*l*, *m*, *theta*, *phi*]	spherical harmonics $Y_l^m(\theta, \phi)$
HermiteH[*n*, *x*]	Hermite polynomials $H_n(x)$
LaguerreL[*n*, *x*]	Laguerre polynomials $L_n(x)$
LaguerreL[*n*, *a*, *x*]	generalized Laguerre polynomials $L_n^a(x)$
ClebschGordan[{*j1*, *m1*}, {*j2*, *m2*}, {*j*, *m*}]	Clebsch-Gordan coefficient
BesselJ[*n*, *z*] and BesselY[*n*, *z*]	Bessel functions $J_n(z)$ and $Y_n(z)$
Hypergeometric1F1[*a*, *b*, *z*]	confluent hypergeometric function $_1F_1(a; b; z)$

(There are several notations for the associated Laguerre polynomials in quantum mechanics texts. For the relations between the generalized Laguerre polynomials in *Mathematica* and these associated Laguerre polynomials, see Example 2.2.6 of this book and p. 439 of [Lib92].)

Consider, for example, the evaluation of the Clebsch–Gordan coefficient $\langle 1\,0\,1\,0 \,|\, 2\,0 \rangle$:

In[30] := **ClebschGordan[{1,0},{1,0},{2,0}]**

Out[30] = Sqrt[$\frac{2}{3}$]

2.1.11   *Matrices*

Mathematica represents vectors and matrices by lists and nested lists, respectively:

$$\text{vector}: \quad \begin{pmatrix} a \\ b \\ c \end{pmatrix} \qquad \text{list: \{a, b, c\}}$$

$$\text{matrix}: \quad \begin{pmatrix} a & b & c \\ d & e & f \\ g & h & i \end{pmatrix} \qquad \text{nested list: \{\{a, b, c\}, \{d, e, f\}, \{g, h, i\}\}}$$

Some functions for vectors are

$c\,v$	product of scalar c and vector v
$u.v$	dot product of vectors u and v

❋ 3.0   **Cross** $[u, v]$ cross product of vectors u and v

Here are some functions for matrices:

$c\,m$	product of scalar c and matrix m
$m.n$	product of matrices m and n
Inverse $[m]$	inverse of matrix m
MatrixPower $[m, k]$	kth power of matrix m
Det $[m]$	determinant of matrix m
Transpose $[m]$	transpose of matrix m
Eigenvalues $[m]$	eigenvalues of matrix m
Eigenvectors $[m]$	eigenvectors of matrix m
MatrixForm $[list]$	$list$ displayed in matrix form

EXAMPLE 2.1.1   Find the inverse of the matrix

$$\begin{pmatrix} 16 & 0 & 0 \\ 0 & 14 & -6 \\ 0 & -6 & -2 \end{pmatrix}$$

In terms of a nested list, the matrix takes the form

In[31] := **m = {{16, 0, 0},{0, 14, -6},{0, -6, -2}}**

Out[31] = {{16, 0, 0}, {0, 14, -6}, {0, -6, -2}}

where we have assigned the matrix to the variable **m** so that we can refer to it later. The function **MatrixForm** displays the matrix in familiar two-dimensional form:

In[32] := **MatrixForm[m]**

Out[32]//MatrixForm=

```
16    0     0
0     14   -6
0     -6   -2
```

The function **Inverse** gives the inverse of the matrix:

In[33] := **mInv = Inverse[m]**

Out[33] = $\{\{\frac{1}{16}, 0, 0\}, \{0, \frac{1}{32}, -(\frac{3}{32})\}, \{0, -(\frac{3}{32}), -(\frac{7}{32})\}\}$

where we have assigned the inverse matrix to the variable **mInv**. To verify that **mInv** is the inverse of **m**, we multiply the matrices together:

In[34] := **m.mInv**

Out[34] = {{1, 0, 0}, {0, 1, 0}, {0, 0, 1}}

In standard two-dimensional matrix form, the result can be displayed as

In[35] := **MatrixForm[%]**

Out[35]//MatrixForm=

```
1    0    0
0    1    0
0    0    1
```

which is the identity matrix. Again, it is a good practice to remove unneeded values assigned to variables:

In[36] := **Clear[m, mInv]**   ∎

2.1.12 *Double Square Brackets*

Double square brackets allow us to pick out elements of lists and nested lists:

list[[*i*]]	the *i*th element of *list*
list[[{*i*, *j*, *k*, ...}]]	a list of the *i*th, *j*th, *k*th, ... elements of *list*
list[[*i*, *j*]]	the *j*th element in the *i*th sublist of a nested list

EXAMPLE 2.1.2 Assign the name *ourlist* to the list {1, 3, 5, 7, 9, 11, 13, 15}, extract the second element, and create a list of the first, fourth, and fifth elements.

We begin by naming the list:

In[37] := **ourlist = {1, 3, 5, 7, 9, 11, 13, 15}**

Out[37] = {1, 3, 5, 7, 9, 11, 13, 15}

We then pick out the second element:

In[38] := **ourlist[[2]]**

Out[38] = 3

Finally, we generate a list of the first, fourth, and fifth elements:

In[39] := **ourlist[[{1, 4, 5}]]**

Out[39] = {1, 7, 9}   ∎

EXAMPLE 2.1.3   Give the name *mymatrix* to the matrix

$$\begin{pmatrix} 1 & 2 & 3 \\ 4 & 5 & 6 \\ 7 & 8 & 9 \end{pmatrix}$$

Pick out (a) the second row, and (b) the second element in the third row.

In *Mathematica*, a matrix is represented as a nested list. We begin by assigning the nested list to the variable **myMatrix**:

In[40] := **myMatrix = {{1,2,3},{4,5,6},{7,8,9}}**

Out[40] = {{1, 2, 3}, {4, 5, 6}, {7, 8, 9}}

Here is the second row of the matrix:

In[41] := **myMatrix[[2]]**

Out[41] = {4, 5, 6}

Here is the second element in the third row:

In[42] := **myMatrix[[3, 2]]**

Out[42] = 8

In[43] := **Clear[ourlist, myMatrix]**   ∎

2.1.13   Least-Squares Fit

Fit[*data, funs, vars*] finds a least-squares fit to a list of data in terms of a linear combination of the functions in list *funs* of the variables in list *vars*. The argument *funs* can be any list of functions that depend only on the variables in list *vars*. When list *vars* has only one variable, **Fit**[*data, funs, vars*] takes the form

Fit[{{x1, y1, {x2, y2}, ...}, {f1, f2, ...}, {x}]

where the curly brackets in the third argument {x} are optional. If $x1 = 1$, $x2 = 2$, ..., that is, $xi = i$, this can be written as

Fit[{{y1, y2, ...}, {f1, f2, ...}, x]

EXAMPLE 2.1.4   Given here is the data of distance versus time for a hot Volkswagen, where *d* is in meters and *t* in seconds. Find the equation that gives *d* as a function of *t*.

t	d
0	0
1	1.5
2	6.0
3	13.5
4	24.0
5	37.5
6	54.0
7	73.5
8	96.0
9	121.5

We begin by defining `time` and `distance`:

```
In[44] := time = Table[i, {i, 0, 9}]
```

```
Out[44] = {0, 1, 2, 3, 4, 5, 6, 7, 8, 9}
```

```
In[45] := distance = {0, 1.5, 6.0, 13.5, 24.0, 37.5,
                      54.0, 73.5, 96.0, 121.5}
```

```
Out[45] = {0, 1.5, 6., 13.5, 24., 37.5, 54., 73.5, 96., 121.5}
```

where the function `Table` will be discussed in Section 2.1.18. We then generate the nested list for the data and name it `vwdata`:

```
In[46] := vwdata = Transpose[{time, distance}]
```

```
Out[46] = {{0, 0}, {1, 1.5}, {2, 6.}, {3, 13.5}, {4, 24.}, {5, 37.5},
          {6, 54.}, {7, 73.5}, {8, 96.}, {9, 121.5}}
```

The function `MatrixForm` allows us to see the data in familiar two-dimensional form:

```
In[47] := MatrixForm[vwdata]
```

```
Out[47]////MatrixForm=
  0      0
  1      1.5
  2      6.
  3      13.5
  4      24.
  5      37.5
  6      54.
  7      73.5
  8      96.
  9      121.5
```

Let us fit the data with a linear combination of functions, $1, t, t^2$, and t^3:

In[48] := **Fit[vwdata, {1, t, t^2, t^3}, t]**

Out[48] = $1.3059 \ 10^{-14} + 2.77556 \ 10^{-15} \ t + 1.5 \ t^2 - 1.97758 \ 10^{-16} \ t^3$

We can use the function Chop to remove terms that are close to zero. Chop[*expr*] replaces in *expr* all approximate real numbers with magnitude less than 10^{-10} by the exact integer 0:

In[49] := **Chop[%]**

Out[49] = $1.5 \ t^2$

Thus, the formula for distance versus time is $d = 1.5 \ t^2$. Again, it is a good idea to clear unneeded values assigned to variables as soon as possible:

In[50] := **Clear[time, distance, vwdata]**   ∎

2.1.14 Complex Numbers

Mathematica works with complex numbers as well as real numbers. What follow are some complex number operations:

Abs[z]	absolute value
Arg[z]	the argument
Re[z]	real part
Im[z]	imaginary part
Conjugate[z]	complex conjugate

Consider, for example, putting the expression

$$\sqrt{-7} + \ln(2 + 8i)$$

in the form $a + ib$, where a and b are approximate real numbers, and finding the complex conjugate:

In[51] := **N[Sqrt[-7] + Log[2 + 8 I]]**

Out[51] = $2.10975 + 3.97157 \ I$

In[52] := **Conjugate[%]**

Out[52] = $2.10975 - 3.97157 \ I$

2.1.15 Numerical Solution of Polynomial Equations

NSolve[*eqns*, *vars*] finds the numerical approximations to the roots of a polynomial equation or a system of polynomial equations. In *Mathematica,* an equation is written as *lhs* == *rhs*. It is important to distinguish between the two *Mathematica* operators "=" and "==". The operator "=" is for assignments, whereas the operator "==" is for specifying equations.

For example, we can determine the roots of the polynomial equation

$$x^5 + 4x^3 + 3x^2 + 2x = 10$$

In[53] := **NSolve[x^5 + 4 x^3 + 3 x^2 + 2 x == 10, x]**

Out[53] = {{x -> -1. - 1. I}, {x -> -1. + 1. I}, {x -> 0.5 - 2.17945 I},
        {x -> 0.5 + 2.17945 I}, {x -> 1.}}

NSolve gives solutions as rules of the form x -> *sol*. Section 2.2.4 will introduce *Mathematica* transformation rules.

We can also find the roots of the set of simultaneous polynomial equations

$$x^2 - xy - y^2 = 1$$
$$x^3 + 3xy - y^3 = 9$$

With a system of equations, the arguments of **NSolve** are lists:

In[54] := **NSolve[{x^2 - x y - y^2 == 1, x^3 + 3 x y - y^3 == 9},
            {x, y}]**

Out[54] = {{x -> -2.07194 - 1.87447 I, y -> -1.16444 - 1.26905 I},
        {x -> -2.07194 + 1.87447 I, y -> -1.16444 + 1.26905 I},
        {x -> -0.43195 - 0.816763 I, y -> 0.552813 + 1.71762 I},
        {x -> -0.43195 + 0.816763 I, y -> 0.552813 - 1.71762 I},
        {x -> 1.74511, y -> 0.802781}, {x -> 1.76268, y -> -2.57952}}

2.1.16 Numerical Integration

NIntegrate[*f*, {*x, xmin, xmax*}**]** gives a numerical approximation to the integral

$$\int_{xmin}^{xmax} f(x)\,dx$$

Let us evaluate, for example, the integral

$$\int_0^2 \sin[\,\sqrt{1 + x^4 + \sin(x^3)}\,]\,dx$$

In[55] := **NIntegrate[Sin[Sqrt[1 + x^4 + Sin[x^3]]], {x, 0, 2}]**

Out[55] = 1.31859

NIntegrate[*f*, {*x, xmin, xmax*}, {*y, ymin, ymax*}, ...**]** finds a numerical approximation to the multidimensional integral

$$\int_{xmin}^{xmax} dx \int_{ymin}^{ymax} dy \ldots f(x, y, \ldots)$$

EXAMPLE 2.1.5 A uniformly charged disk of radius a and charge density σ is placed on the xy plane with its center at the origin. At the point (x, y, z), the electric potential due to the disk is

$$V(x, y, z) = \frac{\sigma}{4\pi\varepsilon_0} \int_0^a dr \int_0^{2\pi} d\varphi \frac{r}{\sqrt{(x - r\cos\varphi)^2 + (y - r\sin\varphi)^2 + z^2}}$$

in which ε_0 is the permittivity constant. Determine V at $(a, 2a, 5a)$.

In units of $\sigma a / 4\pi\varepsilon_0$, the electric potential can be expressed as

$$V(x, y, z) = \int_0^1 dr \int_0^{2\pi} d\varphi \frac{r}{\sqrt{(x - r\cos\varphi)^2 + (y - r\sin\varphi)^2 + z^2}}$$

where the unit of x, y, and z is a. For $x = 1$, $y = 2$, and $z = 5$, **NIntegrate** evaluates numerically the double integral:

```
In[56] := NIntegrate[r/Sqrt[(1 - r Cos[phi])^2 +
                      (2 - r Sin[phi])^2 + 5^2],
              {r, 0, 1}, {phi, 0, 2Pi}]
```

```
Out[56] = 0.570012
```

Thus, the electric potential at $(a, 2a, 5a)$ equals $0.570012\sigma a / 4\pi\varepsilon_0$.   ■

2.1.17   *Numerical Solution of Differential Equations*

NDSolve[*eqns, y, {x, xmin, xmax}*] numerically solves *eqns*, which is a differential equation together with the necessary initial conditions, for the function y with independent variable x in the range *xmin* to *xmax*.

EXAMPLE 2.1.6 The equation of motion for a damped harmonic oscillator is

$$\frac{d^2x}{dt^2} + \gamma\frac{dx}{dt} + \omega_0^2 x = 0$$

Let $\omega_0^2 = 9\text{ s}^{-2}$, $\gamma = 0.5\text{ s}^{-1}$, $x(0) = 1\text{ m}$, and $x'(0) = 0$. Determine $x(t)$.

Let us numerically determine, in SI units, $x(t)$ for t in the range 0 to 12 and assign the result to the variable **sol**:

```
In[57] := sol = NDSolve[{x''[t] + 0.5 x'[t] + 9 x[t] == 0,
                    x'[0] == 0, x[0] == 1},
               x, {t, 0, 12}]
```

```
Out[57] = {{x -> InterpolatingFunction[{{0., 12.}}, <>]}}
```

The numerical solution to the differential equation is a list of pairs of numbers (t_i, x_i). *Mathematica* returns the solution as an interpolating function, which is an approximate function interpolating these pairs.

We can find the value of x at any time from 0 to 12. For $t = 2$,

```
In[58] := x[2] /. sol
```

```
Out[58] = {0.563522}
```

We can also plot $x(t)$ with t in the range 0 to 12:

```
In[59] := Plot[Evaluate[x[t] /. sol], {t, 0, 12},
              AxesLabel -> {" t (s)", " x (m)"}];
```

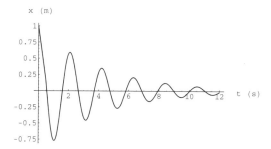

The operator `/.` and the functions `Evaluate` and `Plot` will be discussed in Sections 2.2.4 and 2.3.1.1, respectively.

To see how well the numerical solution approximates the exact solution, we substitute the numerical solution into the *lhs* of the differential equation, plot the result, and observe the deviation of the curve from zero:

```
In[60] := Plot[Evaluate[(x''[t] + 0.5 x'[t] + 9 x[t]) /. sol],
            {t, 0, 12}, PlotRange -> {-0.004, 0.004},
            AxesLabel -> {" t (s)", "lhs (m/s^2)"}];
```

With the numerical solution, the *lhs* of the differential equation is zero to within ±0.0037.

```
In[61] := Clear[sol]
```

`NDSolve[eqns, {y1, y2, ...}, {x, xmin, xmax}]` numerically solves *eqns*, which is a set of differential equations together with the necessary initial conditions, for the functions *y1, y2,* ... with independent variable *x* in the range *xmin* to *xmax*.

EXAMPLE 2.1.7 The magnitude of the force on a planet due to the Sun is

$$F = \frac{GMm}{r^2}$$

where *G*, *r*, *M*, and *m* are the gravitational constant, the distance of the planet from the Sun, the mass of the Sun, and the mass of the planet, respectively. The direction of the force is indicated in Figure 2.1.1.

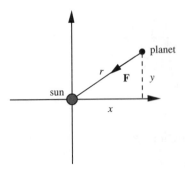

Figure 2.1.1   A planet under the gravitational attraction of the Sun

Newton's second law gives

$$\frac{d^2x}{dt^2} + \frac{GMx}{(x^2 + y^2)^{3/2}} = 0$$

$$\frac{d^2y}{dt^2} + \frac{GMy}{(x^2 + y^2)^{3/2}} = 0$$

where we have assumed that the Sun is stationary at the origin of the Cartesian coordinate system. In a system of units where the unit of length is the astronomical unit (AU), that is, the length of the semimajor axis of the Earth's orbit, and the unit of time is one year, $GM = 4\pi^2$ (see [GT88]). Solve these equations of motion for $x(t)$ and $y(t)$ with t ranging from 0 to 1.6 and plot the trajectory of the planet. Let $x(0) = 1$, $y(0) = 0$, $x'(0) = -\pi$, and $y'(0) = 2\pi$.

NDSolve finds a numerical solution to the equations of motion with the specified initial conditions:

```
In[62] := orbit =
          NDSolve[
              {x''[t] + (4 Pi^2) x[t]/((x[t]^2 + y[t]^2)^(3/2)) == 0,
               y''[t] + (4 Pi^2) y[t]/((x[t]^2 + y[t]^2)^(3/2)) == 0,
               x[0] == 1,
               y[0] == 0,
               x'[0] == -Pi,
               y'[0] == 2 Pi
              },
              {x, y}, {t, 0, 1.6}
             ]

Out[62] = {{x -> InterpolatingFunction[{{0., 1.6}}, <>],
            y -> InterpolatingFunction[{{0., 1.6}}, <>]}}
```

Let us plot the path of the planet:

```
In[63] := ParametricPlot[Evaluate[{x[t], y[t]} /. orbit],
                {t, 0, 1.6},
                AspectRatio -> Automatic,
                AxesLabel -> {"x (AU)", " y (AU)"}];
```

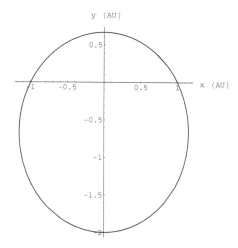

The trajectory is an ellipse with the Sun at one focus in agreement with Kepler's first law. The function **ParametricPlot** will be discussed in Section 2.3.1.6.

```
In[64] := Clear[orbit]
```

∎

2.1.18 Iterators

Iterators instruct *Mathematica* to perform certain tasks a number of times. The iterators notation is

{*imax*}	repeat *imax* times
{*i*, *imax*}	repeat with *i* running from 1 to *imax* in steps of 1
{*i*, *imin*, *imax*}	repeat with *i* running from *imin* to *imax* in steps of 1
{*i*, *imin*, *imax*, *di*}	repeat with *i* running from *imin* to *imax* in steps of *di*
{*i*, *imin*, *imax*},	repeat with *i* running from *imin* to *imax* in steps of 1; for each *i*,
{*j*, *jmin*, *jmax*}, ...	let *j* go from *jmin* to *jmax* in steps of 1; and so forth

Table and **Do** are examples of functions that use iterators. **Table**[*expr*, *iterator*] generates a list of *expr*. **Do**[*expr*, *iterator*] evaluates *expr* as many times as indicated by the *iterator*.

For example, we can generate the list $\{1, 1/4, 1/8, 1/16, 1/32, \ldots, 1/4096\}$ with the function
Table:

In[65] := **Table[1/2^(i - 1), {i, 13}]**

Out[65] = $\{1, \dfrac{1}{2}, \dfrac{1}{4}, \dfrac{1}{8}, \dfrac{1}{16}, \dfrac{1}{32}, \dfrac{1}{64}, \dfrac{1}{128}, \dfrac{1}{256}, \dfrac{1}{512}, \dfrac{1}{1024}, \dfrac{1}{2048}, \dfrac{1}{4096}\}$

Here, i goes from 1 to 13 in steps of 1.

We can plot $\sin(2x)$, $\sin(3x)$, and $\sin(4x)$ with the function **Do**:

In[66] := **Do[Plot[Sin[k x], {x, -2Pi, 2Pi},**
AxesLabel -> {x, Sin[k x]}
],
{k, 2, 4}
]

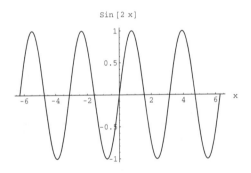

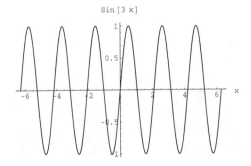

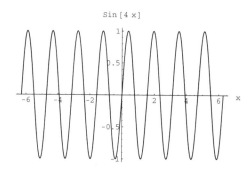

In this example, k runs from 2 to 4 in steps of 1.

2.1.19   Exercises

1. Evaluate

$$\left[10 \times \left(\frac{10.8 \times 10^3}{300} \right)^{1/2} + 4 \right]^{1/3}$$

   Answer: 4.

2. Evaluate

$$\frac{(10^{-24} \times 10^{12})}{10^{-14}} \times \sqrt{\frac{32,000}{2^3}}$$

   Answer: 6.32×10^3

3. Determine the length a of the triangle shown.

   Answer: 2.29×10^2

4. Determine the length b of the triangle shown.

Answer: 3.28×10^2

5. Given the matrices A and B, determine AB.

$$A = \begin{pmatrix} 1 & 0 & -1 \\ 2 & 4 & 7 \\ 5 & 3 & 0 \end{pmatrix} \qquad B = \begin{pmatrix} 6 & 1 \\ 0 & 4 \\ -2 & 3 \end{pmatrix}$$

Answer: 8   -2
            -2   39
            30   17

6. Given the matrix C, determine its inverse.

$$C = \begin{pmatrix} -2 & 1 & 3 \\ 0 & -1 & 1 \\ 1 & 2 & 0 \end{pmatrix}$$

Answer: $-(\frac{1}{4})$ $\frac{3}{4}$ $\frac{1}{2}$

$\frac{1}{8}$ $-(\frac{3}{8})$ $\frac{1}{4}$

$\frac{1}{8}$ $\frac{5}{8}$ $\frac{1}{4}$

7. Luap Yllek, a Martian physicist, has discovered a very peculiar gas. He measured the volume and pressure of this gas and obtained the following data (V, P), where V is in Martian liters and P in Martian atmospheres: (0.608, 0.05), (0.430, 0.10), (0.304, 0.20), (0.248, 0.30), (0.215, 0.40), (0.192, 0.50). Determine the equation of the curve that best fits this set of data. Use a combination of 1, $1/V$, and $1/V^2$.

Answer: $-0.00120577 + \dfrac{0.0182999}{V^2} + \dfrac{0.000988148}{V}$

8. Use the method of least-squares to find the best straight line for the four points (4, 5), (6, 8), (8, 10), (9, 12).

Answer: $-0.288136 + 1.33898\,x$

9. Determine the real and imaginary parts of the expression

$$\frac{4 + i}{2 + 3i}$$

Answers: $\dfrac{11}{13}$

$-(\dfrac{10}{13})$

10. Express the following complex number in the form $p + qi$, where p and q are floating-point numbers with 15 digits of accuracy:

$$(3\pi + 7i) \cos 37° + (2 + 8i)e^{-3i+2}$$

Answer: 1.23869141386657 - 55.0159195013496 I

11. Solve the equation

$$x^7 + x^5 + 2x^3 + x^2 + 1 = 0$$

Answers: {{x -> -0.812432}, {x -> -0.640787 - 1.07931 I},
           {x -> -0.640787 + 1.07931 I},
           {x -> 0.254825 - 0.700968 I},
           {x -> 0.254825 + 0.700968 I},
           {x -> 0.792178 - 0.881387 I},
           {x -> 0.792178 + 0.881387 I}}

12. Solve the system of equations

$$x^2 + xy + y^2 = 1$$
$$x^3 + x^2y + xy^2 + y^3 = 4$$

Answers: {{x -> -1. - 1.41421 I, y -> -1. + 1.41421 I},
           {x -> -1. + 1.41421 I, y -> -1. - 1.41421 I},
           {x -> -0.683802 - 1.13355 I,
            y -> 1.6838 + 0.133552 I},
           {x -> -0.683802 + 1.13355 I,
            y -> 1.6838 - 0.133552 I},
           {x -> 1.6838 - 0.133552 I,
            y -> -0.683802 + 1.13355 I},
           {x -> 1.6838 + 0.133552 I,
            y -> -0.683802 - 1.13355 I}}

13. Evaluate the integral

$$\int_0^1 \frac{x^3}{e^x - 1} dx$$

Answer: 0.224805

14. Evaluate the integral

$$\int_{-\infty}^{\infty} dx\, H_4(x)^2 e^{-x^2}$$

where $H_n(x)$ are the Hermite polynomials.

Answer: 680.622

15. Solve the differential equation

$$\frac{d^2y}{dx^2} = 2x + y + 3\frac{dy}{dx}$$

with $y(2) = 1$ and $y'(2) = -1$. Find $y(2.2)$ and also plot $y(x)$ with x ranging from 2.0 to 2.3.

Answers: `{{y -> InterpolatingFunction[{{2.1, 2.3}}, <>]}}`

`{0.851096}`

16. Solve the differential equation

$$\frac{d^2y}{dx^2} + \sin^2 x \frac{dy}{dx} + 3y^2 = e^{-x^2}$$

with $y(0) = 1$ and $y'(0) = 0$. Determine $y(1)$, $y(2)$, and $y(3)$ and also plot $y(x)$ with x in the range 0 to 3.

Answers: `{{y -> InterpolatingFunction[{{0., 3.}}, <>]}}`

`{{0.345833}, {-0.155487}, {-0.465167}}`

17. Solve the differential equation

$$\frac{d^2y}{dx^2} + 5x\frac{dy}{dx} - (1 - x^3)y = 0$$

with $y(0) = 0$ and $y'(0) = 1$. Determine $y(1.8)$ and also plot $y(x)$ with x ranging from 0 to 4.

18. Consider again Example 2.1.7 on the motion of a planet.
  (a) Plot the trajectory for the initial conditions $x(0) = 1$, $y(0) = 0$, $x'(0) = 0$, and $y'(0) = 8$. Since $x'(0) = 0$ and $y'(0) = 8$, should the orbit be a circle? Explain.
  (b) Plot the trajectory for $x(0) = 1$, $y(0) = 0$, $x'(0) = -2.5$, and $y'(0) = 8$. Should the foci of the ellipse be located on the x or y axis? Explain.
  (c) Find a set of initial conditions for which the orbit is a parabola. (*Hint:* Trial and error is too difficult; thinking is easier.)
  (d) Find a set of initial conditions for which the orbit is a hyperbola.

19. The equation of motion of the Van der Pol oscillator can be written as

$$\frac{d^2x}{dt^2} - \varepsilon(1 - x^2)\frac{dx}{dt} + x = 0$$

where ε is a positive parameter. This second-order differential equation can be reduced to two first-order equations:

$$\frac{dx}{dt} = v$$

$$\frac{dv}{dt} = \varepsilon(1 - x^2)v - x$$

Consider the case in which $\varepsilon = 1$. Solve these coupled differential equations numerically and plot the phase diagrams (i.e., plot $v(t)$ versus $x(t)$) with t ranging from 0 to 7π and for five sets of initial conditions: $v(0) = 0$ and $x(0)$ varies from 1 to 3 in steps of 0.5. Repeat with several other sets of randomly selected initial conditions. From these phase diagrams, infer the meaning of "limit cycle."

20. A uniform ring of radius a and total charge q is placed on the xy plane with its center at the origin. Determine the electric potential at the point $(2a, 3a, 5a)$, in units of $q/4\pi\varepsilon_0 a$, where ε_0 is the permittivity constant.

2.2 Symbolic Capabilities

Section 2.1 considered the numerical capabilities of *Mathematica*. *Mathematica* can work with variables or symbols without assigned values as well as with numbers. This section discusses its symbolic capabilities. Most concepts and techniques developed for numerical computations can be easily generalized to symbolic computations.

2.2.1 Transforming Algebraic Expressions

Mathematica has many functions for transforming algebraic expressions. Some of them are

`Apart[`*expr, var*`]`	rewrite a rational expression as a sum of terms with minimal denominators, treating all variables other than *var* as constants
`Cancel[`*expr*`]`	reduce a fraction to lowest terms
`Collect[`*expr, x*`]`	collect together terms involving the same power of x
`Expand[`*expr*`]`	distribute multiplications over additions
`ExpandAll[`*expr*`]`	apply `Expand` to all subexpressions
`ExpandDenominator[`*expr*`]`	apply `Expand` to the denominator
`ExpandNumerator[`*expr*`]`	apply `Expand` to the numerator
`Factor[`*expr*`]`	factor *expr* completely, that is, till no factor can be factored further
`FactorTerms[`*expr, x*`]`	pull out factors that do not depend on x
`Simplify[`*expr*`]`	return the simplest form by applying various standard algebraic transformations
`Together[`*expr*`]`	reduce a sum of fractions to a single fraction with the least common denominator as denominator

`ComplexExpand[`*expr*`]`	expand *expr* assuming that all variables are real
`PowerExpand[`*expr*`]`	expand nested powers, powers of products, logarithms of powers, and logarithms of products; that is, expand out $(ab)^c$, $(a^b)^c$, etc. (Use `PowerExpand` with caution because it is oblivious to issues of branches of multivalued functions.)

❋ **3.0**

`FullSimplify[`*expr*`]`	return the simplest form by applying a wide range of transformations involving elementary and special functions

---

Let us illustrate the use of these functions with several examples.

Using the function `Apart`, we can find the partial fraction decomposition of the rational expression

$$\frac{5x^2 - 4x + 16}{(x^2 - x + 1)^2(x - 3)}$$

In[1] := `Apart[(5 x^2 - 4 x + 16)/(((x^2 - x + 1)^2)(x - 3)), x]`

Out[1] $= \dfrac{1}{-3 + x} + \dfrac{-3 - 2 x}{(1 - x + x^2)^2} + \dfrac{-2 - x}{1 - x + x^2}$

Applying the function `Cancel` to the fraction

$$\frac{y^2 - 5y + 4}{y - 1}$$

reduces it to lowest terms:

In[2] := `Cancel[(y^2 - 5 y + 4)/(y - 1)]`

Out[2] $= -4 + y$

Let us apply `Expand`, `ExpandAll`, `ExpandDenominator`, and `ExpandNumerator` to

$$\frac{(x + 3)(x - 1)^2}{(x^2 + 1)(x + 5)^2}$$

and observe the differences among the results:

In[3] := `myexpr = ((x + 3)(x - 1)^2)/((x^2 + 1)(x + 5)^2)`

Out[3] $= \dfrac{(-1 + x)^2 \ (3 + x)}{(5 + x)^2 \ (1 + x^2)}$

In[4] := **Expand[myexpr]**

$$Out[4] = \frac{3}{(5 + x)^2\,(1 + x^2)} - \frac{5\,x}{(5 + x)^2\,(1 + x^2)} + \frac{x^2}{(5 + x)^2\,(1 + x^2)} +$$

$$\frac{x^3}{(5 + x)^2\,(1 + x^2)}$$

In[5] := **ExpandAll[myexpr]**

$$Out[5] = \frac{3}{25 + 10\,x + 26\,x^2 + 10\,x^3 + x^4} - \frac{5\,x}{25 + 10\,x + 26\,x^2 + 10\,x^3 + x^4} +$$

$$\frac{x^2}{25 + 10\,x + 26\,x^2 + 10\,x^3 + x^4} + \frac{x^3}{25 + 10\,x + 26\,x^2 + 10\,x^3 + x^4}$$

In[6] := **ExpandDenominator[myexpr]**

$$Out[6] = \frac{(-1 + x)^2\,(3 + x)}{25 + 10\,x + 26\,x^2 + 10\,x^3 + x^4}$$

In[7] := **ExpandNumerator[myexpr]**

$$Out[7] = \frac{3 - 5\,x + x^2 + x^3}{(5 + x)^2\,(1 + x^2)}$$

The results, which are equivalent algebraic expressions, have different forms; that is, the four expanding functions produce different effects.

In[8] := **Clear[myexpr]**

The function **Factor** factors

$$16y^2 - 25y^4$$

completely:

In[9] := **Factor[16 y^2 - 25 y^4]**

$$Out[9] = -(y^2\,(-4 + 5\,y)\,(4 + 5\,y))$$

Applying **Expand** to this result gives back the original expression:

In[10] := **Expand[%]**

$$Out[10] = 16\,y^2 - 25\,y^4$$

Finally, consider simplifying

$$\frac{1}{x^2 - 16} - \frac{x + 4}{x^2 - 3x - 4}$$

The obvious function to use is `Simplify`:

    In[11] := Simplify[1/(x^2 - 16) - (x + 4)/(x^2 - 3 x - 4)]

    Out[11] = $\dfrac{4 + x}{4 + 3\ x\ -\ x^2}\ +\ \dfrac{1}{-16\ +\ x^2}$

Contrary to expectation, `Simplify` does not always return the "simplest" form. Let us try the function `Together`:

    In[12] := Together[1/(x^2 - 16) - (x + 4)/(x^2 - 3 x - 4)]

    Out[12] = $\dfrac{-15\ -\ 7\ x\ -\ x^2}{(-4 + x)\ (1 + x)\ (4 + x)}$

One can argue that the result is still not in the "simplest" form because the denominator is factored. That is just a matter of preference. We can apply `ExpandDenominator` to the output:

    In[13] := ExpandDenominator[%]

    Out[13] = $\dfrac{-15\ -\ 7\ x\ -\ x^2}{-16\ -\ 16\ x\ +\ x^2\ +\ x^3}$

In general, to get an expression into a desired form, experiment with different combinations of transforming functions.

❦ **3.0**   `FullSimplify` can transform, slowly at times, some expressions that elude `Simplify`:

    In[14] := Simplify[BesselJ[-n, z] BesselJ[n - 1, z] +
                        BesselJ[-n + 1, z] BesselJ[n, z]]

    Out[14] = BesselJ[-1 + n, z] BesselJ[-n, z] +
              BesselJ[1 - n, z] BesselJ[n, z]

    In[15] := FullSimplify[BesselJ[-n, z] BesselJ[n - 1, z] +
                            BesselJ[-n + 1, z] BesselJ[n, z]]

    Out[15] = $\dfrac{2\ Sin[n\ Pi]}{Pi\ z}$

where `BesselJ[n, z]` is the Bessel function of the first kind $J_n(z)$.

2.2.2   *Transforming Trigonometric Expressions*

2.2.2.1   *MATHEMATICA* VERSION 3.0 SPECIFICS

With the exception of `Simplify` and `ComplexExpand`, the algebraic transformation functions introduced in Section 2.2.1 treat trigonometric functions as indivisible objects and leave them unchanged in algebraic manipulations. For instance, `Expand` does not rewrite the trigonometric expression $\sin(x)\cos(x)$ as $\sin(2x)/2$:

In[16] := **Expand[Sin[x] Cos[x]]**

Out[16] = Cos[x] Sin[x]

Expand left the trigonometric functions unaltered. However, **Simplify** transforms, for example, $\sin^2(x) + \cos^2(x)$ into 1:

In[17] := **Simplify[Sin[x]^2 + Cos[x]^2]**

Out[17] = 1

Also, **ComplexExpand** expands, for instance, $\cos(x + iy)$, assuming that x and y are real:

In[18] := **ComplexExpand[Cos[x + I y]]**

Out[18] = Cos[x] Cosh[y] - I Sin[x] Sinh[y]

Mathematica provides several other functions for manipulating trigonometric expressions:

TrigExpand[*expr*]	expand trigonometric expressions out into a sum of terms
TrigFactor[*expr*]	factor trigonometric expressions into products of terms
TrigReduce[*expr*]	apply trigonometric multiple angle identities
TrigToExp[*expr*]	write trigonometric functions in terms of exponentials
ExpToTrig[*expr*]	write exponentials in terms of trigonometric functions

Let us apply **TrigReduce** to the trigonometric expression $\sin^2(x) \cos(y) \sin(z)$:

In[19] := **TrigReduce[Sin[x]^2 Cos[y] Sin[z]]**

Out[19] = (Sin[2 x - y - z] - 2 Sin[y - z] + Sin[2 x + y - z] -

Sin[2 x - y + z] + 2 Sin[y + z] - Sin[2 x + y + z]) / 8

TrigReduce turns products and powers of trigonometric functions into sums of trigonometric functions with combined arguments. Now apply **TrigFactor** to the result generated by **TrigReduce**:

In[20] := **TrigFactor[%]**

Out[20] = Cos[y] Sin[x]² Sin[z]

We get back the original expression! In this case, **TrigFactor** is the inverse of **TrigReduce**.

Applying **TrigExpand**, for instance, to $\cos(x+y+z)$ gives a sum of products of trigonometric functions:

In[21] := **TrigExpand[Cos[x + y + z]]**

Out[21] = Cos[x] Cos[y] Cos[z] - Cos[z] Sin[x] Sin[y] -

Cos[y] Sin[x] Sin[z] - Cos[x] Sin[y] Sin[z]

Applying the function **ExpToTrig**, for example, to e^{iz} verifies Euler's formula:

In[22] := **ExpToTrig[Exp[I z]]**

Out[22] = Cos[z] + I Sin[z]

2.2.2.2   *MATHEMATICA* VERSION 2.2 SPECIFICS

With the exception of `Simplify` and `ComplexExpand`, the algebraic transformation functions introduced in Section 2.2.1 treat trigonometric functions as indivisible objects and leave them unchanged in algebraic manipulations. For instance, `Expand` does not rewrite the trigonometric expression $\sin(x)\cos(x)$ as $\sin(2x)/2$:

```
Expand[Sin[x] Cos[x]]
```

```
Cos[x] Sin[x]
```

`Expand` left the trigonometric functions unaltered. However, `Simplify` transforms, for example, $\sin^2(x) + \cos^2(x)$ into 1:

```
Simplify[Sin[x]^2 + Cos[x]^2]
```

```
1
```

Also, `ComplexExpand` expands, for instance, $\cos(x + iy)$, assuming that x and y are real:

```
ComplexExpand[Cos[x + I y]]
```

```
Cos[x] Cosh[y] - I Sin[x] Sinh[y]
```

We can empower the algebraic transformation functions to manipulate trigonometric expressions if "`Trig -> True`" is given as additional arguments, called options, in the functions. (There must not be any space between the "`-`" and "`>`" in the symbol "`->`".) The default setting is `Trig -> False` for all algebraic functions except `Simplify`, which has the default `Trig -> True`. Applying `Expand` with the additional argument to $\sin(x)\cos(x)$ now gives $\sin(2x)/2$:

```
Expand[Sin[x] Cos[x], Trig -> True]
```

$$\frac{\text{Sin}[2\ x]}{2}$$

The specification `Trig -> False` disables `Simplify`; it can no longer transform, for example, $\sin^2(x) + \cos^2(x)$ into 1:

```
Simplify[Sin[x]^2 + Cos[x]^2, Trig -> False]
```

$$\text{Cos}[x]^2 + \text{Sin}[x]^2$$

Let us apply `Expand` to the trigonometric expression $\sin^2(x)\cos(y)\sin(z)$:

```
Expand[Sin[x]^2 Cos[y] Sin[z], Trig -> True]
```

$$\frac{\text{Sin}[2\ x\ -\ y\ -\ z]}{8} - \frac{\text{Sin}[y\ -\ z]}{4} + \frac{\text{Sin}[2\ x\ +\ y\ -\ z]}{8} -$$
$$\frac{\text{Sin}[2\ x\ -\ y\ +\ z]}{8} + \frac{\text{Sin}[y\ +\ z]}{4} - \frac{\text{Sin}[2\ x\ +\ y\ +\ z]}{8}$$

`Expand` turns products and powers of trigonometric functions into sums of trigonometric functions with combined arguments. Now apply `Factor` to the result generated by `Expand`:

```
Factor[%, Trig -> True]
```

$$Cos[y] \, Sin[x]^2 \, Sin[z]$$

We get back the original expression!

When we specify `Trig -> True`, what does *Mathematica* actually do? With the inclusion of `Trig -> True` as the second argument of an algebraic transformation function, *Mathematica* no longer treats trigonometric functions as indivisible objects; instead, it writes them as rational functions of exponentials, that is, as quotients of polynomial functions of exponentials. It then performs the specified algebraic operations and finally combines the exponentials into trigonometric functions before returning the result.

The package `Algebra`Trigonometry`` contains three additional functions for trigonometric manipulations:

`TrigReduce[`*expr*`]`	apply trigonometric multiple angle identities
`TrigToComplex[`*expr*`]`	write trigonometric functions in terms of complex exponentials
`ComplexToTrig[`*expr*`]`	write complex exponentials as trigonometric functions

To use these functions, first load the package:

```
Needs["Algebra`Trigonometry`"]
```

Applying `TrigReduce`, for instance, to $\cos(x + y + z)$

```
TrigReduce[Cos[x + y + z]]
```

```
-(Sin[x] (Cos[z] Sin[y] + Cos[y] Sin[z])) +
   Cos[x] (Cos[y] Cos[z] - Sin[y] Sin[z])
```

and then `Expand` to the result

```
Expand[%]
```

```
Cos[x] Cos[y] Cos[z] - Cos[z] Sin[x] Sin[y] -
   Cos[y] Sin[x] Sin[z] - Cos[x] Sin[y] Sin[z]
```

gives a sum of products of trigonometric functions. Applying the function `ComplexToTrig`, for example, to e^{iz}

```
ComplexToTrig[Exp[I z]]
```

```
Cos[z] + I Sin[z]
```

verifies Euler's formula.

2.2.2.3   TRIGONOMETRIC IDENTITIES

There is a simple method for proving trigonometric identities: To prove the identity *lhs* == *rhs*, apply `Simplify` to the equivalent identity *lhs* - *rhs* == 0 and show that it returns `True`.

Let us prove several trigonometric identities:

(a) $(1 - \cot x)^2 + (1 - \tan x)^2 = (\sec x - \csc x)^2$

(b) $2 \csc 4x + 2 \cot 4x = \cot x - \tan x$

(c) $\tan \dfrac{x}{2} \cot \dfrac{y}{2} - \cot \dfrac{x}{2} \tan \dfrac{y}{2} = \dfrac{2(\cos y - \cos x)}{\sin x \sin y}$

```
In[23] := Simplify[
              (1 - Cot[x])^2 + (1 - Tan[x])^2 -
              (Sec[x] - Csc[x])^2 == 0
              ]
```

Out[23] = True

```
In[24] := Simplify[
              (2 Csc[4x] + 2 Cot[4x]) -
              (Cot[x] - Tan[x]) == 0
              ]
```

Out[24] = True

```
In[25] := Simplify[
              (Tan[x/2] Cot[y/2] - Cot[x/2] Tan[y/2]) -
              2(Cos[y] - Cos[x])/(Sin[x] Sin[y]) == 0
              ]
```

Out[25] = True

2.2.3   *Units, Conversion of Units, and Physical Constants*

Units can be tagged to numbers in calculations.

EXAMPLE 2.2.1 A race car accelerates along a straight line from 6.50 m/s at a rate of 5.00 m/s^2. What is the speed of the car after it has traveled 29.5 m?

An equation of kinematics is

$$v^2 = v_0^2 + 2ax$$

In this problem, $v_0 = 6.50$ m/s, $a = 5.00$ m/s^2, and $x = 29.5$ m. Thus, v equals

```
In[26] := N[Sqrt[(6.50 m/s)^2 + 2(5.00 m/s^2)(29.5 m)], 3]
```

Out[26] = $18.4 \; \text{Sqrt}[\dfrac{m^2}{s^2}]$

where N[*expr*, *n*] evaluates *expr* numerically with *n*-digit precision. *Mathematica* does not automatically convert Sqrt[z^2] to z nor Sqrt[a b] to Sqrt[a] Sqrt[b] because these conversions are certain to be correct only if z, a, and b are real and nonnegative. These conversions can be done, at our discretion, with the function PowerExpand:

In[27] := **PowerExpand[%]**

Out[27] = $\dfrac{18.4 \text{ m}}{\text{s}}$ ■

The package **Miscellaneous`Units`** contains the functions **Convert**, **ConvertTemperature**, and **SI** for conversion of units. To use these functions, we must first load the package:

In[28] := **Needs["Miscellaneous`Units`"]**

Convert[*old, new*] converts *old* to a form involving the combination of units *new*. The names of units must be spelled out in full. What follow are some unit names.

- Electrical Units
  Amp, Coulomb, Farad, Henry, Ohm, Statampere, Statcoulomb, Statfarad, Stathenry, Statohm, Statvolt, Volt

- Units of Length
  AU, Centimeter, Fathom, Feet, Fermi, Inch, LightYear, Meter, Micron, Mil, Mile, NauticalMile, Parsec, Yard

- Units of Time
  Day, Hour, Minute, Month, Second, Year

- Units of Mass
  AtomicMassUnit, Gram, Kilogram, MetricTon, Slug, TroyOunce

- Magnetic Units
  BohrMagneton, Gauss, Oersted, Tesla, Weber

- Units of Pressure
  Atmosphere, Bar, MillimeterMercury, Pascal, Torr

- Units of Energy
  BTU, Calorie, ElectronVolt, Erg, Joule, Rydberg, Therm

- Others
  Newton, Dyne, HorsePower, Watt, Liter, Curie, Rad

For instance, we can convert the density of water 1.00×10^3 kg/m^3 at STP to gram/cm^3:

In[29] := **Convert[1.00 10^3 Kilogram/Meter^3, Gram/Centimeter^3]**

Out[29] = $\dfrac{1. \text{ Gram}}{\text{Centimeter}^3}$

Prefixes can be given for units. They must be entered as separate words. Some prefixes used in the SI system are **Centi, Deci, Femto, Giga, Kilo, Mega, Micro, Milli, Nano,** and **Pico**. As an example, convert 55 mi/h to km/h:

In[30] := **N[Convert[55 Mile/Hour, Kilo Meter/Hour], 2]**

Out[30] = $\dfrac{89. \text{ Kilo Meter}}{\text{Hour}}$

Temperature conversions require `ConvertTemperature`[*temp, old, new*] that converts *temp* from the *old* scale to the *new* scale. The function `Convert` cannot be used because temperature conversions are not multiplicative. Some temperature scales are `Celsius`, `Fahrenheit`, and `Kelvin`. Let us, for example, convert 50°F to Celsius:

In[31] := `ConvertTemperature[50, Fahrenheit, Celsius]`

Out[31] = 10.

`SI`[*expr*] converts *expr* to the SI system. As an example, a superconducting magnet generates a magnetic field of 3×10^5 G. What is the field in the SI system?

In[32] := `SI[3 10^5 Gauss]`

Out[32] = 30 Tesla

The package `Miscellaneous`PhysicalConstants`` contains the values of many physical constants. Again, full names of the constants must be invoked. Here are some available constants:

```
AccelerationDueToGravity
AgeOfUniverse
AvogadroConstant
BohrRadius
BoltzmannConstant
EarthMass
EarthRadius
ElectronCharge
ElectronMass
FineStructureConstant
GravitationalConstant
HubbleConstant
PlanckConstant
PlanckConstantReduced
ProtonMass
RydbergConstant
SpeedOfLight
```

After loading the package

In[33] := `Needs["Miscellaneous`PhysicalConstants`"]`

we can determine, for example, the rest energy of an electron:

In[34] := `ElectronMass SpeedOfLight^2`

$$Out[34] = \frac{8.18711 \ 10^{-14} \ \text{Kilogram Meter}^2}{\text{Second}^2}$$

2.2.4 Assignments and Transformation Rules

We can assign symbolic as well as numerical values to variables. To make an assignment, enter

variable = *value*

For example, let us expand $(a + b)(a - b)$ and assign the value to the variable t:

In[35] := **t = (a + b)(a - b)//Expand**

Out[35] = $a^2 - b^2$

Now, t has the value $a^2 - b^2$. Thus, adding b^2 to t gives a^2:

In[36] := **t + b^2**

Out[36] = a^2

Names of variables can consist of any number of alphanumeric characters. *Mathematica* version 3.0 (alas, not version 2.2) permits the use of Greek letters, subscripts, and other letter-like forms to be discussed in Section 2.5. The only restriction is that a name cannot begin with a number. For instance, **3x** is not an acceptable name (**3x** stands for **3** times **x**), whereas **x3** is. By convention, user-defined variables should begin with lowercase letters, because names of built-in *Mathematica* objects start with uppercase letters.

When new values are assigned to variables, old values are discarded. For an example, assign the value d^2 to t in the previous example and then add b^2 to it:

In[37] := **t = d^2**

Out[37] = d^2

In[38] := **t + b^2**

Out[38] = $b^2 + d^2$

The answer for $t + b^2$ is no longer a^2, since t now has the value d^2.

Assignments remain in a *Mathematica* session until the values are changed or removed. Forgetting that certain variables were assigned values earlier in a session and proceeding to use them as formal variables is a common mistake. Let us, for example, solve numerically the differential equation

$$\frac{d^2x}{dt^2} - \frac{dx}{dt} - x = 0$$

with $x = 1$ and $dx/dt = 0$ at $t = 0$:

In[39] := **NDSolve[{x''[t] + x'[t] + x[t] == 0,**
                  x'[0] == 0, x[0] == 1}, x, {t, 0, 5}]

   General::ivar: d^2 is not a valid variable.

   NDSolve::dsvar: d^2 cannot be used as a variable.

Out[39] = NDSolve[{x[d^2] + x'[d^2] + x''[d^2] == 0, x'[0] == 0, x[0] == 1}, x,
                  {d^2, 0, 5}]

We get an error message! What happens is that t was assigned the value d^2 in the previous example.

To avoid the mistake of forgetting previous assignments, we should develop the habit of removing unneeded values as soon as possible with either the deassignment operator =. or the function `Clear`. Let us apply `Clear` to `t` and solve the differential equation again:

In[40] := **Clear[t]**

In[41] := **NDSolve[{x''[t] + x'[t] + x[t] == 0,**
            x'[0] == 0, x[0] == 1}, x, {t, 0, 5}]

Out[41] = {{x -> InterpolatingFunction[{{0., 5.}}, <>]}}

The equation is solved!

Clearing the declared values as they are no longer needed is definitely a good habit. Should we also routinely apply at the beginning of a calculation the function `Clear` to all the variables to be used later just in case they possess preassigned values? That is a good practice if we know ahead of time the names of the variables. Unlike other languages such as C and Pascal, *Mathematica* does not require commencing calculations with variable declarations. Therefore, we often do not know all the variable names till the end of the computation. Beginning a new *Mathematica* session, like starting with a clean slate, for each problem is perhaps a better, albeit time-consuming practice. (An alternative, almost as good as starting with a clean slate, is to execute the command `Clear["Global`*"]`, which clears the values for all symbols in the `Global`` context, except those that are protected. Section 3.3.5 will discuss symbol protection and Section 3.7.1 will introduce *Mathematica* contexts.)

A simple way to avoid forgetting earlier assignments is by refraining from making assignments whenever possible. Replacements of variables with values can be made with transformation rules. A transformation rule is written as *variable -> value*, with no space between "`-`" and "`>`". The transformation rule is applied to an expression with the replacement operator `/.`, with no space between the slash "`/`" and the dot "`.`". To make a replacement, write

   expr `/.` *variable* `->` *value*

We can read "`/.`" as "given that" and "`->`" as "goes to." For example, "`2x + 10 /. x -> 10`" is interpreted as `2x + 10` given that `x` goes to `10`:

In[42] := **2x + 10 /. x -> 10**

Out[42] = 30

The advantage of a replacement over an assignment is that it only affects the specific expression. The variable `x` in the preceding example can still be used as a formal variable. Let us ask for `x`:

In[43] := **x**

Out[43] = x

Indeed, `x` has no global value.

We can also replace a subpart of an expression with a value by writing

   expr `/.` *lhs* `->` *rhs*

where the value *rhs* is substituted for the subpart *lhs* in the expression *expr*. For example, we can determine the rest energy of an electron in Joules. Since the package `Miscellaneous`PhysicalConstants`` was already loaded in Section 2.2.3, we have, for the rest energy of an electron,

In[44] := **ElectronMass SpeedOfLight^2**

$$Out[44] = \frac{8.18711\ 10^{-14}\ \text{Kilogram Meter}^2}{\text{Second}^2}$$

In[45] := **% /. Kilogram Meter^2/Second^2 -> Joule**

Out[45] = $8.18711\ 10^{-14}$ Joule

where the subexpression

$$\frac{\text{Kilogram Meter}^2}{\text{Second}^2}$$

is replaced by Joule.

2.2.5 Equation Solving

Section 2.1.15 discussed solving equations numerically. This section considers exact solution of equations. `Solve[`*lhs* == *rhs*, *var*`]` attempts to solve exactly an equation for the variable *var*. For example, consider solving the equation

$$x^2 - 3x - 10 = 0$$

In[46] := **Solve[x^2 - 3 x - 10 == 0, x]**

Out[46] = {{x -> -2}, {x -> 5}}

`Solve` returns a nested list of transformation rules stating that the equation is satisfied if x is replaced by -2 or 5. We can verify the solutions with the replacement operator `/.` (again, with no space between the slash and the dot):

In[47] := **x^2 - 3 x - 10 == 0 /. %**

Out[47] = {True, True}

Mathematica reports that the equation is true if x is replaced by -2 or 5. To obtain a list of the solutions, apply the transformation rules to **x** in the form

In[48] := **x /. %%**

Out[48] = {-2, 5}

Though `Solve` is intended primarily for solving polynomial equations, it can solve some other equations. As an example, consider a one-dimensional relativistic particle of rest mass m moving at a velocity v. Its momentum is

$$p = \frac{mv}{\sqrt{1 - v^2/c^2}}$$

where c is the speed of light. Let us solve for v:

In[49] := **Solve[p == (m v)/Sqrt[1 - (v^2/c^2)], v]**

Out[49] = $\{\{v \rightarrow -(\dfrac{c\ p}{\text{Sqrt}[c^2\ m^2 + p^2]})\}, \{v \rightarrow \dfrac{c\ p}{\text{Sqrt}[c^2\ m^2 + p^2]}\}\}$

The first solution is, of course, physically unacceptable because it implies that the velocity and momentum have opposite directions. For another example, solve the equation

$$9\cos^2 x + 8\cos x = 1$$

In[50] := **Solve[9 Cos[x]^2 + 8 Cos[x] == 1, x]**

Solve::ifun: Inverse functions are being used by Solve, so some
     solutions may not be found.

Out[50] = $\{\{x \rightarrow -Pi\}, \{x \rightarrow Pi\}, \{x \rightarrow -\text{ArcCos}[\frac{1}{9}]\}, \{x \rightarrow \text{ArcCos}[\frac{1}{9}]\}\}$

The warning message alerts us to the fact that for every solution to an equation of the form $\cos(x) = a$, there are an infinite number of solutions differing by multiples of 2π.

Mathematically, complicated equations such as transcendental equations and polynomial equations with degrees higher than 4 do not always have exact solutions. Consider, for example, the transcendental equation

$$-5 + x^2 e^{-x} + x = 0$$

In[51] := **Solve[-5 + x^2 Exp[-x] + x == 0, x]**

Solve::tdep: The equations appear to involve transcendental functions
     of the variables in an essentially non-algebraic way.

Out[51] = $\text{Solve}[-5 + x + \dfrac{x^2}{E^x} == 0, x]$

Mathematica cannot find exact solutions for the equation and returns the input unevaluated. For another example, consider solving the polynomial equation

$$x^5 - 3x^4 + 2x^3 + x = 4$$

In[52] := **Solve[x^5 - 3x^4 + 2x^3 + x == 4, x]**

Out[52] = $\{\{x \rightarrow \text{Root}[-4 + \#1 + 2\ \#1^3 - 3\ \#1^4 + \#1^5\ \&,\ 1]\},$
              $\{x \rightarrow \text{Root}[-4 + \#1 + 2\ \#1^3 - 3\ \#1^4 + \#1^5\ \&,\ 2]\},$
              $\{x \rightarrow \text{Root}[-4 + \#1 + 2\ \#1^3 - 3\ \#1^4 + \#1^5\ \&,\ 3]\},$
              $\{x \rightarrow \text{Root}[-4 + \#1 + 2\ \#1^3 - 3\ \#1^4 + \#1^5\ \&,\ 4]\},$
              $\{x \rightarrow \text{Root}[-4 + \#1 + 2\ \#1^3 - 3\ \#1^4 + \#1^5\ \&,\ 5]\}\}$

(The output has a different appearance in *Mathematica* version 2.2.) The result indicates that **Solve** fails to produce explicit solutions. We can, of course, determine approximate solutions to the equation numerically:

In[53] := **%//N**

Out[53] = {{x -> 2.15819}, {x -> -0.637669 - 0.748825 I},

{x -> -0.637669 + 0.748825 I}, {x -> 1.05857 - 0.89183 I},

{x -> 1.05857 + 0.89183 I}}

Solve[{*lhs1* == *rhs1*, *lhs2* == *rhs2*, ...}, {*var1*, *var2*, ...}**]** attempts to solve a set of equations exactly for the variables *var1*, *var2*, etc. For instance, we can solve the system of equations

$$x + y = z$$

$$10 - 6x - 2z = 0$$

$$6x - 24 - 4y = 0$$

In[54] := **Solve[**{x + y == z, 10 - 6x - 2z == 0,

6x - 24 - 4y == 0},

{x, y, z}**]**

Out[54] = {{x -> 2, y -> -3, z -> -1}}

We conclude this section with a problem from direct current circuits.

EXAMPLE 2.2.2   For the circuit in Figure 2.2.1, find all currents in terms of the resistances and the emf. Assume that the internal resistance of the battery is negligible.

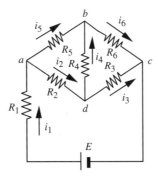

Figure 2.2.1   A multiloop circuit

We have arbitrarily assigned directions to the currents. If we assumed a wrong direction for a current, its resulting value will be negative but the magnitude will be correct. Kirchhoff's rules applied to the circuit give

junction a : $i_1 = i_2 + i_5$

junction b : $i_6 = i_4 + i_5$

junction d : $i_2 = i_3 + i_4$

loop *abda* : $-i_5 R_5 + i_4 R_4 + i_2 R_2 = 0$

loop *dbcd* : $-i_4 R_4 - i_6 R_6 + i_3 R_3 = 0$

loop *adca* : $-i_2 R_2 - i_3 R_3 + E - i_1 R_1 = 0$

Since **E** is a built-in *Mathematica* object representing the exponential constant *e*, let **emf** be the alias of *E*. **Solve** solves the system of linear equations:

```
In[55] := Timing[
            Simplify[
            sol = Solve[
                    {i1 == i2 + i5,
                     i6 == i4 + i5,
                     i2 == i3 + i4,
                     -i5 R5 + i4 R4 + i2 R2 == 0,
                     -i4 R4 - i6 R6 + i3 R3 == 0,
                     -i2 R2 - i3 R3 + emf - i1 R1 == 0
                    },
                    {i1, i2, i3, i4, i5, i6}
                    ]
                ]
            ]


{226.367 Second, {{i1 ->
    (emf (R4 R5 + R3 (R4 + R5) + R4 R6 + R5 R6 + R2 (R3 + R4 + R6))) /
    (R1 (R4 R5 + R3 (R4 + R5) + R4 R6 + R5 R6 + R2 (R3 + R4 + R6)) +
     R2 (R4 R5 + R4 R6 + R5 R6 + R3 (R5 + R6)) + R3 (R5 R6 + R4
     (R5 + R6)))
    , i6 -> (emf (R2 (R3 + R4) + R3 (R4 + R5))) /
    (R1 (R4 R5 + R3 (R4 + R5) + R4 R6 + R5 R6 + R2 (R3 + R4 + R6)) +
     R2 (R4 R5 + R4 R6 + R5 R6 + R3 (R5 + R6)) + R3 (R5 R6 + R4
     (R5 + R6)))
    , i5 -> (emf (R3 R4 + R2 (R3 + R4 + R6))) /
    (R1 (R4 R5 + R3 (R4 + R5) + R4 R6 + R5 R6 + R2 (R3 + R4 + R6)) +
     R2 (R4 R5 + R4 R6 + R5 R6 + R3 (R5 + R6)) + R3 (R5 R6 + R4
     (R5 + R6)))
    , i2 -> (emf (R3 R5 + R5 R6 + R4 (R5 + R6))) /
    (R1 (R4 R5 + R3 (R4 + R5) + R4 R6 + R5 R6 + R2 (R3 + R4 + R6)) +
     R2 (R4 R5 + R4 R6 + R5 R6 + R3 (R5 + R6)) + R3 (R5 R6 + R4
     (R5 + R6)))
    , i3 -> (emf ((R2 + R5) R6 + R4 (R5 + R6))) /
    (R1 (R4 R5 + R3 (R4 + R5) + R4 R6 + R5 R6 + R2 (R3 + R4 + R6)) +
     R2 (R4 R5 + R4 R6 + R5 R6 + R3 (R5 + R6)) + R3 (R5 R6 + R4
     (R5 + R6)))
    , i4 -> (emf (R3 R5 - R2 R6)) /
    (R1 (R4 R5 + R3 (R4 + R5) + R4 R6 + R5 R6 + R2 (R3 + R4 + R6)) +
     R2 (R4 R5 + R4 R6 + R5 R6 + R3 (R5 + R6)) + R3 (R5 R6 + R4
     (R5 + R6)))
    }}}
```

`Timing[expr]` evaluates *expr* and returns a list of time used, together with the result obtained. The evaluation takes 227 seconds on a Macintosh IIci and 31 seconds on a Gateway 2000 4DX2-66V. *Mathematica* spends most of the time in simplifying rather than finding the solution. Finding the solution takes only 2.5 seconds on a Macintosh IIci and 0.28 second on a Gateway 2000 4DX2-66V. For this book, all timings are performed with *Mathematica* version 2.2.

Checking answers with special cases is a good practice because computer algebra systems are known to have bugs and because we can gain insight into the problems. Let us consider the case where $R_2 = R_3 = R_5 = R_6 = R$. Symmetry of the circuit requires that $i_4 = 0$ and $i_2 = i_3 = i_5 = i_6 = i_1/2$:

$In[56] := $ `Simplify[sol /. {R2 -> R, R3 -> R, R5 -> R, R6 -> R}]`

$Out[56] = \left\{\left\{ i1 \to \dfrac{emf}{R + R1}, \quad i6 \to \dfrac{emf}{2\ (R + R1)}, \quad i3 \to \dfrac{emf}{2\ (R + R1)}, \right.\right.$

$\left.\left. i2 \to \dfrac{emf}{2\ (R + R1)}, \quad i4 \to 0, \quad i5 \to \dfrac{emf}{2\ (R + R1)} \right\}\right\}$

Indeed, it is so.

$In[57] := $ `Clear[sol]`   ∎

2.2.6 Differentiation

`D[f, x]` gives

$$\frac{\partial f}{\partial x}$$

the partial derivative of f with respect to x. It simply gives the derivative of f with respect to x if f is a function of x only. For instance, we can differentiate the function

$$\frac{\cos 4x}{1 - \sin 4x}$$

simplify the result, and evaluate it at, say, $x = 0$:

$In[58] := $ `D[Cos[4x]/(1 - Sin[4x]), x]`

$Out[58] = \dfrac{4\ Cos[4\ x]^2}{(1\ -\ Sin[4\ x])^2} - \dfrac{4\ Sin[4\ x]}{1\ -\ Sin[4\ x]}$

$In[59] := $ `Simplify[%]`

$Out[59] = \dfrac{-4}{-1\ +\ Sin[4\ x]}$

$In[60] := $ `% /. x -> 0`

$Out[60] = 4$

`D[f, {x, n}]` gives

$$\frac{\partial^n f}{\partial x^n}$$

the nth partial derivative of f with respect to x. For example, determine the fourth derivative of the function

$$4x^2 - 5x + 8 - \frac{3}{x}$$

```
In[61] := D[4x^2 - 5x + 8 -3/x, {x, 4}]
```
$$Out[61] = \frac{-72}{x^5}$$

`D[f, x₁, x₂, ... ]` gives

$$\frac{\partial}{\partial x_1} \frac{\partial}{\partial x_2} \cdots f$$

the multiple partial derivative of f with respect to $x_1, x_2, \ldots$. Let us find the second partial derivative

$$\frac{\partial^2 f}{\partial x \partial y}$$

for

$$f(x, y) = x^3 y^2 - 2x^2 y + 3x$$

```
In[62] := D[x^3 y^2 - 2 x^2 y + 3 x, x, y]
```

$$Out[62] = -4 x + 6 x^2 y$$

2.2.7   *Integration*

`Integrate[f, x]` gives the indefinite integral

$$\int f\, dx$$

For example, we can evaluate

$$\int \frac{x^4\, dx}{a^2 + x^2}$$

```
In[63] := Integrate[x^4/(a^2 + x^2), x]
```

$$Out[63] = -(a^2\ x) + \frac{x^3}{3} + a^3\ \mathrm{ArcTan}[\frac{x}{a}]$$

Note that *Mathematica* does not include an arbitrary constant of integration in the result. Differentiating the result should, of course, recover the integrand:

```
In[64] := D[%, x]
```

$$Out[64] = -a^2 + x^2 + \frac{a^2}{1 + \frac{x^2}{a^2}}$$

To show that this expression is indeed the integrand, use the function Simplify:

$In[65] :=$ **Simplify[%]**

$Out[65] = \dfrac{x^4}{a^2 + x^2}$

Integrate$[f, \{x, xmin, xmax\}]$ gives the definite integral

$$\int_{xmin}^{xmax} f\, dx$$

Let us evaluate

$$\int_1^4 (5x - 2\sqrt{x} + \frac{32}{x^3})\, dx$$

$In[66] :=$ **Integrate[(5 x - 2 Sqrt[x] + 32/x^3), {x, 1, 4}]**

$Out[66] = \dfrac{259}{6}$

There are integrals that *Mathematica* cannot do. For instance, it does not know how to evaluate the integral

$$\int_0^1 \tan(\cos x)\, dx$$

$In[67] :=$ **Integrate[Tan[Cos[x]], {x, 0, 1}]**

$Out[67] =$ Integrate[Tan[Cos[x]], {x, 0, 1}]

Mathematica returns the input unevaluated. We can, of course, obtain an approximate value for the integral:

$In[68] :=$ **N[%]**

$Out[68] = 1.16499$

Integrate$[f, \{x, xmin, xmax\}, \{y, ymin, ymax\}]$ gives the multiple integral

$$\int_{xmin}^{xmax} dx \int_{ymin}^{ymax} dy f$$

For an example, evaluate

$$\int_0^1 dx \int_0^{2-2x} dy(x^2 + y^2 + 1)$$

$In[69] :=$ **Integrate[x^2 + y^2 + 1, {x, 0, 1}, {y, 0, 2 - 2 x}]**

$Out[69] = \dfrac{11}{6}$

We end this section with an example from quantum physics.

EXAMPLE 2.2.3 Planck's blackbody radiation (also known as cavity radiation) formula is

$$R(\lambda, T) = \frac{2\pi hc^2}{\lambda^5} \frac{1}{e^{hc/\lambda kT} - 1}$$

where $R(\lambda, T)$ is the spectral radiancy as a function of wavelength λ and temperature T, k is Boltzmann's constant, c is the speed of light, and h is Planck's constant. Derive the Stefan-Boltzmann law that relates the radiant intensity $R(T)$ and temperature T by

$$R(T) = \sigma T^4$$

in which

$$\sigma = \frac{2\pi^5 k^4}{15 c^2 h^3}$$

is called the Stefan-Boltzmann constant.

$R(\lambda, T) d\lambda$ is defined as the energy radiated with wavelengths in the interval λ to $\lambda + d\lambda$ per unit time from unit area of the surface. Thus, the radiant intensity $R(T)$, which is the total energy radiated at all wavelengths per unit time from unit area, is given by

$$R(T) = \int_0^\infty R(\lambda, T) \, d\lambda$$

● *Mathematica* **Version 2.2 Specifics**
Let **x** be the alias of λ. **Integrate** evaluates the integral and validates the Stefan-Boltzmann law:

```
Integrate[(2 Pi h c^2)/((x^5)(Exp[(h c)/(x k T)] - 1)),
        {x, 0, Infinity}]
```

$$\frac{2 \; k^4 \; Pi^5 \; T^4}{15 \; c^2 \; h^3}$$

● *Mathematica* **Version 3.0 Specifics**
Let **x** be the alias of λ. **Integrate** evaluates the integral and validates the Stefan-Boltzmann law:

```
In[70] := Integrate[(2Pi h c^2)/((x^5)(Exp[(h c)/(x k T)] - 1)),
              {x, 0, Infinity}]
```

$$Out[70] = If[Re[\frac{c \; h}{k \; T}] > 0, \; \frac{2 \; k^4 \; Pi^5 \; T^4}{15 \; c^2 \; h^3},$$

$$Integrate[\frac{2 \; c^2 \; h \; Pi}{(-1 + E^{(c \; h)/(k \; T \; x)}) \; x^5}, \; \{x, \; 0, \; \infty\}]]$$

where the output indicates that the result is

$$\frac{2 \; k^4 \; Pi^5 \; T^4}{15 \; c^2 \; h^3}$$

if $\mathrm{Re}(ch/kT) > 0$ is true, and that *Mathematica* cannot do the integral if it is false. For a definite integral, we can include a trailing argument, called an option, in the form **Assumptions -> {**relation1, relation2, …**}** to declare assumptions about the parameters:

```
In[71] := Integrate[(2Pi h c^2)/((x^5)(Exp[(h c)/(x k T)] - 1)),
                     {x, 0, Infinity}, Assumptions -> {Re[(c h)/(k T)] > 0}]
```

$$Out[71] = \frac{2\ k^4\ \mathrm{Pi}^5\ T^4}{15\ c^2\ h^3}$$

Mathematica returns the result that is consistent with the specified relation. (Section 2.3.1.2 will discuss *Mathematica* options.) ∎

2.2.8 Sums

Sum[*f*, {*i*, *imin*, *imax*}] evaluates the sum

$$\sum_{i=imin}^{imax} f$$

The second argument of the function **Sum** is an example of iterators introduced in Section 2.1.18. Let us, for example, generate the sum

$$1 + \frac{1}{x} + \frac{1}{x^2} + \frac{1}{x^3} + \frac{1}{x^4} + \frac{1}{x^5} + \frac{1}{x^6}$$

```
In[72] := Sum[1/x^i, {i, 0, 6}]
```

$$Out[72] = 1 + x^{-6} + x^{-5} + x^{-4} + x^{-3} + x^{-2} + \frac{1}{x}$$

We can also compute the sum

$$1 + \frac{1}{2^4} + \frac{1}{3^4} + \frac{1}{4^4} + \frac{1}{5^4} + \frac{1}{6^4} + \frac{1}{7^4}$$

```
In[73] := Sum[1/i^4, {i, 1, 7}]
```

$$Out[73] = \frac{33654237761}{31116960000}$$

The result is exact. To determine the approximate numerical value, use the function **N**:

```
In[74] := N[%]
```

$$Out[74] = 1.08154$$

Sum[*f*, {*i*, *imin*, *imax*, *di*}] evaluates the sum with *i* increasing in steps of *di*. To generate the sum

$$1 + \frac{1}{x^2} + \frac{1}{x^4} + \frac{1}{x^6} + \frac{1}{x^8} + \frac{1}{x^{10}}$$

we enter

```
In[75] := Sum[1/x^i, {i, 0, 10, 2}]
```

$$Out[75] = 1 + x^{-10} + x^{-8} + x^{-6} + x^{-4} + x^{-2}$$

Here, **i** increases in steps of 2.

Sum[*f*, {*i*, *imin*, *imax*}, {*j*, *jmin*, *jmax*}] evaluates the nested sum

$$\sum_{i=imin}^{imax} \sum_{j=jmin}^{jmax} f$$

For instance, we can evaluate the sum

$$\sum_{i=1}^{3} \sum_{j=1}^{i} \frac{1}{x^i + y^j}$$

```
In[76] := Sum[1/(x^i + y^j), {i, 1, 3}, {j, 1, i}]
```

$$Out[76] = \frac{1}{x + y} + \frac{1}{x^2 + y} + \frac{1}{x^3 + y} + \frac{1}{x^2 + y^2} + \frac{1}{x^3 + y^2} + \frac{1}{x^3 + y^3}$$

2.2.8.1   *MATHEMATICA* VERSION 2.2 SPECIFICS

`SymbolicSum[f, {i, n}]` in the package `Algebra`SymbolicSum`` performs, if possible, the symbolic or indefinite sum

$$\sum_{i=1}^{n} f$$

For an example, determine the indefinite sum

$$1^3 + 2^3 + 3^3 + 4^3 + 5^3 + \ldots + n^3$$

After loading the package

```
Needs["Algebra`SymbolicSum`"]
```

we have

```
SymbolicSum[i^3, {i, n}]
```

$$\frac{n^2 (1 + n)^2}{4}$$

2.2.8.2   *MATHEMATICA* VERSION 3.0 SPECIFICS

`Sum[f, {i, n}]` performs, if possible, the symbolic or indefinite sum

$$\sum_{i=1}^{n} f$$

For example, consider the indefinite sum

$$1^3 + 2^3 + 3^3 + 4^3 + 5^3 + \ldots + n^3$$

```
In[77] := Sum[i^3, {i, n}]
```

$$Out[77] = \frac{n^2 (1 + n)^2}{4}$$

2.2.9   Power Series

`Series[f, {x, x_0, n}]` finds a power series representation for f in $(x - x_0)$ to order $(x - x_0)^n$. For example, we can determine the power series expansion about $x = 0$ for

$$e^{\sin u}$$

to order u^7:

$In[78] := $ **Series[Exp[Sin[u]], {u, 0, 7}]**

$Out[78] = 1 + u + \frac{u^2}{2} - \frac{u^4}{8} - \frac{u^5}{15} - \frac{u^6}{240} + \frac{u^7}{90} + O[u]^8$

$O[u]^n$ stands for a term of order u^n.

Normal[*series*] converts a power series to an ordinary expression, that is, removes the term $O[u]^n$. Let us approximate the integral

$$\int_0^{0.3} (1 + x^4)^{\frac{1}{3}} dx$$

First, expand the integrand in a series about $x = 0$ up to order x^8:

$In[79] := $ **Series[(1 + x^4)^(1/3), {x, 0, 8}]**

$Out[79] = 1 + \frac{x^4}{3} - \frac{x^8}{9} + O[x]^9$

Second, remove the term $O[x]^9$:

$In[80] := $ **Normal[%]**

$Out[80] = 1 + \frac{x^4}{3} - \frac{x^8}{9}$

Finally, integrate the terms of the series:

$In[81] := $ **Integrate[%, {x, 0, 0.3}]**

$Out[81] = 0.300162$

Let us compare the result with that from numerical integration:

$In[82] := $ **NIntegrate[(1 + x^4)^(1/3), {x, 0, 0.3}]**

$Out[82] = 0.300162$

The results agree to six decimal places.

We conclude this section with an example from introductory physics.

EXAMPLE 2.2.4 The electric potential V on the axis of a uniformly charged disk of radius r and charge density g is given as

$$V = kg\pi \int_0^r \frac{2a\,da}{\sqrt{x^2 + a^2}}$$

where x is the distance from the disk along the axis and k is the Coulomb constant (see [Tip91]). Determine V, and verify that it approaches the electric potential produced by a point charge in the limit $x \gg r$.

The electric potential V is

$In[83] := $ **k g Pi Integrate[2a/Sqrt[x^2 + a^2], {a, 0, r}]//PowerExpand**

$Out[83] = g\ k\ Pi\ (-2\ x + 2\ Sqrt[r^2 + x^2])$

To show that V approaches the electric potential produced by a point charge in the limit $x \gg r$, we make the substitution $x = r/y$:

```
In[84] := % /. x -> r/y
```

$$Out[84] = g \ k \ Pi \ (2 \ Sqrt[r^2 + \frac{r^2}{y^2}] - \frac{2 \ r}{y})$$

Since $y \ll 1$, expand V to the first power of y:

```
In[85] := Series[%, {y, 0, 1}]//PowerExpand
```

$$Out[85] = g \ k \ Pi \ r \ y + O[y]^2$$

To express V as a function of x, apply the function **Normal** to the series and then replace y with r/x:

```
In[86] := (% // Normal) /. y -> r/x
```

$$Out[86] = \frac{g \ k \ Pi \ r^2}{x}$$

The factor $g\pi r^2$ is just the total charge q of the disk. Thus, V equals

```
In[87] := % /. g Pi r^2 -> q
```

$$Out[87] = \frac{k \ q}{x}$$

That is the electric potential produced by a point charge of charge q.   ∎

2.2.10   Limits

Limit[f(x), x -> a] determines the limit of $f(x)$ as x approaches a. For example, find

$$\lim_{x \to \infty} \frac{\sqrt{9x^2 + 2}}{3 - 4x}$$

```
In[88] := Limit[Sqrt[9 x^2 + 2]/(3 - 4 x), x -> Infinity]
```

$$Out[88] = -(\frac{3}{4})$$

Limit[f(x), x -> a, Direction -> -1] determines the right-hand limit of $f(x)$ as x approaches a. We can find

$$\lim_{x \to 3^+} \frac{2x^2}{9 - x^2}$$

```
In[89] := Limit[2 x^2/(9 - x^2), x -> 3, Direction -> -1]
```

$$Out[89] = -\infty$$

Limit[f(x), x -> a, Direction -> 1] determines the left-hand limit of $f(x)$ as x approaches a. We can also find

$$\lim_{x \to 3^-} \frac{2x^2}{9 - x^2}$$

In[90] := **Limit[2 x^2/(9 - x^2), x -> 3, Direction -> 1]**

Out[90] = ∞

The last two results show that the right- and left-hand limits may be different for a discontinuous function.

2.2.11 *Solving Differential Equations*

DSolve[*eqn*, *y[x]*, *x*] solves the differential equation *eqn* for the function *y[x]*, with independent variable *x*. For example, solve

$$\frac{d^2y}{dx^2} - 3\frac{dy}{dx} - 18y = xe^{4x}$$

In[91] := **DSolve[y''[x] - 3 y'[x] - 18 y[x] == x Exp[4 x],**
            y[x], x]

Out[91] = $\{\{y[x] \rightarrow \frac{E^{4\ x}\ (-(\frac{1}{4}) - \frac{x}{2})}{9} - \frac{E^{4\ x}\ (-(\frac{1}{49}) + \frac{x}{7})}{9} + \frac{C[1]}{E^3\ x} + E^6\ ^x\ C[2]\}\}$

C[1] and **C[2]** are the arbitrary constants.

  For another example, determine *x(t)*, the position at time *t*, for a damped harmonic oscillator with the equation of motion

$$\frac{d^2x}{dt^2} = -\omega_0^2 x - \gamma\frac{dx}{dt}$$

In Section 2.1.17, we obtained a numerical solution for *x(t)*. Here, we seek the symbolic solution. While *Mathematica* version 3.0 permits Greek letters and subscripts for input, version 2.2 does not. Let **wo** and **r** be aliases for ω_0 and γ, respectively.

In[92] := **DSolve[x''[t] + r x'[t] + wo^2 x[t] == 0, x[t], t]**

Out[92] = $\{\{x[t] \rightarrow E^{(t\ (-r\ -\ Sqrt[r^2\ -\ 4\ wo^2]))/2}\ C[1] +$
            $E^{(t\ (-r\ +\ Sqrt[r^2\ -\ 4\ wo^2]))/2}\ C[2]\}\}$

Again, **C[1]** and **C[2]** are the arbitrary constants. (Section 2.5 will discuss the use of Greek letters and subscripts in *Mathematica* version 3.0. For more information on the damped harmonic oscillator, refer to [MT95].)

  DSolve[{*eqn1*, *eqn2*, ...}, {*y1[x1*, ...], ...}, {*x1*, ...}] solves the list of differential equations {*eqn1*, *eqn2*, ...} for the functions *y1[x1*, ...], ..., with independent variables *x1*,.... Boundary conditions may be included in the list of equations. Let us solve

$$\frac{d^2y}{dx^2} + 8\frac{dy}{dx} + 16y = 0$$

with *y* = 2 and *dy/dx* = 1 when *x* = 0:

In[93] := **DSolve[{y''[x] + 8 y'[x] + 16 y[x] == 0,**
              y[0] == 2, y'[0] == 1}, y[x], x]

Out[93] = $\{\{y[x] \rightarrow \frac{2\ +\ 9\ x}{E^4\ x}\}\}$

What follows is an example from classical mechanics.

EXAMPLE 2.2.5 Two simple harmonic oscillators are coupled together by a spring, as in Figure 2.2.2. While m_1 is held fixed, m_2 is displaced to the right by a distance a from its equilibrium position. At time $t = 0$, both masses are released. Determine the subsequent motion. Consider only the case where $k_1 = k_2 = k_{12} = k$ and $m_1 = m_2 = m$.

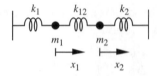

Figure 2.2.2 Two coupled harmonic oscillators

Let x_1 and x_2 be the displacements of m_1 and m_2, respectively. Each displacement is measured from the equilibrium position, and the direction to the right is positive. The force on m_1 is $-kx_1 - k(x_1 - x_2)$, and the force on m_2 is $-kx_2 - k(x_2 - x_1)$. Thus, the equations of motion are

$$m\frac{d^2 x_1}{dt^2} + 2kx_1 - kx_2 = 0$$

$$m\frac{d^2 x_2}{dt^2} + 2kx_2 - kx_1 = 0$$

At $t = 0$,

$$\frac{dx_1}{dt} = 0$$

$$\frac{dx_2}{dt} = 0$$

$$x_1 = 0$$

$$x_2 = a$$

Let $w^2 = k/m$. DSolve solves the equations of motion subject to the initial conditions:

```
In[94] := DSolve[{x1''[t] + 2 w^2 x1[t] - w^2 x2[t] == 0,
            x2''[t] + 2 w^2 x2[t] - w^2 x1[t] == 0,
            x1'[0] == 0, x2'[0] == 0,
            x1[0] == 0, x2[0] == a},
            {x1[t], x2[t]}, t]
```

$Out[94]$ = $\{\{x1[t]$ -> $(\frac{-I}{3}$ E$^{-I\ t\ w}$ - I Sqrt[3] t w

$(\frac{-3\ I}{4}$ a E$^{I\ t\ w}$ w + $\frac{3\ I}{4}$ a E^{I} Sqrt[3] t w w +

$$\frac{3\ I}{4}\ a\ E^{2\ I\ t\ w\ +\ I\ Sqrt[3]\ t\ w}\ _{w}\ -$$

$$\frac{3\ I}{4}\ a\ E^{I\ t\ w\ +\ 2\ I\ Sqrt[3]\ t\ w}\ _{w}))\ /\ w,$$

$$x2[t]\ ->\ (\frac{-I}{3}\ E^{-I\ t\ w\ -\ I\ Sqrt[3]\ t\ w}$$

$$(\frac{3\ I}{4}\ a\ E^{I\ t\ w}\ _{w}\ +\ \frac{3\ I}{4}\ a\ E^{I\ Sqrt[3]\ t\ w}\ _{w}\ +$$

$$\frac{3\ I}{4}\ a\ E^{2\ I\ t\ w\ +\ I\ Sqrt[3]\ t\ w}\ _{w}\ +$$

$$\frac{3\ I}{4}\ a\ E^{I\ t\ w\ +\ 2\ I\ Sqrt[3]\ t\ w}\ _{w}))\ /\ w\}\}$$

■ *Mathematica* **Version 2.2 Specifics**

We can simplify the solution:

ExpandAll[ComplexToTrig[%]]

$$\{\{x1[t]\ ->\ \frac{a\ Cos[t\ w]}{2}\ -\ \frac{a\ Cos[Sqrt[3]\ t\ w]}{2},$$

$$x2[t]\ ->\ \frac{a\ Cos[t\ w]}{2}\ +\ \frac{a\ Cos[Sqrt[3]\ t\ w]}{2}\}\}$$

There are two normal or characteristic frequencies: $\sqrt{k/m}$ and $\sqrt{3k/m}$. The function **ComplexToTrig** is defined in the package **Algebra`Trigonometry`** that was loaded in Section 2.2.2.

■ *Mathematica* **Version 3.0 Specifics**

We can simplify the solution:

In[95] := **ExpandAll[FullSimplify[ExpToTrig[%]]]**

Out[95] = $\{\{x1[t]\ ->\ \frac{a\ Cos[t\ w]}{2}\ -\ \frac{a\ Cos[Sqrt[3]\ t\ w]}{2},$

$$x2[t]\ ->\ \frac{a\ Cos[t\ w]}{2}\ +\ \frac{a\ Cos[Sqrt[3]\ t\ w]}{2}\}\}$$

There are two normal or characteristic frequencies: $\sqrt{k/m}$ and $\sqrt{3\ k/m}$. ■

2.2.12 *Immediate versus Delayed Assignments and Transformation Rules*

As discussed in Section 2.2.4, we make an immediate assignment by writing *variable* = *value*. We can also make a delayed assignment by typing *variable* := *value* (with no space between : and =). What is the difference between these assignments? With an immediate assignment, *value* is evaluated as soon as the assignment is made; for a delayed assignment, *value* is evaluated later when it is requested. Let us illustrate the difference between the two assignments with the **Random** function that generates, upon request, a random number between zero and one. To make an immediate assignment to the variable **imasm**, we enter

In[96] := **imasm = Random[]**

Out[96] = 0.232502

The Random function is evaluated immediately, and the variable imasm is assigned a permanent value until it is changed or removed. To make a delayed assignment to the variable dlasm, we write

In[97] := **dlasm := Random[]**

In this case, the Random function is not evaluated, and no value is assigned to the variable dlasm until it is called. Now call these variables several times with the **Table** function introduced in Section 2.1.18:

In[98] := **Table[imasm, {4}]**

Out[98] = {0.232502, 0.232502, 0.232502, 0.232502}

In[99] := **Table[dlasm, {4}]**

Out[99] = {0.9211, 0.534473, 0.21089, 0.750662}

While the value for imasm remains the same, that for dlasm varies from call to call.

As shown in Section 2.2.4, we can specify an immediate transformation rule in the form *expr* /. *lhs* -> *rhs*. We can also specify a delayed transformation rule in the form *expr* /. *lhs* :> *rhs* (with no space between : and >). The difference between these rules rests on when *rhs* is evaluated. Whereas *rhs* of an immediate rule is evaluated upon specification of the rule, *rhs* of a delayed rule is evaluated afresh when the rule is invoked. Let us use the Random function to illustrate the difference between these rules:

In[100] := **{f[a], f[a], f[a], f[a]} /. f[a] -> Random[]**

Out[100] = {0.810566, 0.810566, 0.810566, 0.810566}

In[101] := **{f[a], f[a], f[a], f[a]} /. f[a] :> Random[]**

Out[101] = {0.739093, 0.335429, 0.3537, 0.934887}

In the first case, the same transformation rule is used for all replacements because Random[] is evaluated right away. In the second case, the delayed rule varies from replacement to replacement since Random[] is evaluated afresh each time.

2.2.13   Defining Functions

An obvious definition for a function is *name[arg1, arg2, ...]* := *body*. To illustrate the flaw of this definition, we define the function $f(x) = x^3 - 1$:

In[102] := **f[x] := x^3 -1**

Let us call f[x]:

In[103] := **f[x]**

Out[103] = $-1 + x^3$

As expected, it gives $x^3 - 1$. Now, call f[y] and f[2]:

In[104] := **f[y]**

Out[104] = f[y]

In[105] := **f[2]**

Out[105] = f[2]

Contrary to expectation, *Mathematica* does not return $y^3 - 1$ and 7 for **f[y]** and **f[2]**, respectively. The problem is that while *Mathematica* knows **f[x]**, it does not recognize **f[y]** and **f[2]**.

To define a function, we need to write

name[*arg1_*, *arg2_*, ...] := *body*

where "_", called a blank, stands for "any expression." The symbol *arg_* means any expression to be named *arg* in the *body* of the definition. To show that this definition works, we return to the previous example:

In[106] := **f[x_] := x^3 - 1**

In[107] := **f[y]**
Out[107] = $-1 + y^3$

In[108] := **f[2]**
Out[108] = 7

The results for **f[y]** and **f[2]** are indeed correct.

We can use the immediate assignment operator = instead of the delayed assignment operator := in a definition if we prefer immediate evaluation of the body of the function upon definition. Yet, in most cases, we would rather have the bodies of the functions evaluated when they are called. When in doubt, use the delayed assignment operator " := ".

In Section 2.2.4, we discussed the advisability of clearing as often as possible unneeded values assigned to variables. It is also a good practice to remove unneeded definitions for functions as soon as possible.

In[109] := **Clear[f]**

Let us illustrate the writing of functions with a physics problem.

EXAMPLE 2.2.6 The normalized hydrogenic eigenfunctions are given as

$$\psi_{nlm}(r, \theta, \phi) = R_{nl}(r)Y_{lm}(\theta, \phi)$$

where R_{nl} and Y_{lm} are the radial functions and spherical harmonics, respectively. In terms of the associated Laguerre polynomials L_p^q and the Bohr radius a, R_{nl} can be expressed as

$$a^{-3/2} \frac{2}{n^2} \sqrt{\frac{(n-l-1)!}{[(n+l)!]^3}} F_{nl}\left(\frac{2r}{na}\right)$$

with

$$F_{nl}(x) = x^l e^{-x/2} L_{n-l-1}^{2l+1}(x)$$

Find the radial functions for the 1*s*, 2*s*, 2*p*, and 3*s* states.

According to convention, all user-defined function names should start with lowercase letters because built-in _Mathematica_ functions begin with uppercase letters. Let f[n, l, x] and radialFunction[n, l, r] represent $F_{nl}(x)$ and $R_{nl}(r)$, respectively.

In terms of the generalized Laguerre polynomials LaguerreL[p, q, x] in _Mathematica_, the associated Laguerre polynomials $L_p^q(x)$ are given by

$$L_p^q(x) = (p + q)!\texttt{LaguerreL[p, q, x]}$$

([Lib92] compares three common notations for the associated Laguerre polynomials. The notation adopted here is that used in [Lib92] and [Mes62].) Thus, the function f can be defined as

```
In[110] := f[n_, l_, x_] := (x^l Exp[-x/2] (n + l)! *
                             LaguerreL[n - l - 1, 2 l + 1, x])
```

The function radialFunction can be written as

```
In[111] := radialFunction[n_, l_, r_] :=
                       (a^(-3/2) (2/(n^2)) *
                        Sqrt[(n - l - 1)!/(n + l)!^3] *
                        f[n, l, 2r/(n a)])
```

For $R_{10}(r)$, $R_{20}(r)$, $R_{21}(r)$, and $R_{30}(r)$, we have

```
In[112] := radialFunction[1, 0, r]
```

$$Out[112] = \frac{2}{a^{3/2}\,E^{r/a}}$$

```
In[113] := radialFunction[2, 0, r]
```

$$Out[113] = \frac{2 - \frac{r}{a}}{2\,\text{Sqrt}[2]\,a^{3/2}\,E^{r/(2\,a)}}$$

```
In[114] := radialFunction[2, 1, r]
```

$$Out[114] = \frac{r}{2\,\text{Sqrt}[6]\,a^{5/2}\,E^{r/(2\,a)}}$$

```
In[115] := radialFunction[3, 0, r]
```

$$Out[115] = \frac{2\,(27\,a^2 - 18\,a\,r + 2\,r^2)}{81\,\text{Sqrt}[3]\,a^{7/2}\,E^{r/(3\,a)}}$$

```
In[116] := Clear[f, radialFunction]
```
    ■

Section 2.1.8 pointed out that N[_expr_] has a special input form, namely, the postfix form _expr_//N. For example, Sin[50 Degree]//N and N[Sin[50 Degree]] are identical:

```
In[117] := N[Sin[50 Degree]] == Sin[50 Degree]//N
```

```
Out[117] = True
```

In general, functions with a single argument have two special input forms:

$$func \ @ \ arg \qquad \text{prefix form of } func\,[arg]$$
$$arg \ // \ func \qquad \text{postfix form of } func\,[arg]$$

We can, for example, determine `Sqrt[3.2]` with the prefix form `Sqrt@3.2` or the postfix form `3.2//Sqrt`:

```
In[118] := Sqrt@3.2
```

```
Out[118] = 1.78885
```

```
In[119] := 3.2//Sqrt
```

```
Out[119] = 1.78885
```

If the argument contains *Mathematica* operators, we must be careful in using the prefix form. For instance, `f@x y + z` equals `f[x] y + z` rather than `f[x y + z]`. To get `f[x y + z]`, we must write `f@(x y + z)`. On the other hand, the postfix form is usually forgiving; for example, `x y + z//f` gives `f[x y + z]`. When in doubt, parenthesize the argument.

To illustrate the use of prefix form, consider a problem from quantum mechanics.

EXAMPLE 2.2.7 The time-independent Schrödinger equation is

$$H\psi = E\psi$$

where H is the Hamiltonian operator, E is the energy eigenvalue, and ψ is the corresponding eigenfunction. For a particle with mass m moving in a one-dimensional potential $V(x)$, the Hamiltonian operator is given by

$$H = -\frac{\hbar^2}{2m}\frac{d^2}{dx^2} + V(x)$$

Solve the time-independent Schrödinger equation for a free particle in one dimension.

The prefix special input form is most suited for working with operators. If the Hamiltonian is defined in *Mathematica* by

```
In[120] := hamiltonian[V_] @ psi_ := (-hb^2/(2m) D[psi, {x, 2}]
                                        + V psi)
```

the time-independent Schrödinger equation in one dimension is

```
In[121] := hamiltonian[V[x]] @ psi[x] == energy psi[x]
```

$$Out[121] = psi[x] \ V[x] \ - \ \frac{hb^2 \ psi''[x]}{2 \ m} \ == \ energy \ psi[x]$$

with the aliases `hamiltonian[V[x]]` for H, `energy` for E, `psi` for ψ, and `hb` for $\hbar$, that is, Planck's constant divided by 2π. (Section 2.5 will discuss the use of ψ and $\hbar$ for input in *Mathematica* version 3.0.) With a free particle, $V = 0$. Let us show that e^{ikx} is a free-particle eigenfunction by substituting

```
In[122] := psi[x] = Exp[I k x]
```

$$Out[122] = E^{I \ k \ x}$$

into

$In[123] := $ **hamiltonian[0] @ psi[x] == energy psi[x]**

$$Out[123] = \frac{E^{I\ k\ x}\ hb^2\ k^2}{2\ m} == E^{I\ k\ x}\ energy$$

We see that e^{ikx} is indeed an eigenfunction, and the energy eigenvalue is given by

$In[124] := $ **Solve[%, energy]**

$$Out[124] = \{\{energy \to \frac{hb^2\ k^2}{2\ m}\}\}$$

(For an excellent discourse on using *Mathematica* to do quantum mechanics, see [Fea94].)

$In[125] := $ **Clear[psi]**　　　　　　　　　　　　　　　■

Section 1.3 showed how to access information about built-in *Mathematica* objects with the operators **?** and **??**. We can also obtain information about user-defined functions. For example, let us define a function **f**:

$In[126] := $ **f[x_] := x^3**

Using the operator **?** on **f** yields its definition together with the fact that it is a symbol in the context **Global`**:

$In[127] := $ **?f**

```
Global`f
f[x_] := x^3
```

(Section 3.7.1 will discuss the notion of contexts.) We can specify usage messages for user-defined functions in the form

f **::usage** = "*usage message text*"

For example, let us give **f** a usage statement:

$In[128] := $ **f::usage = "f[x] returns the cube of x."**

$Out[128] = $ f[x] returns the cube of x.

Using **?** on **f** now retrieves the usage statement:

$In[129] := $ **?f**

```
f[x] returns the cube of x.
```

To obtain additional information about **f**, use **??**:

$In[130] := $ **??f**

```
f[x] returns the cube of x.
f[x_] := x^3
```

$In[131] := $ **Clear[f]**

2.2.14　*Relational and Logical Operators*

Relational operators compare two expressions and return **True** or **False** whenever comparisons are possible. *Mathematica* has several relational operators or functions:

lhs == *rhs* or **Equal**[*lhs*, *rhs*]	return **True** if *lhs* and *rhs* are identical
lhs != *rhs* or **Unequal**[*lhs*, *rhs*]	return **False** if *lhs* and *rhs* are identical
lhs > *rhs* or **Greater**[*lhs*, *rhs*]	yield **True** if *lhs* is determined to be greater than *rhs*
lhs >= *rhs* or **GreaterEqual**[*lhs*, *rhs*]	yield **True** if *lhs* is determined to be greater than or equal to *rhs*
lhs < *rhs* or **Less**[*lhs*, *rhs*]	yield **True** if *lhs* is determined to be less than *rhs*
lhs <= *rhs* or **LessEqual**[*lhs*, *rhs*]	yield **True** if *lhs* is determined to be less than or equal to *rhs*

For example, consider

In[132] := **f[x_] := x^2**

In[133] := **g[x_] := x^3**

In[134] := **g[2] > f[3] + 1**

Out[134] = False

We can combine relational operators:

In[135] := **8 - 5 == 3 == 6/2**

Out[135] = True

In[136] := **4 > 3 <= 2**

Out[136] = False

When *Mathematica* is unable to compare the expressions, it returns the input:

In[137] := **1 + x > 2 y**

Out[137] = 1 + x > 2 y

Mathematica supports several logical operators or functions:

!*expr* or **Not**[*expr*]	give **False** if *expr* is **True**, and **True** if it is **False**
expr1 && *expr2* && ... or **And**[*expr1*, *expr2*, ...]	evaluate the expressions in order from left to right, giving **False** immediately if any of them are **False**, and **True** if they are all **True**

expr1 || *expr2* || ... or Or[*expr1*, *expr2*, ...]	evaluate the expressions in order from left to right, giving True immediately if any of them are True, and False if they are all False
Xor[*expr1*, *expr2*, ...]	give True if an odd number of the *expri* are True and the rest are False; give False if an even number of the *expri* are True and the rest are False

For example, consider

In[138] := **4 > 3 || 5 <= 10**

Out[138] = True

In[139] := **Xor[4 > 3, 5 <= 10]**

Out[139] = False

Whereas || is inclusive, Xor is exclusive. In other words, || returns True when one or both expressions are True, but Xor gives True only when one expression is True and the other is False.

With *expr1* && *expr2* && ..., *Mathematica* evaluates the expressions from left to right, stopping at the first expression that is False:

In[140] := **EvenQ[g[2] + 1] && 1/0 < 10**

Out[140] = False

EvenQ[*expr*] gives True if *expr* is an even integer, and False otherwise. The second expression was not evaluated; otherwise, *Mathematica* would have given a warning message. This can be seen by reversing the expressions:

In[141] := **1/0 < 10 && EvenQ[g[2] + 1]**

$$\frac{1}{0}$$
  Power::infy: Infinite expression - encountered.

  Less::nord: Invalid comparison with ComplexInfinity attempted.

Out[141] = False

When *Mathematica* cannot ascertain the truth of the expressions, it returns the input:

In[142] := **a || b || c**

Out[142] = a || b || c

In[143] := **Clear[f, g]**

2.2.15 Fourier Transforms

The Fourier transform, if it exists, of the function $f(x)$ is

$$g(u) = \left(\frac{a}{2\pi}\right)^{\frac{1}{2}} \int_{-\infty}^{\infty} f(x)e^{-iaux}dx \tag{2.2.1}$$

and the inverse Fourier transform of $g(u)$ is

$$f(x) = \left(\frac{a}{2\pi}\right)^{\frac{1}{2}} \int_{-\infty}^{\infty} g(u)e^{iaux}du \tag{2.2.2}$$

where a is a positive real constant. With $a = 1$, the variables x and u usually represent, respectively, position and wave number or time and angular frequency. With $a = 2\pi/h$ where h is Planck's constant, x and u often stand for position and momentum or time and energy. (For a discussion of the Fourier transform and an introduction to the theory of generalized functions, see [DK67].)

The standard *Mathematica* package `Calculus`FourierTransform`` contains several functions useful for evaluating Fourier transforms:

`FourierTransform[`
  f `[x], x, u,`
    `FourierOverallConstant -> ` A,
    `FourierFrequencyConstant -> ` B
      `]`
	give A `* Integrate[Exp[` B `I` u x `]`
	f `[x], {x, -Infinity, Infinity}]`

`InverseFourierTransform[`
  g `[u], u, x,`
    `FourierOverallConstant -> ` A,
    `FourierFrequencyConstant -> ` B
      `]`
	give $(B$ `/(2 Pi` A`))` `* Integrate[`
	`Exp[-` B `I` u x `]` g `[u],`
	`{u, -Infinity, Infinity}]`

`DiracDelta[x - ` $x0$ `]`	Dirac delta function with singularity located at $x = x0$
`UnitStep[x]`	a function that is 1 for $x > 0$ and 0 for $x < 0$

The last two arguments of `FourierTransform` and `InverseFourierTransform` are named optional arguments or, simply, options that are specified in terms of transformation rules introduced in Section 2.2.4. The order in which these arguments are entered is irrelevant as long as they are trailing arguments of the function. Both A and B have a default value of 1. That is, if the rule `FourierOverallConstant -> ` A is omitted, A assumes the value of 1, and if the rule `FourierFrequencyConstant -> ` B is absent, B also takes on the value of 1.

Comparing the definition of `FourierTransform` and Equation 2.2.1 reveals that the Fourier transform of $f(x)$ is given by

```
FourierTransform[
      f[x], x, -u,
      FourierOverallConstant -> Sqrt[a/(2 Pi)],
      FourierFrequencyConstant -> a
                  ]
```

Similarly, the inverse Fourier transform of $g(u)$ is given by

```
InverseFourierTransform[
          g[u], u, -x,
          FourierOverallConstant -> Sqrt[a/(2 Pi)],
          FourierFrequencyConstant -> a
                      ]
```

Let us set $a = 1$ and $u = k$ and determine the Fourier transforms of the several functions. First, load the package `Calculus`FourierTransform`:

In[144] := `Needs["Calculus`FourierTransform`"]`

Consider

$$f(x) = \sqrt{b}\, e^{-b|x|}$$

with $b > 0$. The Fourier transform of $f(x)$ is

In[145] := `FourierTransform[`
`                   Sqrt[b](Exp[-b x] UnitStep[x] +`
`                           Exp[b x] UnitStep[-x]),`
`                   x, -k,`
`                   FourierOverallConstant -> Sqrt[1/(2 Pi)]`
`                          ] // Simplify`

Out[145] = $\dfrac{b^{3/2}\, \mathrm{Sqrt}[\frac{2}{\mathrm{Pi}}]}{b^2 + k^2}$

The inverse Fourier transform is

In[146] := `InverseFourierTransform[`
`                   %, k, -x,`
`                   FourierOverallConstant -> Sqrt[1/(2 Pi)]`
`                      ]`

Out[146] = $\dfrac{\mathrm{Sqrt}[b]}{\mathrm{E}^{b\ \mathrm{Abs}[x]}}$

which is, of course, $f(x)$ itself. Now find the Fourier transform of the Dirac delta function:

```
In[147] := FourierTransform[
              DiracDelta[x - x0], x, -k,
              FourierOverallConstant -> Sqrt[1/(2 Pi)]
              ]
```

$$Out[147] = \frac{1}{E^{I \ k \ x0} \ Sqrt[2 \ Pi]}$$

Finally, determine the Fourier transform of

$$f(x) = \begin{cases} i \sqrt{2b}\, e^{-bx} & x > 0 \\ 0 & x < 0 \end{cases}$$

with $b > 0$:

```
In[148] := FourierTransform[
              I Sqrt[2 b] Exp[-b x] UnitStep[x],
              x, -k,
              FourierOverallConstant -> Sqrt[1/(2 Pi)]
              ]
```

$$Out[148] = \frac{I \ Sqrt[b]}{(b + I \ k) \ Sqrt[Pi]}$$

Consider a particle in one dimension. If $\psi(x)$ and $\varphi(p)$ are, respectively, its wave functions in configuration space and momentum space, then $\psi^*(x)\psi(x)dx$ is the probability of finding the particle's position between x and $x + dx$ and $\varphi^*(p)\varphi(p)dp$ is the probability of finding the particle's momentum between p and $p + dp$. Furthermore, the wave functions are transforms of each other:

$$\varphi(p) = \left(\frac{1}{2\pi\hbar}\right)^{\frac{1}{2}} \int_{-\infty}^{\infty} \psi(x)\, e^{\frac{-(ipx)}{\hbar}}\, dx \tag{2.2.3}$$

$$\psi(x) = \left(\frac{1}{2\pi\hbar}\right)^{\frac{1}{2}} \int_{-\infty}^{\infty} \varphi(p)\, e^{\frac{(ipx)}{\hbar}}\, dp \tag{2.2.4}$$

Comparing Equations 2.2.3 and 2.2.4 with Equations 2.2.1 and 2.2.2 establishes that these transforms are just Fourier and inverse transforms if we identify u as momentum p and a as $2\pi/h$ where h is, again, Planck's constant. For an example, a wave function in configuration space has the form

$$\psi(x) = \begin{cases} A\, e^{ip_0 x/\hbar} & |x| < b \\ 0 & |x| > b \end{cases}$$

where A and b are positive real constants. We can plot the probability density $P(x) = \psi(x)^*\psi(x)$:

```
In[149] := Plot[UnitStep[x + 1] - UnitStep[x - 1], {x, -2, 2},
              AxesLabel -> {" x (b)", " P(x) (A^2)"}
              ];
```

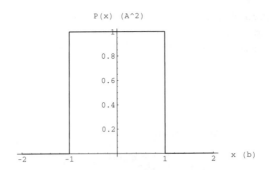

with $P(x)$ and x expressed in units of A^2 and b, respectively. (The function `Plot` will be discussed in Section 2.3.1.1.) The uncertainty Δx in the particle's position is approximately $2b$. The wave function in momentum space is the Fourier transform of $\psi(x)$:

```
In[150] := FourierTransform[
               A Exp[(2 Pi I p0 x)/h](UnitStep[x + b] -
                                      UnitStep[x - b]),
               x, -p,
               FourierOverallConstant -> Sqrt[1/h],
               FourierFrequencyConstant -> ((2 Pi)/h)
                   ]//Simplify//PowerExpand
```

$$Out[150] = \frac{A \; \text{Sqrt}[h] \; \text{Sin}[\frac{2 \; b \; (p - p0) \; \text{Pi}}{h}]}{(p - p0) \; \text{Pi}}$$

Let us plot the probability density $P(p) = \varphi^*(p)\varphi(p)$:

```
In[151] := Plot[(Sin[p - 10]/(p - 10))^2, {p, -5, 20},
               PlotRange -> All,
               AxesLabel ->
                  {" p (h/2Pib)", " P(p) ((1/h)(2Ab)^2)"}
               ];
```

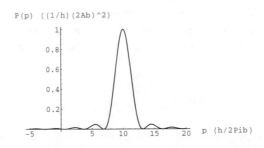

with $P(p)$ and p expressed in units of $4A^2b^2/h$ and $h/2\pi b$, respectively. We have also set $p_0 = 10$ in the graph. The probability density has a sharp peak at p_0 and falls off rapidly on both sides to zero at $p = p_0 \pm h/2b$. Therefore, a reasonable estimate of the uncertain Δp in the particle's momentum is h/b. The product of the uncertainties in position and momentum is

$$\Delta x \Delta p \approx 2h = 4\pi\hbar$$

This is consistent with the Heisenberg's uncertainty relations.

2.2.16 Exercises

1. Factor the expression

$$x^4 + 2x^3 - 3x - 6$$

   Answer: $(2 + x)\ (-3 + x^3)$

2. Factor the expression

$$6t^3 + 9t^2 - 15t$$

   Answer: $3\ (-1 + t)\ t\ (5 + 2\ t)$

3. Factor the expression

$$ax^2 + ay + bx^2 + by$$

   Answer: $(a + b)\ (x^2 + y)$

4. Factor the expression

$$2x^4y^6 + 6x^2y^3 - 20$$

   Answer: $2\ (-2 + x^2\ y^3)\ (5 + x^2\ y^3)$

5. Simplify

$$\frac{1}{x^2 - 16} - \frac{x + 4}{x^2 - 3x - 4}$$

   Answer: $\dfrac{-15\ -\ 7\ x\ -\ x^2}{(-4 + x)\ (1 + x)\ (4 + x)}$

6. Add

$$\frac{3}{x^3 - x} + \frac{4}{x^2 + 2x + 1}$$

   Answer: $\dfrac{3\ -\ x\ +\ 4\ x^2}{(-1 + x)\ x\ (1 + x)^2}$

7. Solve the simultaneous equations

$$4x + 5y = 5$$
$$6x + 7y = 7$$

   Answer: $\{\{x\ ->\ 0,\ y\ ->\ 1\}\}$

8. Solve the equation

$$t^2 + 4t - 8 = 0$$

Answers: `{{t -> 2 (-1 + Sqrt[3])}, {t -> -2 (1 + Sqrt[3])}}`

9. Solve for *x*:

$$\sqrt{x + 2} + 4 = x$$

Answer: `{{x -> 7}}`

10. Solve for *y*:

$$\sqrt{y - 5} - \sqrt{y} = 1$$

Answer: `{}`

11. Calculate the derivative of the function

$$y(x) = \sin(e^{x^2})$$

Answer: `2 E`$^{x^2}$` x Cos[E`$^{x^2}$`]`

12. Determine the second derivative of the function

$$y(x) = e^x \sin x$$

Answer: `2 E`^{X}` Cos[x]`

13. Let the functions *f(t)* and *r(t)* be related by the equation

$$f(t) = \frac{dr(t)}{dt}$$

To within an additive constant, *r(t)* is given in terms of *f(t)* by

$$r(t) = \int f(t) \, dt$$

Determine *r(t)* for

$$f(t) = 5 \ln t - e^{-3t}$$

Answer: $\dfrac{1}{3 \text{ E}^3 \text{ t}}$ ` - 5 t + 5 t Log[t]`

   or

   `{{r[t] -> ` $\dfrac{1}{3 \text{ E}^3 \text{ t}}$ ` - 5 t + C[1] + 5 t Log[t]}}`

14. Let the functions *f(t)* and *r(t)* be related by the equation

$$f(t) = \frac{dr(t)}{dt}$$

Determine $r(t)$ for

$$f(t) = \frac{t}{\sqrt{d^2 + t^2}}$$

where d is a constant.

Answer: `Sqrt[d^2 + t^2]`

or

`{{r[t] -> Sqrt[d^2 + t^2] + C[1]}}`

15. Evaluate

$$\int_0^1 \frac{4}{1 + x^2} dx$$

What is special about this integral?

Answer: `Pi`

The integral evaluates to π exactly.

16. Consider the integral

$$\int \frac{x}{a^3 + x^3} dx$$

(a) Evaluate the integral.   (b) Differentiate the result and recover the integrand.

Answers:

$$\frac{\text{ArcTan}[\frac{-a + 2x}{\text{Sqrt}[3] a}]}{\text{Sqrt}[3] a} - \frac{\text{Log}[a + x]}{3 a} + \frac{\text{Log}[a^2 - a x + x^2]}{6 a}$$

$$\frac{-1}{3 a (a + x)} + \frac{-a + 2x}{6 a (a^2 - a x + x^2)} + \frac{2}{3 a^2 (1 + \frac{(-a + 2x)^2}{3 a^2})}$$

$$\frac{x}{a^3 + x^3}$$

17. Find a power series approximation, up to terms of order x^8, for

$$\sqrt[3]{1 + x^4}$$

Answer: $1 + \dfrac{x^4}{3} - \dfrac{x^8}{9} + O[x]^9$

18. Find a power series approximation for $\sec(x)$ about $x = \pi/3$, accurate to order $(x - \pi/3)^3$.

Answer: $2 + 2 \text{ Sqrt}[3] \ (\frac{-Pi}{3} + x) + 7 \ (\frac{-Pi}{3} + x)^2 +$

$2 \ (\frac{-1}{2 \text{ Sqrt}[3]} + 4 \text{ Sqrt}[3]) \ (\frac{-Pi}{3} + x)^3 +$

$O[\frac{-Pi}{3} + x]^4$

19. Compute the limit as x goes to infinity of

$$\frac{\log(x - a)}{(a - b)(a - c)} + \frac{\log 2(x - b)}{(b - c)(b - a)} + \frac{\log(x - c)}{(c - a)(c - b)}$$

Answer:
$$\frac{\text{Log}[2]}{-(a\ b)\ +\ b^2\ +\ a\ c\ -\ b\ c}$$

20. Compute the limit as x goes to infinity of

$$\frac{1}{e^x - e^{(x - x^{-2})}}$$

Hint: Load the package `Calculus`Limit`.

Answer: 0

21. In the bimolecular reaction $A + B \rightarrow M$, α moles per liter of A and β moles per liter of B are combined. If x denotes the number of moles per liter that have reacted after time t, the rate of reaction is given by

$$\frac{dx}{dt} = k(\alpha - x)(\beta - x)$$

where k is the velocity constant. For the case $\alpha = \beta$:

(a) Show

$$x = \frac{\alpha^2 kt}{1 + \alpha kt}$$

(b) Find

$$\lim_{t \to \infty} x(t)$$

Answers: $\{\{x[t] \rightarrow \dfrac{a^2\ k\ t}{1\ +\ a\ k\ t}\}\}$

with **a** as the alias of α.

22. Solve the differential equation

$$\frac{d^2y}{dx^2} - 6\frac{dy}{dx} + 13y = e^x \cos x$$

Answer: $\{\{y[x] \rightarrow (E^x\ (65\ E^{2\ x}\ C[2]\ \text{Cos}[2\ x]\ -\ 4\ \text{Sin}[x]\ +$
$\text{Cos}[x]\ (7\ -\ 130\ E^{2\ x}\ C[1]\ \text{Sin}[x]))) /\ 65\}$
$\}$

23. For the circuit shown, (a) find all currents in terms of the emf and the resistances and (b) obtain numerical values for the currents with $E = 12V$, $R_1 = 2.0\ \Omega$, $R_2 = 4.0\ \Omega$, $R_3 = 6.0\ \Omega$, $R_4 = 2.0\ \Omega$, and $R_5 = 3.0\ \Omega$. Assume that the internal resistance of the battery is negligible.

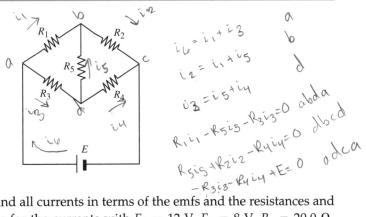

24. For the circuit shown, (a) find all currents in terms of the emfs and the resistances and (b) obtain numerical values for the currents with $E_1 = 12$ V, $E_2 = 8$ V, $R_1 = 20.0\ \Omega$, $R_2 = 15.0\ \Omega$, $R_3 = 3.0\ \Omega$, $R_4 = 6.0\ \Omega$, $R_5 = 12.0\ \Omega$, and $R_6 = 2.0\ \Omega$. Assume that the internal resistances of the batteries are negligible.

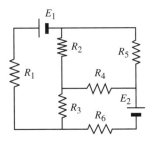

Answer: {{i5 -> 0.481 amperes, i6 -> 1.15 amperes,

i1 -> 0.596 amperes, i2 -> - 0.116 amperes,

i3 -> 0.556 amperes, i4 -> - 0.671 amperes}}

25. Prove that
$$\frac{\sec x - \csc x}{\tan x + \cot x} = \frac{\tan x - \cot x}{\sec x + \csc x}$$

26. Prove that
$$\frac{2\sin 2x - \sin 3x}{\cos x} = \sin x(8\sin^2 \frac{x}{2} + \sec x)$$

27. Find the Fourier transform of the function
$$\psi(x) = \begin{cases} 1 - |x| & |x| < 1 \\ 0 & |x| \geq 1 \end{cases}$$

In Equation 2.2.1, let $a = 2\pi/h$ and $u = p$.

Answer: $\dfrac{h^{3/2}\ \mathrm{Sin}[\frac{p}{h}\frac{Pi}{}]^2}{p^2\ Pi^2}$

28. Determine the Fourier transform of the function

$$\psi(x) = \begin{cases} 0 & |x| \geq b \\ A & 0 < x < b \\ -A & 0 > x > -b \end{cases}$$

where *A* and *b* are positive real constants. In Equation 2.2.1, let *a* = 1 and *u* = *k*. For *Mathematica* version 2.2, use the function `ComplexToTrig` in the package `Algebra`Trigonometry` to simplify the result.

Answer:
$$\frac{\text{I A Sqrt}[\frac{2}{\text{Pi}}] \ (-1 + \text{Cos}[b \ k])}{k}$$

2.3   *Graphical Capabilities*

Mathematica has won many advocates because of its graphical capabilities. This section discusses numerical equation solving and animation as well as two- and three-dimensional plotting.

2.3.1   *Two-Dimensional Graphics*

2.3.1.1   BASIC PLOTS

`Plot[f, {x, xmin, xmax}]` plots *f* as a function of *x* from *xmin* to *xmax*.

EXAMPLE 2.3.1   The normalized eigenfunction for the first excited state of the one-dimensional harmonic oscillator is given as

$$u_1(x) = \left(\frac{4}{\pi}\right)^{1/4} x \ e^{-x^2/2}$$

where *x* is in units of $(h/2\pi m\omega)^{1/2}$. Plot $u_1(x)$ for *x* from −4 to 4.

In[1] := `Plot[(4/Pi)^(1/4) x Exp[-(x^2)/2], {x, -4, 4}]`

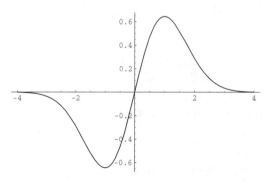

Out[1] = -Graphics-

(The function `Plot` returns `-Graphics-`, representing a graphics object; the graph is the result of its rendering. Section 3.5.1 will elaborate on *Mathematica* graphics objects. For a discussion of the one-dimensional harmonic oscillator in quantum mechanics, see [Gos92].)

■

With `Plot[`*f*`, {`*x*`,` *xmin*`,` *xmax*`}]`, *Mathematica* evaluates *f* from scratch for each needed value of *x*. In some cases, it is better to evaluate *f* first into a more explicit symbolic expression before it is subsequently evaluated numerically for those values of *x*. For the latter approach, we can use `Plot[Evaluate[`*f*`], {`*x*`,` *xmin*`,` *xmax*`}]`.

EXAMPLE 2.3.2 Plot the radial probability density $P(r) = r^2 R_{nl}^2$ for the 1*s* state of the hydrogen atom. The radial functions R_{nl} are given in Example 2.2.6 in Section 2.2.13.

From Example 2.2.6, we have

```
In[2] := f[n_, l_, x_] := (x^l Exp[-x/2] (n + l)! *
                           LaguerreL[n - l - 1, 2 l + 1, x])
```

```
In[3] := radialFunction[n_, l_, r_] :=
                       (a^(-3/2) (2/(n^2)) *
                        Sqrt[(n - l - 1)!/(n + l)!^3] *
                        f[n, l, 2r/(n a)])
```

with `radialFunction` as the alias of *R*. In what follows, let $a = 1$; that is, *r* is in units of the Bohr radius *a*.

In this case, it is better to evaluate $r^2 R_{10}^2$ first into a more explicit symbolic expression. Otherwise, for each needed value of *r*, *Mathematica* must invoke the function `radialFunction`, which calls the function `f`, which in turn calls the function `LaguerreL`. We now plot the probability distribution $r^2 R_{10}^2$:

```
In[4] := Plot[Evaluate[((r radialFunction[1, 0, r])^2) /.
                        a -> 1], {r, 0, 6}];
```

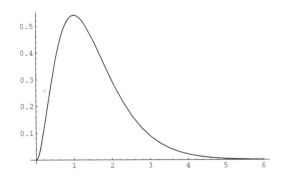

Due to space limitations of this book, we put a semicolon at the end of the graphical input to suppress the output, `Out[`*n*`]` = `-Graphics-`.

■

In numerical solution of differential equations, results are given as transformation rules in terms of interpolating functions. For plotting the solutions, it is better to use `Plot[Evaluate[y[x]/.solution], {x, xmin, xmax}]` so that the `InterpolatingFunction` objects can be set up first. For an illustrative instance, refer to Example 2.1.6 in Section 2.1.17. (To see the `InterpolatingFunction` object in the example, enter `InputForm[x/.sol[[1]]]` after the differential equation is solved. Section 3.5.1 will introduce the function `InputForm`.)

2.3.1.2   OPTIONS

When *Mathematica* generates a graph, it makes many choices such as the ratio of height to width for the plot, the range of coordinates to include in the plot, the number of points at which to sample the function, and so on. In other words, *Mathematica* specifies the values of the options in the plotting function. The command `Options[f]` gives the list of options for the function *f* and their default values. For the function `Plot`, `Options[Plot]` shows its default options:

```
In[5] := Options[Plot]

Out[5] = {AspectRatio -> ──────────, Axes -> Automatic, AxesLabel -> None,
                          GoldenRatio
            AxesOrigin -> Automatic, AxesStyle -> Automatic,
            Background -> Automatic, ColorOutput -> Automatic,
            Compiled -> True, DefaultColor -> Automatic, Epilog -> {},
            Frame -> False, FrameLabel -> None, FrameStyle -> Automatic,
            FrameTicks -> Automatic, GridLines -> None,
            ImageSize -> Automatic, MaxBend -> 10., PlotDivision -> 30.,
            PlotLabel -> None, PlotPoints -> 25, PlotRange -> Automatic,
            PlotRegion -> Automatic, PlotStyle -> Automatic, Prolog -> {},
            RotateLabel -> True, Ticks -> Automatic,
            DefaultFont :> $DefaultFont, DisplayFunction :> $DisplayFunction,
            FormatType :> $FormatType, TextStyle :> $TextStyle}
```

(The list of default options is somewhat different in *Mathematica* version 2.2.) Options are specified as transformation rules in the form *OptionName -> OptionValue*. Default values are quite satisfactory for most plots. Yet, at times, it is necessary to specify different values for some options. That can be done by giving additional arguments to the `Plot` function:

```
Plot[f, {x, xmin, xmax}, OptionName1 -> OptionValue1,
    OptionName2 -> OptionValue2, ...]
```

Options can be specified in any order after the second argument. If an option is not included, *Mathematica* uses the default value.

    EXAMPLE 2.3.3   Plot the Lennard-Jones potential

$$u(r) = 4\varepsilon \left[\left(\frac{\sigma}{r} \right)^{12} - \left(\frac{\sigma}{r} \right)^{6} \right]$$

where *r* is the distance between atoms while σ and ε are the empirical parameters.

Let $u(r)$ and r be measured in units of 4ε and σ, respectively.

In[6] := **Plot[(1/r^12) - (1/r^6), {r, 0.001, 3.0}];**

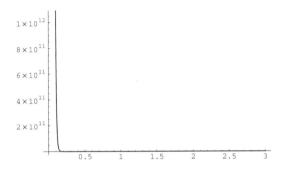

The graph does not show the essential features of the potential. The option **PlotRange** allows us to specify the range of values for $u(r)$ to include in the plot:

In[7] := **Plot[(1/r^12) - (1/r^6), {r, 0.001, 3.0},**
         PlotRange -> {-0.3, 0.6}];

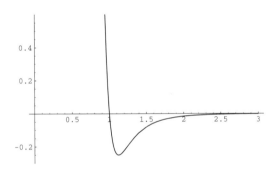

This graph is more interesting. It shows that a minimum occurs at $r/\sigma \approx 1.12$, that the curve is very steep inside the minimum, and that it is flat outside the minimum. (For a discussion of the Lennard-Jones potential, see [AM76].) ■

EXAMPLE 2.3.4 Two waves of slightly different frequencies travel along a line in the same direction. At a point in space, the displacements, y_1 and y_2, of the waves as functions of time are

$$y_1 = A\cos 20t$$
$$y_2 = A\cos 24t$$

where the time t is in seconds. Plot the resultant displacement at the point as a function of time. (For *Mathematica* version 2.2, let $y_1 = A\cos 22t$. Of course, the plot will look somewhat different.)

Let the displacements be measured in units of A. The resultant displacement, $y_1 + y_2$, equals $\cos 20t + \cos 24t$ (or, for *Mathematica* version 2.2, $\cos 22t + \cos 24t$).

In[8] := **Plot[Cos[20t] + Cos[24t], {t, Pi, 3Pi}];**

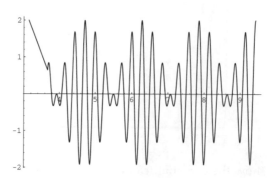

The graph is incorrect. The problem is that *Mathematica* does not sample the function at a sufficient number of points. The default value of the option **PlotPoints**, which gives the minimum number of points at which to sample the function, is 25. Let us increase the value to 70:

In[9] := **Plot[Cos[20t] + Cos[24t],**
              {t, Pi, 3Pi}, PlotPoints -> 70];

This is the familiar beat pattern.   ∎

Show[*plot, options*] redraws a plot with the specified options. Since **Show** redraws a plot with the original set of points, some options, such as **PlotPoints**, cannot be used.

Let us label the axes of the previous plot with **Show** by giving the option **AxesLabel**, which has the form **AxesLabel** -> {"*xlabel*", "*ylabel*"}:

In[10] := **Show[%, AxesLabel -> {" t (s)", " y (A)"}];**

To remove the axes and give the plot a label and also a frame with labels on its two edges but without tick marks, we specify the options **Axes**, **PlotLabel**, **Frame**, **FrameLabel**, and **FrameTicks**:

```
In[11] := Show[%, Axes -> False,
                PlotLabel -> "The Beat Phenomenon",
                Frame -> True,
                FrameLabel -> {"t", "y"},
                FrameTicks -> None
           ];
```

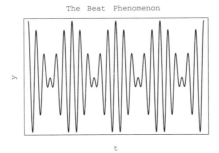

2.3.1.3 MULTIPLE PLOTS

Plot[{*f1***, ***f2***, ...}, {***x***, ***xmin***, ***xmax***}]** and **Plot[Evaluate[{***f1***, ***f2***, ...}], {***x***, ***xmin***, ***xmax***}]** plot functions ***f1***, ***f2***, ... together.

EXAMPLE 2.3.5 Plot the radial probability density $P(r) = r^2 R_{nl}^2$ for the $1s$, $2s$, and $3s$ states of the hydrogen atom. The radial functions R_{nl} are given in Example 2.3.2 in Section 2.3.1.1.

```
In[12] := Plot[Evaluate[Table[(r radialFunction[i, 0, r])^2,
                    {i, 3}
                  ] /. a -> 1
            ],
```

```
    {r, 0, 25}, PlotRange -> {0, 0.55},
    AxesLabel -> {" r (a)", " P(r) (1/a)"}
];
```

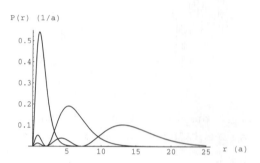

To make the plot easier to comprehend, we can assign a different style to each curve with the option **PlotStyle**. The option has the form **PlotStyle** -> {{_styles for curve1_}, {_styles for curve2_}, ...}. Some styles are

GrayLevel[_i_]	The argument _i_ varies between 0 (black) and 1 (white).
Thickness[_r_]	The argument _r_ specifies the thickness of the curve as a fraction of the width of the entire plot.
Dashing[{_r1_, _r2_, ...}]	The arguments $r1, r2, \ldots$ are the lengths of the successive drawn and undrawn segments of the dashed curve. The arguments are cyclical; that is, $\{r1, r2\}$ has the same effect as $\{r1, r2, r1, r2, \ldots\}$.

Let us specify different styles for the curves in the preceding graph:

```
In[13] := Plot[Evaluate[Table[(r radialFunction[i, 0, r])^2,
                    {i, 3}
                  ] /. a -> 1
              ],
          {r, 0, 25}, PlotRange -> {0, 0.55},
          AxesLabel -> {" r (a)", " P(r) (1/a)"},
          PlotStyle -> {
            {Thickness[0.006]},
            {Thickness[0.0075], Dashing[{0.01, 0.02}]},
            {GrayLevel[0.5], Thickness[0.009], Dashing[{0.02}]}
                }
        ];
```

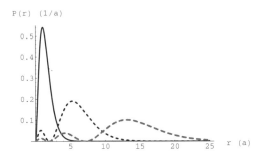

In[14] := **Clear[f, radialFunction]**

2.3.1.4   FINDROOT

Section 2.1.15 discussed numerical equation solving. The function **NSolve** considered there only finds the roots of polynomial equations. It cannot find solutions to more general equations such as transcendental equations.

EXAMPLE 2.3.6   Find the roots of the equation

$$\sin x = x^2$$

In[15] := **NSolve[Sin[x] == x^2, x]**

    Solve::tdep: The equations appear to involve transcendental functions
        of the variables in an essentially non-algebraic way.

    Solve::tdep: The equations appear to involve transcendental functions
        of the variables in an essentially non-algebraic way.

Out[15] = NSolve[Sin[x] == x^2, x]

NSolve fails to find the roots and returns the input unevaluated.   ∎

The function **FindRoot** can find numerical solutions to arbitrary equations. **FindRoot[** *lhs* == *rhs*, {*x*, *x0*}] searches for a numerical solution to the equation *lhs* == *rhs*, starting with *x* = *x0*. The success in finding a particular root depends heavily on the choice of the starting value. If the starting value is sufficiently close to a solution, **FindRoot** always returns that solution. The Macintosh and Windows front ends provide a means for determining the appropriate starting value graphically. The procedure is

1. Plot *lhs* and *rhs* of the equation together. If the curves do not intersect, there is no solution.
2. Click anywhere in the cell other than the cell bracket. A bounding box appears.
3. Enlarge the graphic, if necessary, for increased resolution. To resize the graphic, drag one of the small handles, that is, black squares, at the sides and corners of the bounding box to a desired position.

4. With the pointer inside the bounding box, press the Command key (i.e., the ⌘ key) for the Macintosh or the Control key for Windows. The pointer turns into the coordinates pointer that resembles a cross.

5. Move the center of the pointer over an intersection. The coordinates of the intersection appear in the status line or bar at the bottom of the notebook window.

6. Without moving the pointer, click the mouse. A point is drawn over the intersection.

7. Release the Command or Control key, and select **Copy** in the Edit menu.

8. Place the text insertion point below the graphics cell, and select **Paste** in the Edit menu. The coordinates of the point of intersection are pasted there.

9. Use the x coordinate as the starting value for `FindRoot`.

EXAMPLE 2.3.7   Find the nonzero root of the equation

$$\sin x = x^2$$

Let us follow the procedure just given.

1. Plot sin x and x^2 together:

In[16] := **Plot[{Sin[x], x^2}, {x, -0.5, 1.5}];**

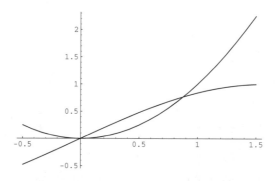

The curves intersect at two points.

2. Click in the cell. A bounding box appears:

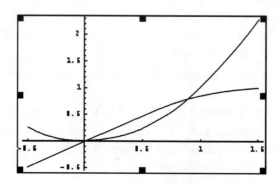

3. Resizing the graphic is unnecessary, as it is sufficiently large.

4. Press the Command or Control key. The coordinates pointer appears:

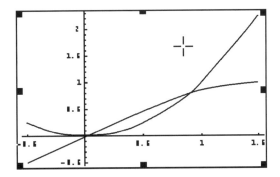

5. For the nonzero root, move the center of the pointer over the intersection that is not the origin:

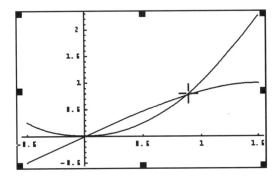

6. Click the mouse. A point is drawn over the intersection.

7. Select **Copy** in the Edit menu.

8. Select **Paste** in the Edit menu. The coordinates of the point of intersection are pasted below the graphics cell:

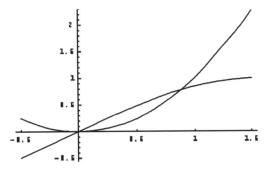

{0.893262, 0.799782}

9. Use 0.893262 as the starting value for `FindRoot`:

In[17] := `FindRoot[Sin[x] == x^2, {x, 0.893262}]`

Out[17] = `{x -> 0.876726}`

The solution is $x = 0.876726$.   ■

EXAMPLE 2.3.8 Consider a particle of mass m in a square well potential

$$V(x) = \begin{cases} V_0 & |x| > a/2 \\ 0 & |x| < a/2 \end{cases}$$

The energy of the particle is given as

$$E = \frac{h^2}{2\pi^2 ma^2} y^2$$

where h is Planck's constant. For energy eigenvalues associated with even eigenfunctions, the values for y are the solutions to the equation

$$y \tan y = \sqrt{d - y^2}$$

with

$$d = \frac{2\pi^2 ma^2}{h^2} V_0$$

For the special case $d = 16$, determine the energy spectrum for the even eigenstates. We begin by plotting $y \tan y$ and $(d - y^2)^{1/2}$ together:

In[18] := `Plot[{y Tan[y], Sqrt[16 - y^2]}, {y, 0, 8},`
                `PlotRange -> {0, 20}];`

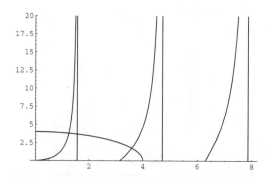

Let us zoom in on the relevant portion of the graph:

```
In[19] := Show[%, PlotRange -> {{0, 4}, {0, 5}}];
```

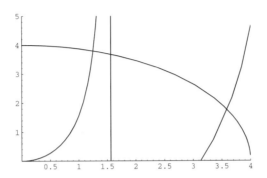

```
{{1.24369, 3.79137}, {3.59878, 1.7715}}
```

The coordinates of the two intersections are obtained with the procedure discussed earlier for determining starting values.

A starting value for **FindRoot** is 1.24369.

```
In[20] := FindRoot[y Tan[y] == Sqrt[16 - y^2], {y, 1.24369}]

Out[20] = {y -> 1.25235}
```

The solution is $y = 1.25235$.

Another starting value for **FindRoot** is 3.59878.

```
In[21] := FindRoot[y Tan[y] == Sqrt[16 - y^2], {y, 3.59878}]

Out[21] = {y -> 3.5953}
```

Therefore, another solution is 3.5953.

In terms of d, the energy E can be expressed as

$$E = \frac{y^2 V_0}{d}$$

With the values 1.25235 and 3.5953 for y, the energies are

```
In[22] := (1.25235^2) Vo/16

Out[22] = 0.0980238 Vo
```

```
In[23] := (3.5953^2) Vo/16

Out[23] = 0.807886 Vo
```

These are the energy eigenvalues of the ground state and the second excited state. (For further discussion of the square well potential, see Section 6.4.) ∎

2.3.1.5 DATA PLOTS

`ListPlot[{{x1, y1}, {x2, y2}, ... }]` plots a list of points with specified x and y coordinates.

EXAMPLE 2.3.9   Consider the hot Volkswagen example in Section 2.1.13. (a) Plot the data. (b) Plot the least-squares fit curve. (c) Superpose the two plots to examine how well the curve fits the data.

Here is the data:

```
In[24] := vwdata = {{0, 0}, {1, 1.5}, {2, 6.0}, {3, 13.5},
                    {4, 24.0}, {5, 37.5}, {6, 54.0},
                    {7, 73.5}, {8, 96.0}, {9, 121.5}};
```

Let us plot the data:

```
In[25] := ListPlot[vwdata, PlotStyle -> PointSize[0.025],
                   AxesLabel -> {" t (s)", " d (m)"}];
```

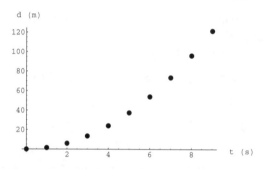

We then plot the least-squares fit curve, namely, $1.5t^2$:

```
In[26] := Plot[1.5 t^2, {t, 0, 10}, PlotRange -> {0, 120},
              AxesLabel -> {" t (s)", " d (m)"}];
```

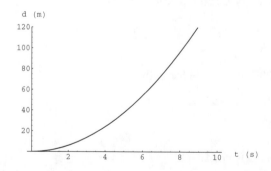

The function **Show** in the form **Show[*plot1*, *plot2*, ... , *options*]** allows us to superpose the two plots to see how well the curve fits the data. (Since **Show** redraws the plots with the original points, some options such as **PlotPoints** and **PlotStyle** cannot be used.)

In[27] := **Show[%, %%];**

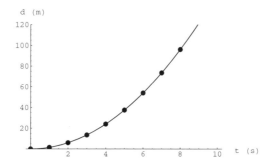

The least-squares fit is excellent.   ■

EXAMPLE 2.3.10 Consider again the hot Volkswagen example. This time, display the last plot without first showing the other two.

Suppose we are only interested in seeing how well the least-squares curve fits the data. There is no reason for taking time to render the plots for the data and the least-squares curve separately. To suppress the display of a plot, set the option **DisplayFunction** to **Identity**.

Let us produce the graphics objects without displaying them:

In[28] := **ListPlot[vwdata, PlotStyle -> PointSize[0.025],**
              DisplayFunction -> Identity];

In[29] := **Plot[1.5 t^2, {t, 0, 10}, PlotRange -> {0, 120},**
              DisplayFunction -> Identity];

To display the two plots together, use the function **Show** with the option **DisplayFunction** returned to its default value **$DisplayFunction**:

In[30] := **Show[%, %%, AxesLabel -> {" t (s)", " d (m)"},**
              DisplayFunction -> $DisplayFunction];

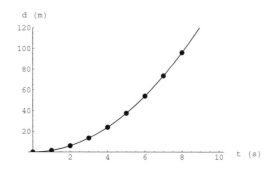

In[31] := **Clear[vwdata]**   ■

2.3.1.6 PARAMETRIC PLOTS

`ParametricPlot[{fx, fy}, {t, tmin, tmax}]` makes a parametric plot with x and y coordinates fx and fy that are functions of the parameter t.

EXAMPLE 2.3.11 Electrons in an oscilloscope are deflected by two mutually perpendicular electric fields in such a way that at any time t the displacement is given by

$$x = A_x \cos(\omega_x t + \phi_x)$$

$$y = A_y \cos(\omega_y t + \phi_y)$$

where $A_x, \omega_x,$ and ϕ_x are, respectively, the amplitude, angular frequency, and phase constant for the horizontal deflection, and $A_y, \omega_y,$ and ϕ_y are those for the vertical deflection. Plot the path of the electrons when (a) $A_x = A_y = A, \omega_x = \omega_y = \omega, \phi_x = 0,$ and $\phi_y = \pi/2;$ (b) $A_x = A_y = A, \omega_x = \omega_y/3 = \omega, \phi_x = 0,$ and $\phi_y = \pi/2;$ (c) $A_x = A_y = A, \omega_x = \omega_y/3 = \omega, \phi_x = 0,$ and $\phi_y = -\pi/4;$ (d) $A_x = A_y = A, \omega_x/5 = \omega_y/4 = \omega, \phi_x = 0,$ and $\phi_y = \pi/2.$
Let the displacements and the time be measured in units of A and $1/\omega$, respectively.

(a) $A_x = A_y = 1, \omega_x = \omega_y = 1, \phi_x = 0,$ and $\phi_y = \pi/2$

```
In[32] := ParametricPlot[{Cos[t], Cos[t + Pi/2]}, {t, 0, 2Pi},
              AxesLabel -> {" x (A)", " y (A)"}];
```

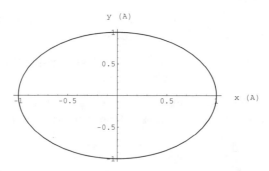

The curve looks like an ellipse rather than a circle because the option `AspectRatio`, which specifies the ratio of height to width for the plot, has the default value of `1/GoldenRatio`. When `AspectRatio` is set to `Automatic`, `ParametricPlot` preserves the true shape of the curve:

```
In[33] := Show[%, AspectRatio -> Automatic,
              AxesLabel -> {" x (A)", " y (A)"}];
```

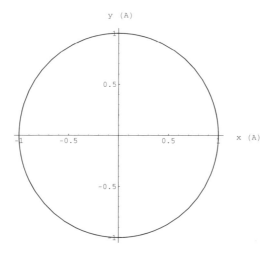

The path of the electrons is a circle, as expected.

(b) $A_x = A_y = 1, \omega_x = \omega_y/3 = 1, \phi_x = 0,$ and $\phi_y = \pi/2$

```
In[34] := ParametricPlot[{Cos[t], Cos[3t + Pi/2]}, {t, 0, 2Pi},
                AspectRatio -> Automatic,
                AxesLabel -> {" x (A)", " y (A)"}];
```

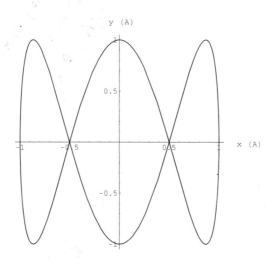

This is a Lissajous curve.

(c) $A_x = A_y = 1, \omega_x = \omega_y/3 = 1, \phi_x = 0,$ and $\phi_y = -\pi/4$

```
In[35] := ParametricPlot[{Cos[t], Cos[3t - Pi/4]}, {t, 0, 2Pi},
                AspectRatio -> Automatic,
                AxesLabel -> {" x (A)", " y (A)"}];
```

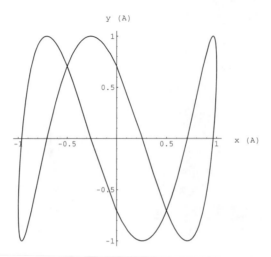

The shape of the Lissajous curve depends on the phase difference $\phi_x - \phi_y$, which is $\pi/4$ in this case whereas it was $-\pi/2$ in part (b).

(d) $A_x = A_y = 1, \omega_x/5 = \omega_y/4 = 1, \phi_x = 0$, and $\phi_y = \pi/2$

```
In[36] := ParametricPlot[{Cos[5t], Cos[4t + Pi/2]}, {t, 0, 2Pi},
                AspectRatio -> Automatic,
                AxesLabel -> {" x (A)", " y (A)"}];
```

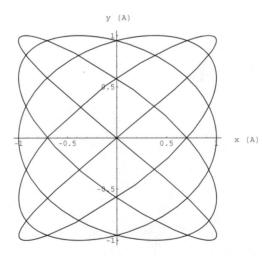

This Lissajous curve is a little more complex than the previous ones.   ■

2.3.2 *Three-Dimensional Graphics*

2.3.2.1 SURFACE PLOTS

`Plot3D[f, {x, xmin, xmax}, {y, ymin, ymax}]` generates a three-dimensional plot of f as a function of the variables x from *xmin* to *xmax* and y from *ymin* to *ymax*.

EXAMPLE 2.3.12 Two point charges, $+Q$ and $-Q$, are placed at $(-b, 0)$ and $(b, 0)$, respectively. For points on the xy plane, the electric potential V produced by these charges is

$$V = kQ \left(\frac{1}{\sqrt{(x + b)^2 + y^2}} - \frac{1}{\sqrt{(x - b)^2 + y^2}} \right)$$

where k is the Coulomb constant. Plot V as a function of x and y.

In units of kQ/b, V can be written as

$$V = \frac{1}{\sqrt{(x + 1)^2 + y^2}} - \frac{1}{\sqrt{(x - 1)^2 + y^2}}$$

where x and y are in units of b.

Let us plot V with the function `Plot3D`:

```
In[1] := Plot3D[1/Sqrt[(x + 1)^2 + y^2] - 1/Sqrt[(x - 1)^2 + y^2],
            {x, -15, 15}, {y, -15, 15},
            Ticks -> {Automatic, Automatic, {-0.04, 0, 0.04}},
            AxesLabel -> {"x (b)", "y (b)", "V (kQ/b)"}
         ];
```

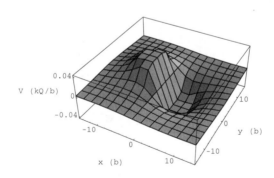

Mathematica has no problem handling the singularities of the potential at $(-b, 0)$ and $(b, 0)$. The option `Ticks` specifies tick marks for axes:

`Ticks -> None`	draw no tick marks
`Ticks -> Automatic`	place tick marks automatically
`Ticks -> {xtick, ytick, ztick}`	specify tick mark options separately for each axis; tick mark option {*coord1, coord2, . . .*} draws tick marks at the specified positions

To show the essential features of the dipole potential, we specify the following options for `Plot3D`:

`PlotRange -> {`*Vmin*`, `*Vmax*`}`	specify the range of values for V
`PlotPoints -> ` *n*	change the number of points in each direction, from the default value of 15 to n, for sampling the function
`BoxRatios -> {`*rx*`, `*ry*`, `*rz*`}`	change the ratios of side lengths for the bounding box from the default values of {1, 1, 0.4} to {rx, ry, rz}
`Boxed -> False`	remove the bounding box
`Axes -> False`	remove the axes

```
In[2] := Plot3D[1/Sqrt[(x + 1)^2 + y^2] - 1/Sqrt[(x - 1)^2 + y^2],
            {x, -15, 15}, {y, -15, 15},
            PlotRange -> {-0.125, 0.125},
            PlotPoints -> 40, BoxRatios -> {1, 1, 1},
            Boxed -> False, Axes -> False
            ];
```

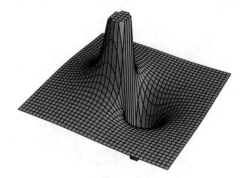

■

2.3.2.2 VIEWPOINT

The option `ViewPoint -> {`*x*`, `*y*`, `*z*`}` specifies the coordinates of the point from which graphic objects are viewed. The origin of the right-handed coordinate system is at the center of the bounding box. The axes are parallel to the edges of the box and scaled so that the longest side of the box equals one. Some special values for `ViewPoint` are

`{1.3, -2.4, 2}`	default
`{0, -2, 0}`	directly in front
`{0, -2, 2}`	in front and up
`{0, 0, 2}`	directly above

The Macintosh and Windows front ends allow us to choose view points interactively. The procedure is

1. Place the text insertion point at a position for including an option in the function `Plot3D` or `Show`.

2. Choose **3D ViewPoint Selector** in the Input menu for *Mathematica* version 3.0 or the Prepare Input submenu of the Action menu for *Mathematica* version 2.2. The following dialog box or a similar one appears:

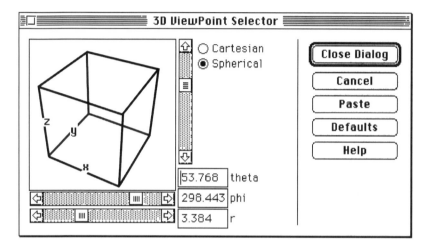

3. Drag the bounding box with the mouse to a desired orientation.

4. Click the **Paste** button in the dialog box. The `ViewPoint` option is pasted in the function `Plot3D` or `Show`.

EXAMPLE 2.3.13 Use the **3D ViewPoint Selector** to specify two viewpoints for looking at the dipole potential in Example 2.3.12.

```
In[3] := Show[%, ViewPoint -> {-1.770, 2.838, 0.514}];
```

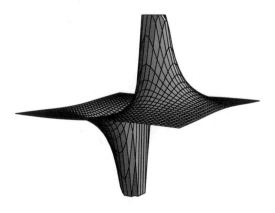

```
In[4] := Show[%, ViewPoint -> {0.806, -3.072, 1.168}];
```

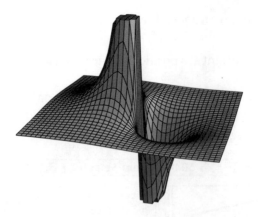

2.3.3 *Animation*

Mathematica simulates motion by displaying a sequence of graphs in rapid succession.

EXAMPLE 2.3.14 Demonstrate the principle of superposition by simulating the interference of two Gaussian pulses traveling along a string in opposite directions.

Let the displacements y_1 and y_2 of the Gaussian pulses be

$$y_1 = A\,e^{-16\alpha(x-vt)^2}$$

$$y_2 = 1.5\,A\,e^{-\alpha(x+vt)^2}$$

These pulses have the same speed v but different amplitudes and widths. The numbers 1.5 and 16 are chosen to make the pulses distinguishable and their interference visible.

If displacement, distance, and time are in units of A, $(1/\alpha)^{1/2}$, and $(1/\alpha v^2)^{1/2}$, respectively, then y_1 and y_2 can be written as

$$y_1 = e^{-16(x-t)^2}$$

$$y_2 = 1.5\,e^{-(x+t)^2}$$

The resultant displacement y equals $y_1 + y_2$. The curves for y_1, y_2, and y are plotted here for a number of values of time t. In each frame, y_1 is at the top and y is at the bottom.

```
In[1] := Do[Plot[{Exp[-16 (x - i)^2] + 1.5 Exp[-(x + i)^2],
               1.5 Exp[-(x + i)^2] + 3,
               Exp[-16 (x - i)^2] + 5},
              {x, -3.0, 3.0},
              PlotStyle -> {{Thickness[0.008]},
                            {Thickness[0.005]},
                            {Thickness[0.005]}},
              PlotRange -> {0, 6.25}, Axes -> False,
              Frame -> True, FrameTicks -> None
              ],
         {i, -2.0, 3.5, 0.5}
         ];
```

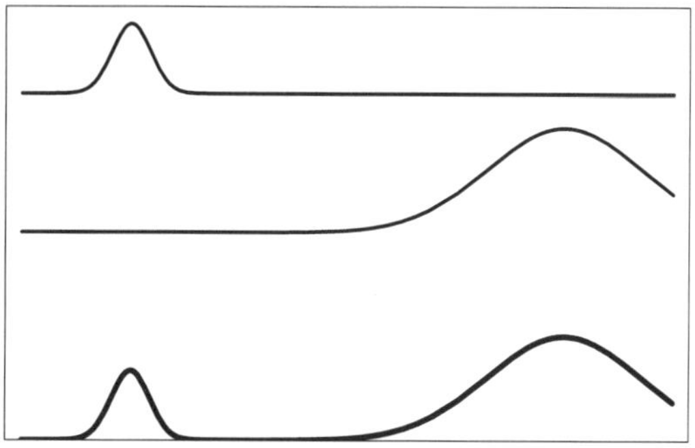

Mathematica displayed the 12 graphics cells in a column. Except for Windows version of *Mathematica* 2.2, these cells belong automatically to a cell group, if the **Automatic Grouping** command from the Cell menu or from the Cell Grouping submenu of the Cell menu is chosen (which is the default for new notebooks). For Windows version of *Mathematica* 2.2, we can manually group these cells by selecting all of them and then choosing the **Group Cells** command from the Cell menu. Only the first cell in the group, called the head cell, is visible because we have closed the group, that is, double-clicked the group bracket.

To simulate the interference of the two pulses, click the group bracket, if it is not already selected, and press Command–Y for the Macintosh or Control+Y for Windows. The panel of buttons at the bottom of the notebook window can be used to adjust the speed and direction of the animation. Slow down or speed up the animation by clicking the two buttons on the right. For Windows version of *Mathematica* 2.2, make sure the **Status Bar** command in the Options menu is selected. To stop the animation, click anywhere in the notebook window.

3.0 When *Mathematica* comes to the end of a sequence of graphs during an animation, it begins immediately again with the first graph of the sequence. To separate the two sequences, we can specify the length of time that the animation will pause at the last frame of the sequence: Open the graphics cell group by double-clicking the group bracket, select the last graphics cell, choose the **Option Inspector** command in the Format menu, open Graphics Options by clicking its button, open Animation, increase the setting of AnimationDisplayTime to about eight times its current value, click Apply, and finally close the dialog box. If **Option Inspector** refuses any changes in AnimationDisplayTime, try another approach: Open the graphics cell group by double-clicking the group bracket, select the last graphics cell, press Command–Y for the Macintosh or Control+Y for Windows, click six or more times the two down arrowheads in the panel of buttons at the bottom of the notebook window.

2.2 When *Mathematica* comes to the end of a sequence of graphs during an animation, it begins immediately again with the first graph of the sequence. To separate the two sequences, we can include two graphs with the words "The End" in a subloop. The procedure for creating the animation subloop is

1. Open the cell group, if it is closed, by double-clicking the group bracket.
2. In a text cell immediately below the last graphics cell of the sequence, type a carriage return, the words "The End," a space, and another carriage return.
3. Click the cell bracket and select **Convert to PICT** in the Graph menu. Click "OK."
4. Copy the cell and paste it immediately below.
5. Select the last graphics cell of the sequence and the two cells with "The End." If the Automatic Grouping feature is turned off, choose twice the **Group Cells** command from the Cell menu. Choose **Aligned Selected Graphics** in the Graph menu. Click "OK."
6. Select the two cells with "The End" and choose **Animation** in the Graph menu.
7. In the dialog box, change the number of times the subloop should be repeated to 4, select "Forward loop," and uncheck "Remove all sub-loops." Click "OK." The dialog box is shown here.

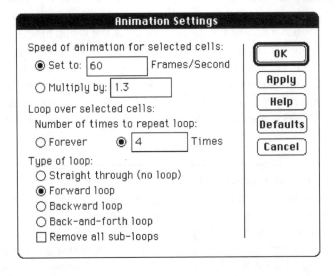

Double-click the group bracket, if the group is open, to close the cell group that includes, for the Macintosh version of *Mathematica* 2.2, the two cells with "The End." To animate, click the group bracket, if it is not already selected, and press Command–Y for the Macintosh or Control+Y for Windows.   ■

EXAMPLE 2.3.15   Create a spinning helix.

The package `Graphics`Shapes`` provides functions for creating the following three-dimensional shapes: cylinder, cone, torus, sphere, möbius strip, helix, and double helix. The package `Graphics`Animation`` contains the function `SpinShow` for rotating three-dimensional graphics objects. `SpinShow[Graphics3D[`*Shape*`[]],` *opts*`]` generates, with options *opts,* a sequence of three-dimensional graphics objects of a *Shape* for an animated rotation. (To display a shape without animation, use the function `Show` instead of `SpinShow`.)

```
In[2] := Needs["Graphics`Shapes`"]
         Needs["Graphics`Animation`"]
         SpinShow[Graphics3D[Helix[]], Boxed -> False,
                Frames -> 16];
```

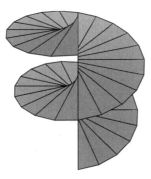

(Only the head cell of the graphics cell group of 16 cells is shown here.) To spin the helix, select all 16 graphics cells, and press Command–Y for the Macintosh or Control+Y for Windows.   ■

2.3.4   Exercises

1. (a) Plot the curve representing the best fit obtained in Exercise 7 of Section 2.1.19. (b) Plot the given points. (c) Superpose the two graphs to see if the fit is acceptable.

Answers:

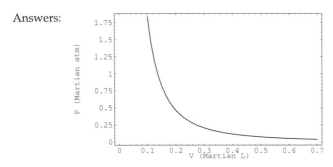

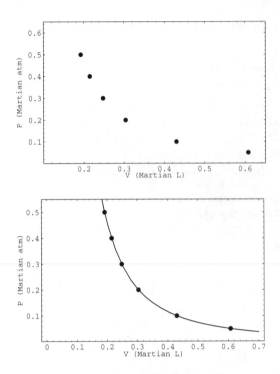

2. (a) Plot the curve representing the best fit obtained in Exercise 8 of Section 2.1.19. (b) Plot the given points. (c) Superpose the two graphs to see if the fit is acceptable.

Answers:

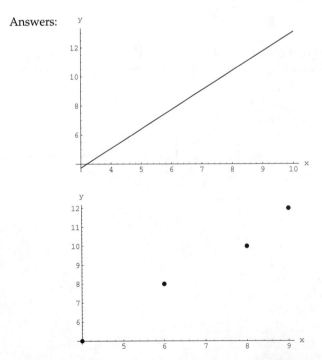

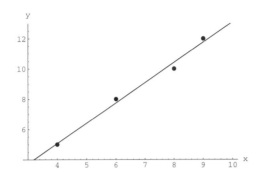

3. Plot the hyperbolic paraboloid given by the equation

$$\frac{x^2}{a^2} - \frac{y^2}{b^2} = 2cz$$

with $a = b = 1$ and $c = 1/2$.

Answer:

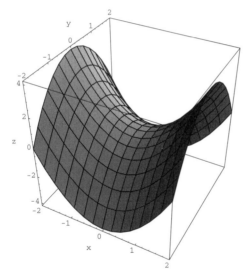

4. Plot the elliptic paraboloid given by the equation

$$\frac{x^2}{a^2} + \frac{y^2}{b^2} = 2cz$$

with $a = b = 1$ and $c = 1/2$.

Answer:

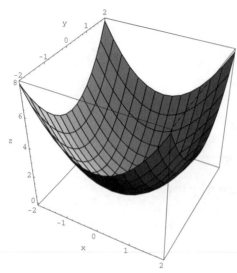

5. Determine the first five positive roots, to six decimal digits of accuracy, of the Bessel function of the first kind $J_0(x)$.

   Answers: {{x -> 2.40483}, {x -> 5.52008}, {x -> 8.65373}, {x -> 11.7915}, {x -> 14.9309}}

6. The normalized eigenfunctions of the one-dimensional harmonic oscillator are

$$u_n(x) = 2^{-n/2}(n!)^{-1/2}\left(\frac{2m\omega}{h}\right)^{1/4} H_n(\sqrt{2\pi m\omega/h}\,x)\exp\left(-\frac{\pi m\omega}{h}x^2\right)$$

   where m, ω, h, n, and H_n are oscillator mass, oscillator frequency, Planck's constant, non-negative integers, and Hermite polynomials, respectively. Plot u_0, u_2, and u_4 together. Give a different style to each curve.

Answer:

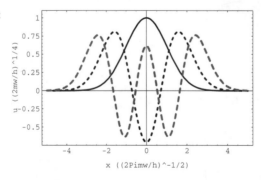

7. The equation of an ellipse is

$$\frac{x^2}{a^2} + \frac{y^2}{b^2} = 1$$

   where $a = 2b$. Plot the ellipse exhibiting its true shape.

Answer:

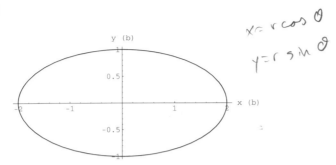

8. Plot the ellipsoid given by the equation

$$\frac{x^2}{a^2} + \frac{y^2}{b^2} + \frac{z^2}{c^2} = 1$$

where $a = 2b = 2c$. *Hint:* Use on-line help to obtain information on the function `ParametricPlot3D`.

Answer:

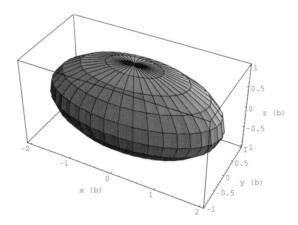

9. In Example 2.3.14, *Mathematica* displayed the graphics cells in a column. With the function `GraphicsArray`, display the graphics in a rectangular array. *Hint:* `Show[GraphicsArray[{{plot11, plot12, ...}, ...},GraphicsSpacing -> {h, v}]]` draws a rectangular array of plots with the specified horizontal spacing h and vertical spacing v between the plots. Define

```
g[i_] := Plot[{Exp[-16 (x - i)^2] + 1.5 Exp[-(x + i)^2],
          1.5 Exp[-(x + i)^2] + 3,
          Exp[-16 (x - i)^2] + 5},
          {x, -3.0, 3.0},
          PlotStyle -> {{Thickness[0.008]},
                        {Thickness[0.005]},
                        {Thickness[0.005]}},
```

```
        PlotRange -> {0, 6.25}, Axes -> False,
        Frame -> True, FrameTicks -> None,
        DisplayFunction -> Identity
    ]
```

Use on-line help or see Section 2.4.6 to obtain information on the function `Partition`.

Answer:

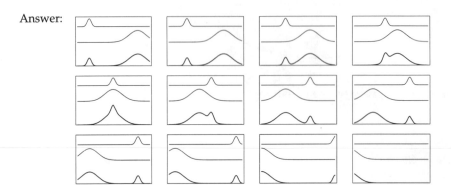

2.4   Lists

Section 2.1.11 discussed representations of vectors and matrices by lists and lists of lists, respectively. *Mathematica* employs lists in many other ways. This section provides a general definition of lists and shows how they are created, used, and manipulated.

2.4.1   Defining Lists

Lists are sequences of *Mathematica* objects separated by commas and enclosed by curly brackets. In Section 2.1.11, the objects were numbers or lists of numbers. In general, the objects can be any expressions such as symbols, character strings, functions, equations, or transformation rules. In a list, the objects are usually, but not necessarily, of the same type.

Built-in *Mathematica* functions provide many examples of lists. They often take lists as arguments and return values in terms of lists. For instance, consider using the function `Solve` to solve the system of equations

$$x^2 - 5x - y + 4 = 0$$

$$x - 4y = 1$$

In[1] := `Solve[{x^2 - 5 x - y + 4 == 0, x - 4 y == 1}, {x, y}]`

Out[1] = $\{\{y \rightarrow 0, x \rightarrow 1\}, \{y \rightarrow \frac{13}{16}, x \rightarrow \frac{17}{4}\}\}$

The first argument of `Solve` is a list of equations; the second argument, a list of variables. The function produces a list of lists of transformation rules for the variables x and y. For another example, `Options[Plot3D]` returns a list of transformation rules for the default options of the function `Plot3D`:

In[2] := **Options[Plot3D]**

Out[2] = {AmbientLight -> GrayLevel[0], AspectRatio -> Automatic,
        Axes -> True, AxesEdge -> Automatic, AxesLabel -> None,
        AxesStyle -> Automatic, Background -> Automatic, Boxed -> True,
        BoxRatios -> {1, 1, 0.4}, BoxStyle -> Automatic,
        ClipFill -> Automatic, ColorFunction -> Automatic,
        ColorOutput -> Automatic, Compiled -> True,
        DefaultColor -> Automatic, Epilog -> {}, FaceGrids -> None,
        HiddenSurface -> True, ImageSize -> Automatic, Lighting -> True,
        LightSources ->
         {{{1., 0., 1.}, RGBColor[1, 0, 0]},
          {{1., 1., 1.}, RGBColor[0, 1, 0]},
          {{0., 1., 1.}, RGBColor[0, 0, 1]}}, Mesh -> True,
        MeshStyle -> Automatic, Plot3Matrix -> Automatic,
        PlotLabel -> None, PlotPoints -> 15, PlotRange -> Automatic,
        PlotRegion -> Automatic, Prolog -> {}, Shading -> True,
        SphericalRegion -> False, Ticks -> Automatic,
        ViewCenter -> Automatic, ViewPoint -> {1.3, -2.4, 2.},
        ViewVertical -> {0., 0., 1.}, DefaultFont :> $DefaultFont,
        DisplayFunction :> $DisplayFunction, FormatType :> $FormatType,
        TextStyle :> $TextStyle}

(The list of default options is somewhat different in *Mathematica* version 2.2.)

2.4.2 *Generating, Displaying, and Sizing Lists*

In addition to the function **Table** introduced in Section 2.1.18, several other functions for
generating lists are

Range[*imax*]	generate the list {1, 2, ..., *imax*}
Range[*imin*, *imax*]	generate the list {*imin*, *imin* + 1, ..., *imax*}
Range[*imin*, *imax*, *di*]	generate the list {*imin*, *imin* + *di*, ..., *imax*}
Array[*a*, *n*]	create the list {*a*[1], *a*[2], ..., *a*[*n*]}
Array[*a*, {*n1*, *n2*, ...}]	create a *n1* $\times$ *n2* $\times$ ... nested list, with elements *a*[*i1*, *i2*, ...]

The functions for displaying lists are

ColumnForm[*list*]	display the elements of *list* in a column
MatrixForm[*list*]	display the elements of *list* in a regular array of square cells of the same size
TableForm[*list*]	display the elements of *list* in an array of rectangular cells that are not necessarily of the same size

The functions for determining the sizes of lists are

Length[*list*]	give the number of elements in *list*
Dimensions[*list*]	give a list of the dimensions of *list*

We can, for example, generate a list of even numbers from 10 to 21 with the function Range, determine the number of elements in the list with the function Length, and display the elements in a column with the function ColumnForm:

In[3] := **Range[10, 21, 2]**

Out[3] = {10, 12, 14, 16, 18, 20}

In[4] := **Length[%]**

Out[4] = 6

In[5] := **ColumnForm[%%]**

Out[5] = 10
10
12
14
16
18
20

For another example, let us use the function Table to create the list $\{1 + \sin x, 1 + \sin^3 x, 1 + \sin^5 x, 1 + \sin^7 x\}$:

In[6] := **Table[1 + Sin[x]^n, {n, 1, 7, 2}]**

Out[6] = {1 + Sin[x], 1 + Sin[x]^3, 1 + Sin[x]^5, 1 + Sin[x]^7}

We can also create a list of the cubes of the first ten integers. First define a function for returning the cube of its argument:

In[7] := **c[x_] := x^3**

The function Array then gives the list:

In[8] := **Array[c, 10]**

Out[8] = {1, 8, 27, 64, 125, 216, 343, 512, 729, 1000}

For the final example, we generate a 3×4 nested list with elements $x^i + y^j$, verify its dimensions, and display the elements in a rectangular array. The function Table generates the nested list:

In[9] := **Table[x^i + y^j, {i, 3}, {j, 4}]**

Out[9] = {{x + y, x + y^2, x + y^3, x + y^4},
{x^2 + y, x^2 + y^2, x^2 + y^3, x^2 + y^4},
{x^3 + y, x^3 + y^2, x^3 + y^3, x^3 + y^4}}

The function Dimensions gives the dimensions of the nested list:

In[10] := **Dimensions[%]**

Out[10] = {3, 4}

The function `TableForm` displays the nested list in tabular form:

```
In[11] := TableForm[%%]
```

```
Out[11]//TableForm=
```

$$
\begin{array}{llll}
x + y & x + y^2 & x + y^3 & x + y^4 \\
x^2 + y & x^2 + y^2 & x^2 + y^3 & x^2 + y^4 \\
x^3 + y & x^3 + y^2 & x^3 + y^3 & x^3 + y^4
\end{array}
$$

```
In[12] := Clear[c]
```

2.4.3   Obtaining List and Sublist Elements

The functions for obtaining elements of lists are

`First`[*list*]	the first element
`Last`[*list*]	the last element
`Part`[*list*, *n*] or *list*[[*n*]]	the *n*th element
`Part`[*list*, {*n1*, *n2*, ...}] or *list*[[{*n1*, *n2*, ...}]]	the list of the *n1*th, *n2*th, ... elements
`Take`[*list*, *n*]	the list of the first *n* elements
`Take`[*list*, {*m*, *n*}]	the list of the *m*th through *n*th elements
`Rest`[*list*]	*list* without the first element
`Drop`[*list*, *n*]	*list* without the first *n* elements
`Drop`[*list*, {*m*, *n*}]	*list* without the *m*th through *n*th elements

A function for selecting elements of sublists is

`Part`[*list*, *i*, *j*, ...] or *list*[[*i*, *j*, ...]]	the element at position $\{i, j, ...\}$ in *list*

❀ **3.0**   Other ways to extract elements of sublists are

`Extract`[*list*, {*i*, *j*, ...}]   the element at position $\{i, j, ...\}$ in *list*, i.e., `Part`[*list*, *i*, *j*, ...]

`Extract`[*list*, {{*i1*, *j1*, ...}, {*i2*, *j2*, ...}, ...}]
the list of elements at positions $\{i1, j1, ...\}, \{i2, j2, ...\}$,
...

To illustrate the use of these functions, consider two examples.

EXAMPLE 2.4.1   (a) Create the list $\{x \cos x, x^2 \cos 2x, x^3 \cos 3x, x^4 \cos 4x, x^5 \cos 5x, x^6 \cos 6x\}$. (b) Pick out the fourth element. (c) Obtain the list of the third and fifth elements. (d) Get the list with the first three elements. (e) Obtain the list without the second through fourth elements.

(a) Let us generate the list with `Table` and name it `mylist`:

In[13] := **mylist = Table[x^n Cos[n x], {n, 6}]**

Out[13] = $\{x \, Cos[x], \, x^2 \, Cos[2 \, x], \, x^3 \, Cos[3 \, x], \, x^4 \, Cos[4 \, x], \, x^5 \, Cos[5 \, x],$
$x^6 \, Cos[6 \, x]\}$

(b) Now select the fourth element:

In[14] := **mylist[[4]]**

Out[14] = $x^4 \, Cos[4 \, x]$

(c) Here is the list of the third and fifth elements:

In[15] := **mylist[[{3, 5}]]**

Out[15] = $\{x^3 \, Cos[3 \, x], \, x^5 \, Cos[5 \, x]\}$

(d) The function `Take` returns the list with the first three elements:

In[16] := **Take[mylist, 3]**

Out[16] = $\{x \, Cos[x], \, x^2 \, Cos[2 \, x], \, x^3 \, Cos[3 \, x]\}$

(e) The function `Drop` gives the list without the second through fourth elements:

In[17] := **Drop[mylist, {2, 4}]**

Out[17] = $\{x \, Cos[x], \, x^5 \, Cos[5 \, x], \, x^6 \, Cos[6 \, x]\}$

In[18] := **Clear[mylist]**

∎

EXAMPLE 2.4.2 (a) Create a 3×4 list of lists with elements c_{ij}. (b) Display the nested list in standard matrix form. (c) Select the third element of the list. (d) Pick out the second element of the first sublist.

(a) The function `Array` generates the nested list with symbolic elements:

In[19] := **nestedlist = Array[c, {3, 4}]**

Out[19] = $\{\{c[1, 1], \, c[1, 2], \, c[1, 3], \, c[1, 4]\},$
$\{c[2, 1], \, c[2, 2], \, c[2, 3], \, c[2, 4]\},$
$\{c[3, 1], \, c[3, 2], \, c[3, 3], \, c[3, 4]\}\}$

(b) The function `MatrixForm` displays the nested list in standard matrix form:

In[20] := **MatrixForm[nestedlist]**

Out[20]//MatrixForm=

c[1, 1]	c[1, 2]	c[1, 3]	c[1, 4]
c[2, 1]	c[2, 2]	c[2, 3]	c[2, 4]
c[3, 1]	c[3, 2]	c[3, 3]	c[3, 4]

(c) Here is the third element of the list:

In[21] := `nestedlist[[3]]`

Out[21] = `{c[3, 1], c[3, 2], c[3, 3], c[3, 4]}`

(d) Finally, select the second element of the first sublist:

In[22] := `nestedlist[[1, 2]]`

Out[22] = `c[1, 2]`

❀ **3.0**   An equivalent command is

In[23] := `Extract[nestedlist, {1, 2}]`

Out[23] = `c[1, 2]`

(For a comparison of the functions `Part` and `Extract`, see Exercise 14 of Section 2.4.10.)

In[24] := `Clear[nestedlist]`   ∎

The functions in this section pick out elements according to their positions in the list. The function `Position` allows us to determine the positions of the elements. `Position[`*list*`,` *elem*`]` gives a list of the positions at which *elem* appears. For example, we can determine the positions of the symbol *a* in the list $\{a, \{a, b, a\}, b, \{\{a, b\}, c\}\}$:

In[25] := `Position[{a, {a, b, a}, b, {{a, b}, c}}, a]`

Out[25] = `{{1}, {2, 1}, {2, 3}, {4, 1, 1}}`

So far we have considered functions that relate to the positions of elements in lists. We now examine a function that selects list elements based on their properties rather than positions:

`Select[`*list*`,` *criterion*`]`	pick out all elements *ei* of *list* for which *criterion*[*ei*] is `True`
`Select[`*list*`,` *criterion*`,` *n*`]`	pick out the first *n* elements for which *criterion*[*ei*] is `True`

Built-in predicates can be used as criteria for the `Select` function. A predicate is a function that returns `True` or `False` in testing an element. Some built-in predicates are

`DigitQ`	`EvenQ`
`IntegerQ`	`LetterQ`
`ListQ`	`LowerCaseQ`
`MachineNumberQ`	`MatrixQ`
`Negative`	`NonNegative`
`NumberQ`	`OddQ`
`OrderedQ`	`Positive`
`PrimeQ`	`StringQ`
`UpperCaseQ`	`ValueQ`
`VectorQ`	

⚜ **3.0**   `NumericQ`

To find out the properties that they test, use on-line help:

> `In[26] := ?EvenQ`

> `EvenQ[expr]` gives True if expr is an even integer, and False
> otherwise.

Let us illustrate the use of `Select` with several examples.

> `In[27] := Select[{1, 2, 3, 4, 5, 6}, EvenQ]`
>
> `Out[27] = {2, 4, 6}`

> `In[28] := Select[{1, 2, 3, 4, 5, 6}, EvenQ, 2]`
>
> `Out[28] = {2, 4}`

`Positive[`x`]` gives `True` if x is a positive number.

> `In[29] := Select[{a, 4 + I 5, x^2, -2, 3, -5, 2/3}, Positive]`
>
> `Out[29] = {3, `$\frac{2}{3}$`}`

`VectorQ[`*expr*`]` gives `True` if *expr* is a list, none of whose elements are themselves lists, and gives `False` otherwise.

> `In[30] := Select[{{a, 2 - 3 I, 7}, x, {-2, {1, -5}}, 2/3}, VectorQ]`
>
> `Out[30] = {{a, 2 - 3 I, 7}}`

The only element that is a list without sublists is `{a, 2 - 3 I, 7}`.

2.4.4   *Changing List and Sublist Elements*

Some functions for inserting, deleting, and replacing list and sublist elements are

`Prepend[`*list*`, `*elem*`]`	insert *elem* at the beginning of *list*
`Append[`*list*`, `*elem*`]`	insert *elem* at the end of *list*
`Insert[`*list*`, `*elem*`, `*i*`]`	insert *elem* at position *i* in *list*
`Insert[`*list*`, `*elem*`, {`*i*`, `*j*`, ...}]`	insert *elem* at position $\{i, j, \ldots\}$ in *list*
`Insert[`*list*`, `*elem*`, {{`*i1*`, `*j1*`, ...}, {`*i2*`, ...}, ...}]`	insert *elem* at positions $\{i1, j1, \ldots\}, \{i2, \ldots\}, \ldots$ in list
`Delete[`*list*`, `*i*`]`	delete the element at position *i* in *list*
`Delete[`*list*`, {`*i*`, `*j*`, ...}]`	delete the element at position $\{i, j, \ldots\}$ in *list*
`Delete[`*list*`, {{`*i1*`, `*j1*`, ...}, {`*i2*`, ...}, ...}]`	delete elements at positions $\{i1, j1, \ldots\}, \{i2, \ldots\}, \ldots$ in *list*
`ReplacePart[`*list*`, `*elem*`, `*i*`]`	replace the element at position *i* in *list* with *elem*

`ReplacePart[list, elem,` `{i, j, ...}]`	replace the element at position $\{i, j, ...\}$ with *elem*
`ReplacePart[list, elem, {{i1,` `j1, ...}, {i2, ...}, ...}]`	replace elements at positions $\{i1, j1, ...\}, \{i2, ...\}, ...,$ with *elem*

Let us illustrate the use of these functions with an example.

EXAMPLE 2.4.3 Consider the list $\{a, \{a, b, a\}, b, \{\{a, b\}, c\}, d\}$. In this list, (a) add x at the beginning, (b) delete d, (c) delete c, (d) insert y at positions $\{2\}, \{2, 2\}$, and $\{4, 1, 2\}$, and (e) replace a with z.

Let us name the list `ourlist`:

```
In[31] := ourlist = {a, {a, b, a}, b, {{a, b}, c}, d}
Out[31] = {a, {a, b, a}, b, {{a, b}, c}, d}
```

(a) Prepend **x** to the list:

```
In[32] := Prepend[ourlist, x]
Out[32] = {x, a, {a, b, a}, b, {{a, b}, c}, d}
```

(b) Delete **d** from the list:

```
In[33] := Delete[ourlist, 5]
Out[33] = {a, {a, b, a}, b, {{a, b}, c}}
```

(c) Delete **c** from the list:

```
In[34] := Delete[ourlist, {4, 2}]
Out[34] = {a, {a, b, a}, b, {{a, b}}, d}
```

(d) Insert **y** at several positions:

```
In[35] := Insert[ourlist, y, {{2}, {2, 2}, {4, 1, 2}}]
Out[35] = {a, y, {a, y, b, a}, b, {{a, y, b}, c}, d}
```

(e) Replace **a** with **z** in the list:

```
In[36] := ReplacePart[ourlist, z, Position[ourlist, a]]
Out[36] = {z, {z, b, z}, b, {{z, b}, c}, d}
```

The function `Position[ourlist, a]` gives a list of positions of the symbol **a** in `ourlist`. Of course, we can also replace **a** with **z** in the list with a transformation rule:

```
In[37] := ourlist /. a -> z
Out[37] = {z, {z, b, z}, b, {{z, b}, c}, d}

In[38] := Clear[ourlist]
```

∎

2.4.5 Rearranging Lists

Some functions for rearranging lists are

Sort [*list*]	sort the elements of *list* into canonical order
Union [*list*]	give a sorted version of *list*, in which all duplicated elements have been dropped
Reverse [*list*]	reverse the order of the elements in *list*
RotateLeft [*list*]	cycle the elements in *list* one position to the left
RotateLeft [*list*, *n*]	cycle the elements in *list* *n* positions to the left
RotateRight [*list*]	cycle the elements in *list* one position to the right
RotateRight [*list*, *n*]	cycle the elements in *list* *n* positions to the right
Permutations [*list*]	generate a list of all possible permutations of the elements in *list*

For example, we can sort the elements of the list $\{abd, abc, ab, a, d, 10, a, 10, 5.5, 1/4, 23/2, ada\}$ into a standard order:

```
In[39] := Sort[{abd, abc, ab, a, d, 10, a, 10, 5.5,
                1/4, 23/2, ada}]
```

$$Out[39] = \{\frac{1}{4}, 5.5, 10, 10, \frac{23}{2}, a, a, ab, abc, abd, ada, d\}$$

Mathematica returns a list in which numbers and symbols are arranged in increasing and alphabetical order, respectively. Furthermore, numbers precede symbols. We can also sort the elements of the list and remove duplicated elements:

```
In[40] := Union[{abd, abc, ab, a, d, 10, a, 10, 5.5,
                 1/4, 23/2, ada}]
```

$$Out[40] = \{\frac{1}{4}, 5.5, 10, \frac{23}{2}, a, ab, abc, abd, ada, d\}$$

For another example, let us rotate the elements of the list of the first ten integers by one position to the left,

```
In[41] := RotateLeft[Range[10]]
```

$$Out[41] = \{2, 3, 4, 5, 6, 7, 8, 9, 10, 1\}$$

or three places to the right,

```
In[42] := RotateRight[Range[10], 3]
```

$$Out[42] = \{8, 9, 10, 1, 2, 3, 4, 5, 6, 7\}$$

For the final example, we give all possible permutations of the list $\{a, b, c\}$:

```
In[43] := Permutations[{a, b, c}]
```

$$Out[43] = \{\{a, b, c\}, \{a, c, b\}, \{b, a, c\}, \{b, c, a\}, \{c, a, b\}, \{c, b, a\}\}$$

2.4.6 Restructuring Lists

Some functions for restructuring lists are

`Partition[list, n]`	partition *list* into nonoverlapping sublists of length n
`Partition[list, n, d]`	generate sublists with offset d
`Transpose[list]`	transpose the first two levels in *list*
`Flatten[list]`	flatten out nested lists
`Flatten[list, n]`	flatten out the top n levels
`FlattenAt[list, i]`	flatten out a sublist that appears as the ith element of *list*
`FlattenAt[list, {i, j, ...}]`	flatten out the element of *list* at position $\{i, j, ...\}$
`FlattenAt[list, {{i1, j1, ...}, {i2, j2, ...}, ...}]`	flatten out elements of *list* at several positions

Let us illustrate the use of these functions with two examples.

EXAMPLE 2.4.4 Consider the list $\{a, b, c, d, e\}$. (a) Partition the list into sublists of two elements. (b) Form sublists of two elements with an offset of one.

(a) Form sublists of two elements:

```
In[44] := Partition[{a, b, c, d, e}, 2]
Out[44] = {{a, b}, {c, d}}
```

The unpaired element at the end is discarded!

(b) Use an offset of one for successive sublists:

```
In[45] := Partition[{a, b, c, d, e}, 2, 1]
Out[45] = {{a, b}, {b, c}, {c, d}, {d, e}}
```

∎

EXAMPLE 2.4.5 (a) Give the list $\{a, \{b, c\}, d, \{a, \{b, \{c, \{e, f\}\}\}\}\}$ a name. (b) Flatten out the list. (c) Flatten out the sublist at position 2. (d) Flatten out the sublist at position $\{4, 2, 2, 2\}$.

(a) Assign the name `demolist` to the list:

```
In[46] := demolist = {a, {b, c}, d, {a, {b, {c, {e, f}}}}}
Out[46] = {a, {b, c}, d, {a, {b, {c, {e, f}}}}}
```

(b) Eliminate nested lists:

```
In[47] := Flatten[demolist]
Out[47] = {a, b, c, d, a, b, c, e, f}
```

The inner brackets of `demolist` are removed.

(c) Flatten out the sublist at position 2 of the list:

```
In[48] := FlattenAt[demolist, 2]
Out[48] = {a, b, c, d, {a, {b, {c, {e, f}}}}}
```

(d) Flatten out the sublist at position $\{4, 2, 2, 2\}$ of the list:

In[49] := **FlattenAt[demolist, {4, 2, 2, 2}]**

Out[49] = {a, {b, c}, d, {a, {b, {c, e, f}}}}

In[50] := **Clear[demolist]**   ∎

2.4.7 *Combining Lists*

Some functions for combining lists are

Join[*list1*, *list2*, ...]	concatenate lists together
Union[*list1*, *list2*, ...]	give a sorted list of all the distinct elements that appear in any of the *listi*
Intersection[*list1*, *list2*, ...]	give a sorted list of the elements common to all the *listi*
Complement[*listall*, *list1*, *list2*, ...]	give a sorted list of the elements in *listall* that are not in any of the *listi*

Let us consider an example.

EXAMPLE 2.4.6   Name the lists $\{1, 2, 3\}$, $\{2, 3, 4, 5\}$, and $\{1, 2, 3, 4, 5, 6, 7, 8\}$ **listA**, **listB**, and **listC**, respectively. Evaluate (a) **Join[listA, listB]**, (b) **Union[listA, listB]**, (c) **Intersection[listA, listB]**, and (d) **Complement[listC, listA, listB]**.
Create and name the lists:

In[51] := **listA = Range[3]**

Out[51] = {1, 2, 3}

In[52] := **listB = Range[2, 5]**

Out[52] = {2, 3, 4, 5}

In[53] := **listC = Range[8]**

Out[53] = {1, 2, 3, 4, 5, 6, 7, 8}

  (a)

In[54] := **Join[listA, listB]**

Out[54] = {1, 2, 3, 2, 3, 4, 5}

The lists are joined and the order is preserved.

  (b)

In[55] := **Union[listA, listB]**

Out[55] = {1, 2, 3, 4, 5}

  (c)

In[56] := **Intersection[listA, listB]**

Out[56] = {2, 3}

(d)

In[57] := **Complement[listC, listA, listB]**

Out[57] = {6, 7, 8}

The functions **Union**, **Intersection**, and **Complement** return sorted lists with duplicated elements removed.

In[58] := **Clear[listA, listB, listC]**   ∎

2.4.8   Operating on Lists

Built-in mathematical functions in *Mathematica* operate separately on each element of a list. For instance, **D[{x, x^2, x^3, x^4}, x]** differentiates each element of the list:

In[59] := **Table[x^n, {n, 4}]**

Out[59] = {x, x^2, x^3, x^4}

In[60] := **D[%, x]**

Out[60] = {1, 2 x, 3 x^2, 4 x^3}

With the list $\{a, b, c\}$ as its argument, the function **Sin** takes the sine of each element:

In[61] := **Sin[{a, b, c}]**

Out[61] = {Sin[a], Sin[b], Sin[c]}

Also, the function **Sqrt** when applied to the list $\{3, 4, 6, 8\}$ returns a list of the square roots:

In[62] := **Sqrt[{3, 4, 6, 8}]**

Out[62] = {Sqrt[3], 2, Sqrt[6], 2 Sqrt[2]}

Functions that operate separately on each element of a list have the attribute **Listable**. Not all built-in functions have this attribute. For example, the function **Print** does not have the attribute **Listable**:

In[63] := **Print[{a, b, c}]**
            {a, b, c}

In[64] := **{Print[a], Print[b], Print[c]}**

            a

            b

            c

Out[64] = {Null, Null, Null}

Null is a *Mathematica* symbol representing "nothing." **Print[{a, b, c}]** does not return the same result as **{Print[a], Print[b], Print[c]}**. Thus, it is not **Listable**.

User-defined functions are, in general, not **Listable** unless they are assigned this attribute, in a manner to be discussed in Section 3.2.5. For instance, define the function

In[65] := **g[x_] := 2**

and evaluate

In[66] := **g[{a, b, c}]**

Out[66] = 2

The function *g* does not act on each element separately; it treats the list as a single object; it returns the number 2 instead of the list {2, 2, 2}.

In[67] := **Clear[g]**

To apply a function that is not **Listable** to each element of a list, we use the function **Map**. **Map[f, list]** or *f* /@ *list* applies *f* to each element of *list*. For example, we can apply *f* to each element of {*a, b, c*}:

In[68] := **Map[f, {a, b, c}]**

Out[68] = {f[a], f[b], f[c]}

We can also use the special input form "/@":

In[69] := **f /@ {a, b, c}**

Out[69] = {f[a], f[b], f[c]}

Sometimes, it is desirable to apply separately a function that is not **Listable** to only some elements of a list. The function **MapAt[f, list, n]** applies *f* to the element at position *n* of *list*. **MapAt[f, list, {{n1}, {n2}, ...}]** applies *f* to elements at positions *n1, n2, ...* of list. For instance, apply *f* separately to elements *a* and *d* of the list {*a, b, c, {d, e}*}:

In[70] := **MapAt[f, {a, b, c, {d, e}}, {{1}, {4, 1}}]**

Out[70] = {f[a], b, c, {f[d], e}}

The positions of **a** and **d** are {1} and {4, 1}, respectively.

It is often necessary to supply the elements of a list as arguments of a function. **Apply[f, {a, b, ...}]** or *f*@@{*a, b, ...*} gives *f*[*a, b, ...*]. For example, let us add the elements of the list {*a, b, c, d*}:

In[71] := **Plus @@ {a, b, c, d}**

Out[71] = a + b + c + d

Plus[a1, a2, ..., an] returns the sum of *a1, a2, ...,* and *an*.

2.4.9   *Using Lists in Computations*

This section illustrates how lists are used in interactive calculations.

EXAMPLE 2.4.7 The change in entropy between states *a* and *b* of a system is given by

$$S_b - S_a = \int_a^b \frac{dQ}{T}$$

where T is the absolute temperature of the system and dQ is the small amount of heat that enters or leaves the system. The integral is evaluated over any reversible path connecting the two states. For a process in which the pressure is held constant and the temperature varies from T_a to T_b, the equation can be written as

$$S_b - S_a = \int_{T_a}^{T_b} \frac{C_p(T)\, dT}{T}$$

where $C_p(T)$ is the heat capacity at constant pressure of the system. The following data were obtained for rhombic sulfur:

T (K)	C_p (cal/K)	T (K)	C_p (cal/K)
15	0.311	160	4.123
20	0.605	170	4.269
25	0.858	180	4.404
30	1.075	190	4.526
40	1.452	200	4.639
50	1.772	210	4.743
60	2.084	220	4.841
70	2.352	230	4.927
80	2.604	240	5.010
90	2.838	250	5.083
100	3.060	260	5.154
110	3.254	270	5.220
120	3.445	280	5.286
130	3.624	290	5.350
140	3.795	298.1	5.401
150	3.964		

Determine the change in entropy and plot C_p versus T.

A reasonable approximation to the preceding integral is given by the trapezoidal rule

$$\sum_i \left(\frac{C_p(T_{i+1})}{T_{i+1}} + \frac{C_p(T_i)}{T_i} \right) \left(\frac{T_{i+1} - T_i}{2} \right)$$

Assign the name t to the list of temperatures in units of K:

```
In[72] := t = {15, 20, 25, 30, 40, 50, 60, 70, 80, 90, 100, 110,
               120, 130, 140, 150, 160, 170, 180, 190, 200, 210,
               220, 230, 240, 250, 260, 270, 280, 290, 298.1};
```

Give the name cp to the corresponding list of heat capacities in cal/K:

```
In[73] := cp = {0.311, 0.605, 0.858, 1.075, 1.452, 1.772,
                2.084, 2.352, 2.604, 2.838, 3.060, 3.254,
                3.445, 3.624, 3.795, 3.964, 4.123, 4.269,
                4.404, 4.526, 4.639, 4.743, 4.841, 4.927,
                5.010, 5.083, 5.154, 5.220, 5.286, 5.350,
                5.401};
```

The change in entropy evaluated with the trapezoidal rule is

```
In[74] := Sum[(cp[[i+1]]/t[[i+1]] + cp[[i]]/t[[i]])
            ((t[[i+1]] - t[[i]])/2),
            {i, Length[t] - 1}
            ]
```

```
Out[74] = 7.50452
```

with the units cal/K. (Instead of using the trapezoidal rule, we can also compute an approximation to the integral with the function ListIntegrate in the package NumericalMath`ListIntegrate`. See Exercise 12 in Section 2.4.10.)

To obtain a list of $\{t_i, cp_i\}$ from the lists t and cp, use the function Transpose:

```
In[75] := Transpose[{t, cp}]
```

```
Out[75] = {{15, 0.311}, {20, 0.605}, {25, 0.858}, {30, 1.075},
           {40, 1.452}, {50, 1.772}, {60, 2.084}, {70, 2.352},
           {80, 2.604}, {90, 2.838}, {100, 3.06}, {110, 3.254},
           {120, 3.445}, {130, 3.624}, {140, 3.795}, {150, 3.964},
           {160, 4.123}, {170, 4.269}, {180, 4.404}, {190, 4.526},
           {200, 4.639}, {210, 4.743}, {220, 4.841}, {230, 4.927},
           {240, 5.01}, {250, 5.083}, {260, 5.154}, {270, 5.22},
           {280, 5.286}, {290, 5.35}, {298.1, 5.401}}
```

ListPlot gives a plot of C_p versus T:

```
In[76] := ListPlot[%, PlotStyle -> PointSize[0.015],
                   AxesLabel -> {" T (K)", " Cp (cal/K)"}];
```

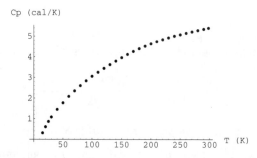

```
In[77] := Clear[t, cp]
```

EXAMPLE 2.4.8 Consider the differential equation

$$\frac{dx}{dt} + x = e^{4t}$$

with $x(0) = 1$. (a) Determine $x(t)$ numerically for t in the range 0 to 0.5. (b) Determine $x(t)$ analytically. (c) Compare the numerical and analytical solutions for t from 0 to 0.5 in steps of 0.1.

(a) To solve the differential equation numerically, use the function **NDSolve** discussed in Section 2.1.17:

```
In[78] := NDSolve[{x'[t] + x[t] == Exp[4 t], x[0] == 1},
              x, {t, 0, 0.5}]

Out[78] = {{x -> InterpolatingFunction[{{0., 0.5}}, <>]}}
```

NDSolve caters to multiple solutions and returns a list of lists of transformation rules. In this case, there is only one solution and one rule. The rule replaces the variable x by the solution that is given in terms of an **InterpolatingFunction**. Let us assign the name **numsol** to the **InterpolatingFunction**:

```
In[79] := numsol = x /. %[[1]]

Out[79] = InterpolatingFunction[{{0., 0.5}}, <>]
```

(b) The function **DSolve** discussed in Section 2.2.11 solves the differential equation analytically:

```
In[80] := DSolve[{x'[t] + x[t] == Exp[4 t], x[0] == 1}, x[t], t]
```

$$Out[80] = \left\{\left\{x[t] \to \frac{4 + E^{5\,t}}{5\,E^t}\right\}\right\}$$

DSolve also allows multiple solutions and returns a nested list of transformation rules. Here again, there is only one solution and one rule. The solution is given as the value of the replacement rule. Let the solution be the body of the definition for a function named **ansol**:

```
In[81] := ansol[t_] = x[t] /. %[[1]]
```

$$Out[81] = \frac{4 + E^{5\,t}}{5\,E^t}$$

(c) Now, generate a list of $\{t_i, \text{numsol}(t_i), \text{ansol}(t_i)\}$ with the function **Table**:

```
In[82] := Table[{t, numsol[t], ansol[t]}, {t, 0, 0.5, 0.1}]

Out[82] = {{0, 1., 1}, {0.1, 1.02224, 1.02223}, {0.2, 1.1001, 1.10009},
          {0.3, 1.25668, 1.25668}, {0.4, 1.52687, 1.52686},
          {0.5, 1.96304, 1.96304}}
```

We can use the functions **Print** and **TableForm** to display the result in a more familiar format:

```
In[83] := Print["\nt      numerical analytical\n"];
          Print[TableForm[%%]]
```

```
t       numerical  analytical
0       1.         1
0.1     1.02224    1.02223
0.2     1.1001     1.10009
0.3     1.25668    1.25668
0.4     1.52687    1.52686
0.5     1.96304    1.96304
```

In[85] := **Clear[numsol, ansol]**   ■

A few words about the function **Print** are in order. **Print** [*expr1, expr2, ...*] prints the expressions *expr1, expr2, ...* with no spaces in between, but with a newline (line feed) at the end. The expressions *expr1, expr2, ...* are evaluated before printing. If an expression has the form "*text*", the string of *text* is printed. The special character \n in a text string indicates a newline. For example, let $y = 2$:

In[86] := **y = 2;**

To print $y + 5 = 7$, enter

In[87] := **Print["y + 5 = ", y + 5]**
              y + 5 = 7

Note that **Print [y + 5 = 7]** differs from **Print ["y + 5 = ", y + 5]**. **Print [y + 5 = 7]** triggers an error message and prints the number to the right of the assignment operator "=":

In[88] := **Print[y + 5 = 7]**

              Set::write: Tag Plus in 2 + 5 is Protected.

              7

In[89] := **Clear[y]**

EXAMPLE 2.4.9 The plane pendulum consists of a particle of mass m constrained by a rigid and massless rod to move in a vertical circle of radius L, as in Figure 2.4.1.

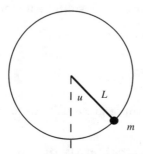

Figure 2.4.1   The plane pendulum

The equation of motion is

$$\frac{d^2u}{dt^2} + \omega_0^2 \sin u = 0 \tag{2.4.1}$$

where u is the angle that the rod makes with the vertical and

$$\omega_0^2 = \frac{g}{L} \tag{2.4.2}$$

In Equation 2.4.2, g is the magnitude of the acceleration due to gravity. If the time t is in units of $1/\omega_0$, Equation 2.4.1 becomes

$$\frac{d^2u}{dt^2} + \sin u = 0 \tag{2.4.3}$$

Let $u(0) = \pi/15$ and $u'(0) = 0$. Plot the positions of the particle for t varying from 0 to π in steps of $\pi/8$.

Let us begin by solving Equation 2.4.3 numerically:

```
In[90] := NDSolve[{u''[t] + Sin[u[t]] == 0,
                u[0] == Pi/15,
                u'[0] == 0},
            u, {t, 0, Pi}
            ]
Out[90] = {{u -> InterpolatingFunction[{{0., 3.14159}}, <>]}}
```

As explained in Example 2.4.8, the function u that is the numerical solution of the differential equation can be obtained with the assignment

```
In[91] := u = u /. First[%]
Out[91] = InterpolatingFunction[{{0., 3.14159}}, <>]
```

If lengths are in units of L and the fixed end of the rod is the origin, the x and y coordinates of the particle for t from 0 to π in steps of $\pi/8$ are

```
In[92] := N[Table[{Sin[u[t]], -Cos[u[t]]}, {t, 0, Pi, Pi/8}], 3]
Out[92] = {{0.208, -0.978}, {0.192, -0.981}, {0.148, -0.989},
        {0.0807, -0.997}, {0.000901, -1.}, {-0.0791, -0.997},
        {-0.147, -0.989}, {-0.192, -0.981}, {-0.208, -0.978}}
```

The function **ListPlot** introduced in Section 2.3.1.5 plots the positions of the particle:

```
In[93] := ListPlot[%, AspectRatio -> Automatic,
                PlotStyle -> PointSize[0.03],
                PlotRange -> {{-0.5, 0.5}, {0, -1.1}},
                AxesLabel -> {" x (L)", " y (L)"}
            ];
```

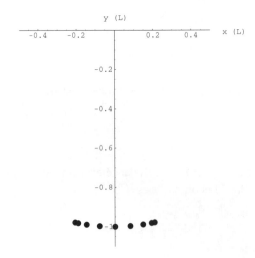

In[94] := Clear[u]

EXAMPLE 2.4.10   Prove that

$$\text{(a)} \qquad [\mathbf{A} \times (\mathbf{B} \times \mathbf{C})] + [\mathbf{B} \times (\mathbf{C} \times \mathbf{A})] + [\mathbf{C} \times (\mathbf{A} \times \mathbf{B})] = 0$$

$$\text{(b)} \qquad \nabla \cdot (f\mathbf{A}) = f(\nabla \cdot \mathbf{A}) + \mathbf{A} \cdot (\nabla f)$$

$$\text{(c)} \qquad \nabla \times (\nabla \times \mathbf{A}) = \nabla(\nabla \cdot \mathbf{A}) - \nabla^2 \mathbf{A}$$

where f is a scalar point function and $\mathbf{A}$, $\mathbf{B}$, and $\mathbf{C}$ are vector point functions.

The package `Calculus`VectorAnalysis`` contains many functions for vector analysis. Some of them are

`CoordinateSystem`	give the name of the default coordinate system
`Coordinates[ ]`	give a list of the default names of the coordinate variables in the default coordinate system
`SetCoordinates[coordsys]`	set the default coordinate system to be *coordsys* with default variable names
`SetCoordinates[coordsys[varnames]]`	
	set the default coordinate system to be *coordsys* with variable names *varnames*
`DotProduct[v1, v2]`	compute the dot product (sometimes called inner product or scalar product) of the three-dimensional vectors *v1* and *v2* in the default coordinate system
`CrossProduct[v1, v2]`	compute the cross product (sometimes called vector product) of the three-dimensional vectors *v1* and *v2* in the default coordinate system
`Div[v]`	give the divergence of the vector point function *v* in the default coordinate system

`Curl[v]`	give the curl of the vector point function v in the default coordinate system
`Grad[f]`	give the gradient of the scalar point function f in the default coordinate system
`Laplacian[f]`	give the Laplacian of the scalar point function f in the default coordinate system

■ *Mathematica* **Version 3.0 Specifics**

In[95] := **Needs["Calculus`VectorAnalysis`"]**

Unless otherwise set, the Cartesian coordinate system is the default system:

In[96] := **CoordinateSystem**

Out[96] = Cartesian

The default names of the coordinate variables are **Xx**, **Yy**, and **Zz**:

In[97] := **Coordinates[ ]**

Out[97] = {Xx, Yy, Zz}

To conform with traditional notation, let us change the names of the coordinate variables to **x**, **y**, and **z**:

In[98] := **SetCoordinates[Cartesian[x, y, z]]**

Out[98] = Cartesian[x, y, z]

We can verify the new names of the coordinate variables:

In[99] := **Coordinates[ ]**

Out[99] = {x, y, z}

■ *Mathematica* **Version 2.2 Specifics**

Needs["Calculus`VectorAnalysis`"]

```
x::shdw: Warning: Symbol x appears in multiple contexts
    {Calculus`VectorAnalysis`, Global`}; definitions in context
    Calculus`VectorAnalysis`
     may shadow or be shadowed by other definitions.

y::shdw: Warning: Symbol y appears in multiple contexts
    {Calculus`VectorAnalysis`, Global`}; definitions in context
    Calculus`VectorAnalysis`
     may shadow or be shadowed by other definitions.

z::shdw: Warning: Symbol z appears in multiple contexts
    {Calculus`VectorAnalysis`, Global`}; definitions in context
    Calculus`VectorAnalysis`
     may shadow or be shadowed by other definitions.
```

```
u::shdw: Warning: Symbol u appears in multiple contexts
    {Calculus`VectorAnalysis`, Global`}; definitions in context
    Calculus`VectorAnalysis`
    may shadow or be shadowed by other definitions.
```

Mathematica indicates that we have invoked the names **x**, **y**, **z**, and **u** before loading the package `Calculus`VectorAnalysis``. As explained in Section 1.5, "Packages," we must now remove these symbols with the function `Remove`:

`Remove[x, y, z, u]`

Unless otherwise set, the Cartesian coordinate system is the default system:

`CoordinateSystem`
Cartesian

The default names of the coordinate variables are **x**, **y**, and **z**:

`Coordinates[ ]`
{x, y, z}

■ Vector Identities

Mathematica represents three-dimensional vectors by three-element lists. Respecting the convention of reserving capitalized names for built-in *Mathematica* objects, we let **a**, **b**, and **c** be aliases of the vector point functions **A**, **B**, and **C**, respectively, and

```
In[100] := a = {ax[x, y, z], ay[x, y, z], az[x, y, z]};
           b = {bx[x, y, z], by[x, y, z], bz[x, y, z]};
           c = {cx[x, y, z], cy[x, y, z], cz[x, y, z]};
           f = func[x, y, z];
```

(a) To prove vector identity (a), show that *lhs* of the identity evaluates to the null vector $\{0, 0, 0\}$:

```
In[104] := CrossProduct[a, CrossProduct[b, c]] +
           CrossProduct[b, CrossProduct[c, a]] +
           CrossProduct[c, CrossProduct[a, b]]
```

`Out[104] = {0, 0, 0}`

We have proved the vector identity for any point (x, y, z). The proof remains valid for the special case when the vector functions **A**, **B**, and **C** are constant vectors independent of the coordinates x, y, and z.

(b) To prove vector identity (b), show that *lhs* − *rhs* evaluates to 0:

`In[105] := (Div[f a] - f Div[a] - DotProduct[a, Grad[f]])//Simplify`

`Out[105] = 0`

(c) To prove vector identity (c), show that *lhs* − *rhs* evaluates to $\{0, 0, 0\}$:

`In[106] := Curl[Curl[a]] - Grad[Div[a]] + Laplacian /@ a`

`Out[106] = {0, 0, 0}`

We have proved the vector identity in Cartesian coordinates. With curvilinear coordinates such as spherical or cylindrical coordinates, the components of the Laplacian of a vector are not, in general, the Laplacians of the corresponding components of the vector. For curvilinear coordinates, vector identity (c) often serves as the definition for the Laplacian of a vector.

In[107] := **Clear[a, b, c, f]**   ∎

EXAMPLE 2.4.11 An uncharged conducting sphere of radius *a* is placed in an initially uniform electric field $\mathbf{E}_0$ aligned in the *z* direction. (a) Show that the potential outside the sphere is

$$V(r, \theta) = -E_0 \left(r - \frac{a^3}{r^2} \right) \cos \theta$$

where *r* is the distance from the origin located at the center of the sphere and θ is the polar angle, that is, the angle down from the *z* axis. (b) Determine the components of the electric field outside the sphere. (c) Find the induced charge density on the surface of the sphere and verify that the total charge on the sphere is zero.

The package **Calculus`VectorAnalysis`** contains many functions for vector analysis. Those relevant to this problem were defined in Example 2.4.10 of this section.

In[108] := **Needs["Calculus`VectorAnalysis`"]**

Let us set the spherical coordinate system to be the default system:

In[109] := **SetCoordinates[Spherical];**

Coordinates[] gives the default names of coordinate variables in the default coordinate system:

❈ **3.0**

In[110] := **Coordinates[]**

Out[110] = {Rr, Ttheta, Pphi}

For convenience, let us change the names of the coordinate variables to **r**, **theta**, and **phi**:

In[111] := **SetCoordinates[Spherical[r, theta, phi]]**

Out[111] = Spherical[r, theta, phi]

❈ **2.2**

Coordinates[]
{r, theta, phi}

(a) To show that $V(r, \theta)$ is the electric potential outside the sphere, we invoke the uniqueness theorem: The solution to Laplace's equation with specified boundary conditions is unique. That is, we must show that V satisfies Laplace's equation and matches the appropriate boundary conditions.

Laplace's equation is

$$\nabla^2 V = 0$$

We can prove that the given potential satisfies this equation. Let **v** be the alias of V.

In[112] := **Clear[a, E0]**

In[113] := **v = - E0 (1 - (a/r)^3) r Cos[theta];**

In[114] := **Laplacian[v] // Simplify**

Out[114] = 0

The boundary conditions are

$$V = 0 \qquad\qquad r = a$$
$$V = -E_0 z = -E_0 r \cos\theta \qquad r \gg a$$

We can verify that V matches these conditions:

In[115] := **v /. r -> a**

Out[115] = 0

In[116] := **v /. ((a/r)^3 -> 0)**

Out[116] = -(E0 r Cos[theta])

where we have translated $r \gg a$ into **(a/r)^3 -> 0**. (For questions raised by this translation, see Exercise 13 of Section 2.4.10.) Hence, the given $V(r, \theta)$ is the potential outside the sphere.

(b) The electric field is given by

$$\mathbf{E} = -\nabla V$$

Thus, **-Grad[v]** yields a list of the radial, polar, and azimuthal components of **E**, namely, $\{E_r, E_\theta, E_\phi\}$:

In[117] := **-Grad[v] // Simplify**

Out[117] = $\{E0 \ (1 + \dfrac{2 \ a^3}{r^3}) \ Cos[theta], \ \dfrac{E0 \ (a^3 - r^3) \ Sin[theta]}{r^3}, \ 0\}$

(c) From Gauss's law, the surface charge density on the sphere is given by

$$\sigma = \varepsilon_0 E_r \big|_{r=a}$$

where ε_0 is the permittivity of free space. Therefore, the induced surface charge density is

In[118] := **e0 %[[1]] /. r -> a**

Out[118] = 3 e0 E0 Cos[theta]

with **e0** being the alias of ε_0. The total charge on the sphere is

$$Q = \int_{\text{sphere}} \sigma \, dS = 2\pi a^2 \int_0^\pi \sigma(\theta) \sin\theta \, d\theta$$

which must be zero, as the sphere was initially uncharged:

In[119] := **(2 Pi a^2) Integrate[% Sin[theta], {theta, 0, Pi}]**

Out[119] = 0

(For a discussion of Laplace's equation as well as this problem of an uncharged metal sphere placed in an otherwise uniform electric field, see [Gri89].)

In[120] := **Clear[v]**   ∎

2.4.10 Exercises

1. Create the list $\{x - 1, x^3 - 1, x^5 - 1, x^7 - 1, x^9 - 1, x^{11} - 1\}$ with the function **Table**. With the function **Drop**, obtain the list without the second and third elements.

   Answer: {-1 + x, -1 + x^7, -1 + x^9, -1 + x^{11}}

2. Determine the length and dimensions of the list $\{\{\{3, 4, 5\}, \{4, 5, 6\}\}, \{\{4, 5, 6\}, \{5, 6, 7\}\}\}$.

   Answers: 2

   {2, 2, 3}

3. Consider the list $\{r, \{s, t\}, \{w, \{x, \{y, z\}\}\}\}$. Pick out the pieces: (a) r, (b) $\{s, t\}$, (c) w, (d) $\{y, z\}$, and (e) z.

   Answers: r

   {s, t}

   w

   {y, z}

   z

4. Consider the list $\{a, \{b, c\}, a, \{d, \{e, \{f, \{a, c, g\}\}\}\}\}$. (a) Flatten out the list. (b) Flatten out sublists $\{b, c\}$ and $\{a, c, g\}$.

   Answers: {a, b, c, a, d, e, f, a, c, g}

   {a, b, c, a, {d, {e, {f, a, c, g}}}}

5. The function f is defined by

$$f(x) = \sin^2(x)$$

Apply f to the elements a in the list $\{a, \{b, c\}, a, \{d, \{e, \{a, c, g\}\}\}\}$.

Answer: {Sin[a]^2, {b, c}, Sin[a]^2,
{d, {e, {Sin[a]^2, c, g}}}}

6. Determine the mean value of the squares of the first eight even integers.

Answer: 102

7. **(a)** Create a list of the cubes of the first four integers.
   (b) Repeat for the next four integers.
   (c) Merge these two lists and call the new list **mylist**.
   (d) Find the average of the elements in **mylist**.
   (e) Create a new list in which each element is the square of the difference between the corresponding element and the average of the elements in **mylist**.

   Answers: {1, 8, 27, 64}

   {125, 216, 343, 512}

   {1, 8, 27, 64, 125, 216, 343, 512}

   162

   {25921, 23716, 18225, 9604, 1369, 2916, 32761, 122500}

8. **(a)** Create a list of the integers from 0 to 10 and call it **xlist**.
   (b) Create a list of the squares of the integers from 0 to 10 and name it **ylist**.
   (c) Generate the list of $\{xlist_i, ylist_i\}$.
   (d) Plot the points $\{xlist_i, ylist_i\}$.

   Answers: {0, 1, 2, 3, 4, 5, 6, 7, 8, 9, 10}

   {0, 1, 4, 9, 16, 25, 36, 49, 64, 81, 100}

   {{0, 0}, {1, 1}, {2, 4}, {3, 9}, {4, 16}, {5, 25},
   {6, 36}, {7, 49}, {8, 64}, {9, 81}, {10, 100}}

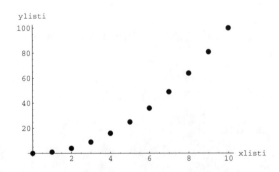

9. Consider the list of the first 100 integers.
   (a) Partition the list into two sublists.
   (b) Reverse the order of the second sublist.
   (c) Add the corresponding elements of the two sublists.
   (d) What can be inferred about the sum and mean value of the first 100 integers?

Answers:
```
{{1, 2, 3, 4, 5, 6, 7, 8, 9, 10, 11, 12, 13, 14,
  15, 16, 17, 18, 19, 20, 21, 22, 23, 24, 25, 26,
  27, 28, 29, 30, 31, 32, 33, 34, 35, 36, 37, 38,
  39, 40, 41, 42, 43, 44, 45, 46, 47, 48, 49, 50},
 {51, 52, 53, 54, 55, 56, 57, 58, 59, 60, 61, 62,
  63, 64, 65, 66, 67, 68, 69, 70, 71, 72, 73, 74,
  75, 76, 77, 78, 79, 80, 81, 82, 83, 84, 85, 86,
  87, 88, 89, 90, 91, 92, 93, 94, 95, 96, 97, 98,
  99, 100}}
```

```
{{1, 2, 3, 4, 5, 6, 7, 8, 9, 10, 11, 12, 13, 14,
  15, 16, 17, 18, 19, 20, 21, 22, 23, 24, 25, 26,
  27, 28, 29, 30, 31, 32, 33, 34, 35, 36, 37, 38,
  39, 40, 41, 42, 43, 44, 45, 46, 47, 48, 49, 50},
 {100, 99, 98, 97, 96, 95, 94, 93, 92, 91, 90, 89,
  88, 87, 86, 85, 84, 83, 82, 81, 80, 79, 78, 77,
  76, 75, 74, 73, 72, 71, 70, 69, 68, 67, 66, 65,
  64, 63, 62, 61, 60, 59, 58, 57, 56, 55, 54, 53,
  52, 51}}
```

```
{101, 101, 101, 101, 101, 101, 101, 101, 101, 101,
 101, 101, 101, 101, 101, 101, 101, 101, 101, 101,
 101, 101, 101, 101, 101, 101, 101, 101, 101, 101,
 101, 101, 101, 101, 101, 101, 101, 101, 101, 101,
 101, 101, 101, 101, 101, 101, 101, 101, 101, 101}
```

5050

$$\frac{101}{2}$$

10. Prove that

$$\nabla \cdot (\mathbf{A} \times \mathbf{B}) = \mathbf{B} \cdot (\nabla \times \mathbf{A}) - \mathbf{A} \cdot (\nabla \times \mathbf{B})$$

where $\mathbf{A}$ and $\mathbf{B}$ are any two vector point functions.

11. Show that

$$\nabla \times (f\mathbf{A}) = f(\nabla \times \mathbf{A}) - \mathbf{A} \times (\nabla f)$$

where f is a scalar point function and $\mathbf{A}$ is a vector point function.

12. Compute an approximation to the integral in Example 2.4.7 with the function `ListIntegrate` in the package ``NumericalMath`ListIntegrate` `` and thus determine the change in entropy of the system. Compare the result with that calculated with the trapezoidal rule. *Hint:* Load the package ``NumericalMath`ListIntegrate` ``; with on-line help, obtain information on the function `ListIntegrate`; refer to [AAB93] or [WR96], if available.

   Answer: `7.51218`

   with the units cal/K.

13. In Example 2.4.11, we evaluated `v /. ((a/r)^3 -> 0)` in order to verify that the potential V matches the boundary condition at $r \gg a$.
   (a) Why don't we apply the transformation rule `(a/r) -> 0` instead of the rule `(a/r)^3 -> 0`?
   (b) Why don't we simply apply the rule `r -> Infinity`?
   (c) Is it more motivating to evaluate `v /. a -> (f r) /. f -> 0` or `v //. {a -> (f r), f -> 0}` rather than `v /. ((a/r)^3 -> 0)`?
   (d) How does the operator "`//.`", which we have not introduced, perform replacements?

3.0 14. Why do we need the function `Extract` when we already have the function `Part`?

2.5 Some New Features in *Mathematica* Version 3.0

In Chapter 1 and earlier sections of Chapter 2, we have seen a number of new features in *Mathematica* version 3.0. This section highlights several more. To reproduce input and output that resemble those in this section, set Default Input FormatType and Default Output FormatType in the Cell menu to **StandardForm** and **TraditionalForm**, respectively.

2.5.1 Special Characters

In addition to ordinary keyboard characters, *Mathematica* recognizes about 700 special characters. We can use the special characters just like ordinary keyboard characters. Section A.12 of *Mathematica* Reference Guide in **The *Mathematica* Book** gives a complete list of these named characters together with their descriptions [Wol96]. To access Section A.12 with on-line help, choose **Help** in the Help menu, click the button for The *Mathematica* Book, and select Reference Guide in the first column and Listing of Named Characters in the second column of Help Browser.

There are three kinds of special characters: letters and letter-like forms, operators, and structural elements. This section shows several ways to enter these characters, gives examples of them, and cautions that there are many similar-looking but different characters.

2.5.1.1 WAYS TO ENTER SPECIAL CHARACTERS

There are three ways to enter special characters in a notebook:

- **Palettes**   Choose **BasicInput**, **BasicTypesetting**, or **CompleteCharacters** in the Palettes submenu of the File menu; click on a character such as β or $\oplus$ to paste it to the notebook.

- **Full Names**   Enter from the keyboard the full name of a character such as \ [Beta] for β or \ [CirclePlus] for $\oplus$; *Mathematica* automatically converts the full name to the character; to display full names, if desired, click the cell bracket, select **Option Inspector** in the Format menu, choose "selection" and "alphabetically" at the top of Option Inspector, and change the setting of ShowSpecialCharacters to False.

- **Aliases**   Enter from the keyboard an alias of a character such as ⎋b⎋ or ⎋beta⎋ for β or ⎋c+⎋ for $\oplus$, where ⎋ stands for the *esc* key.

2.5.1.2 LETTERS AND LETTER-LIKE FORMS

Here are some letters and letter-like forms together with their full names as well as aliases:

Character	Full Name	Aliases
α	\ [Alpha]	⎋a⎋, ⎋alpha⎋
β	\ [Beta]	⎋b⎋, ⎋beta⎋
γ	\ [Gamma]	⎋g⎋, ⎋gamma⎋
δ	\ [Delta]	⎋d⎋, ⎋delta⎋
ϵ	\ [Epsilon]	⎋e⎋, ⎋epsilon⎋
ε	\ [CurlyEpsilon]	⎋ce⎋, ⎋cepsilon⎋
θ	\ [Theta]	⎋q⎋, ⎋th⎋, ⎋theta⎋
λ	\ [Lambda]	⎋l⎋, ⎋lambda⎋
π	\ [Pi]	⎋p⎋, ⎋pi⎋
ϕ	\ [Phi]	⎋f⎋, ⎋ph⎋, ⎋phi⎋
ψ	\ [Psi]	⎋y⎋, ⎋ps⎋, ⎋psi⎋
ω	\ [Omega]	⎋o⎋, ⎋w⎋, ⎋omega⎋
Γ	\ [CapitalGamma]	⎋G⎋, ⎋Gamma⎋
Δ	\ [CapitalDelta]	⎋D⎋, ⎋Delta⎋
Φ	\ [CapitalPhi]	⎋F⎋, ⎋Ph⎋, ⎋Phi⎋
Ψ	\ [CapitalPsi]	⎋Y⎋, ⎋Ps⎋, ⎋Psi⎋
Ω	\ [CapitalOmega]	⎋O⎋, ⎋W⎋, ⎋Omega⎋
ℓ	\ [ScriptL]	⎋scl⎋

$\mathcal{E}$	\[ScriptCapitalE]	:scE:
R	\[DoubleStruckCapitalR]	:dsR:
°	\[Degree]	:deg:
Å	\[Angstrom]	:Ang:
ℏ	\[HBar]	:hb:
∞	\[Infinity]	:inf:
e	\[ExponentialE]	:ee:
i	\[ImaginaryI]	:ii:
j	\[ImaginaryJ]	:jj:
∟	\[Angle]	
•	\[Bullet]	:bu:
†	\[Dagger]	:dg:

In aliases, : stands for the key ESC.

We can use letters and letter-like forms in symbol names just like ordinary keyboard letters. For example, the energy eigenvalues of the one-dimensional harmonic oscillator are

$In[1] := \mathcal{E}[n\_] := (n + 1/2)\, ℏ\, \omega$

where n takes on any nonnegative integer, and the energy for $n = 5$ is

$In[2] := \mathcal{E}[5]$

$Out[2] = \dfrac{11\, \omega\, ℏ}{2}$

$In[3] := Clear[\mathcal{E}]$

While most of the letters and letter-like forms do not have built-in meanings, several do:

Character	Equivalent to
π	Pi
e	E
∞	Infinity
°	Degree
i	I
j	I

For example,

$In[4] := \{\alpha, \psi, \Gamma, \beta\pi, \pi, e\verb|^| - 2, Sin[60°], i\verb|^|2\}\, //\, N$

$Out[4] = \{\alpha, \psi, \Gamma, \beta\pi, 3.14159, 0.135335, 0.866025, -1.\}$

Note that π has built-in meaning, but π in $\beta\pi$ assumes no special meaning.

2.5.1.3   OPERATORS

What follow are some operators grouped according to functionality.

■ **Common Mathematical Operators**

Character	Full Name	Alias	Example	Built-in Meaning
×	\[Times]	:*:	$x \times y$	Times (*)
÷	\[Divide]	:div:	$x \div y$	Divide (/)
√	\[Sqrt]	:sqrt:	$\sqrt{x}$	Sqrt
×	\[Cross]	:cross:	$x \times y$	Cross
ⅆ	\[DifferentialD]	:dd:	$\int f\,\mathrm{d}x$	for use in integrals
∫	\[Integral]	:int:	$\int f\,\mathrm{d}x$	integral sign

The operators ×, ÷, and × are infix operators that go between their operands; the operator √ is a prefix operator that precedes its operand; the characters ⅆ and ∫ are elements of compound operators that evaluate integrals. These operators follow built-in evaluation rules:

$In[5] := \left(3 \times \sqrt{4}\right) \div 6$

$Out[5] = 1$

$In[6] := \{1, 2, 3\} \times \{4, 5, 6\}$

$Out[6] = \{-3, 6, -3\}$

$In[7] := \int \texttt{Sin[x]}\,\mathrm{d}x$

$Out[7] = -\cos(x)$

Note that × is different from × and that the special character ⅆ is different from the ordinary keyboard character *d*.

■ **Logical Operators**

Character	Full Name	Alias	Example	Built-in Meaning
∧	\[And]	:&&:	$x \wedge y$	And (&&)
∨	\[Or]	:||:	$x \vee y$	Or (||)
¬	\[Not]	:!:	$\neg x$	Not (!)
⇒	\[Implies]	:=>:	$x \Rightarrow y$	Implies

The operator ¬ is a prefix operator and the others are infix operators. These operators have built-in meanings:

$In[8] := \neg (5 > 3 \wedge 7 < 5)$

$Out[8] = $ True

■ **Bracketing Operators**

Character	Full Name	Alias	Example	Built-in Meaning
⟦	\[LeftDoubleBracket]	:[[:	$m⟦i⟧$	for use in **Part** ([[)
⟧	\[RightDoubleBracket]	:]]:	$m⟦i, j⟧$	for use in **Part** (]])
⟨	\[LeftAngleBracket]	:<:	$⟨x, y, z⟩$	
⟩	\[RightAngleBracket]	:>:	$⟨x⟩$	
╎	\[LeftBracketingBar]	:l|:	$\|x, y\|$	
╎	\[RightBracketingBar]	:r|:	$\|x\|$	

The first two characters are elements of a compound operator with built-in evaluation rules:

In[9] := **{2, 3, 4, 5} ⟦2⟧**

Out[9] = 3

The other characters do not have built-in meanings and are matchfix operators, which come in matching pairs that enclose or delimit their operands:

In[10] := **{⟨x, y⟩, ╎2 + 3 I╎}**

Out[10] = $\{⟨x,y⟩, |2+3\,i|\}$

The operators ⟨⟩ and ╎╎, formed with pairs of matchfix operators, are named AngleBracket and BracketingBar, respectively. We can assign meanings to these operators. For example,

In[11] := **╎x_╎ := Abs[x]**

In[12] := **╎2 + 3 I╎**

Out[12] = $\sqrt{13}$

For another example, let us define a scalar product of two functions of the variable ζ:

In[13] := **⟨φ_, ψ_⟩ := Integrate[**
                      (φ /. Complex[u_, v_] -> Complex[u, -v]) ψ,
                      {ζ, -∞, ∞}
                      ]

(Section 3.1.2 will explain the necessity of using the rule with the function **Complex** rather than the simpler rule I -> -I to transform φ to $\varphi*$.) The energy eigenfunctions of the one-dimensional harmonic oscillator are

In[14] := **u[n_, ζ_] := (Sqrt[1 / (Sqrt[π] n! 2^n)]***
                      HermiteH[n, ζ] Exp[-(ζ^2) / 2])

where the dimensionless coordinate $\zeta = \sqrt{m\omega/\hbar}\,x$. Thus,

In[15] := **⟨u[1, ζ], u[1, ζ]⟩**

Out[15] = 1

In[16] := ⟨u[1, ζ], ζ^2 u[3, ζ]⟩

Out[16] = $\sqrt{\dfrac{3}{2}}$

In[17] := **Clear[u, BracketingBar, AngleBracket]**

■ **Other Operators**

Character	Full Name	Alias	Example	Built-in Meaning
⊕	\[CirclePlus]	:c+:	$x \oplus y$	
⊗	\[CircleTimes]	:c*:	$\otimes x, x \otimes y$	
·	\[CenterDot]	:.:	$x \cdot y$	
∪	\[Union]	:un:	$x \cup y$	**Union**
∩	\[Intersection]	:inter:	$x \cap y$	**Intersection**
∇	\[Del]	:del:	∇f	
□	\[Square]	:sq:	$\square\, x$	
≡	\[Congruent]	:===:	$x \equiv y$	
~	\[Tilde]	:~:	$x \sim y$	
≈	\[TildeTilde]	:~~:	$x \approx y$	
≃	\[TildeEqual]	:~=:	$x \simeq y$	
∝	\[Proportional]	:prop:	$x \propto y$	
≠	\[NotEqual]	:!=:	$x \neq y$	**Unequal** (!=)
≥	\[GreaterEqual]	:>=:	$x \geq y$	**GreaterEqual** (>=)
≤	\[LessEqual]	:<=:	$x \leq y$	**LessEqual** (<=)
≯	\[NotGreater]	:!>:	$x \ngtr y$	
⊂	\[Subset]	:sub:	$x \subset y$	
∈	\[Element]	:elem:	$x \in y$	
→	\[Rule]	:->:	$x \to y$	**Rule** (->)
:→	\[RuleDelayed]	:>:	$x \rightsquigarrow y$	**RuleDelayed** (:>)
⇀	\[RightVector]	:vec:	$x \rightharpoonup y, \vec{x}$	

These operators are infix operators with the exceptions: □ and ∇ are prefix operators, ⊗ can be a prefix operator, and ⇀ can be an overfix operator that goes over its operand. Several operators have built-in evaluation rules:

In[18] := **{3 ≥ 3, 5 ≠ 5, {a, b, c} ∪ {b, c, d}} /. {a → 1}**

Out[18] = {True, False, {1, *b*, *c*, *d*}}

Many of these operators do not have special meanings:

In[19] := **{x ⊕ y, x ⊗ y, x ∝ y, 3 ≃ 2, ∇x^2}**

Out[19] = $\{x \oplus y,\ x \otimes y,\ x \propto y,\ 3 \simeq 2,\ \nabla x^2\}$

We can assign meanings to them. For instance, let us define the Kronecker product of two matrices:

```
In[20] := ρ_ ⊗ σ_ := Outer[Times, ρ, σ]
```

Consider, for example,

```
In[21] := α = {{α11, α12, α13}, {α21, α22, α23},
               {α31, α32, α33}};
In[22] := β = {{β11, β12}, {β21, β22}};
```

Thus,

```
In[23] := α ⊗ β
```

$$
Out[23] = \begin{pmatrix}
\begin{pmatrix} \alpha11\,\beta11 & \alpha11\,\beta12 \\ \alpha11\,\beta21 & \alpha11\,\beta22 \end{pmatrix} & \begin{pmatrix} \alpha12\,\beta11 & \alpha12\,\beta12 \\ \alpha12\,\beta21 & \alpha12\,\beta22 \end{pmatrix} & \begin{pmatrix} \alpha13\,\beta11 & \alpha13\,\beta12 \\ \alpha13\,\beta21 & \alpha13\,\beta22 \end{pmatrix} \\
\begin{pmatrix} \alpha21\,\beta11 & \alpha21\,\beta12 \\ \alpha21\,\beta21 & \alpha21\,\beta22 \end{pmatrix} & \begin{pmatrix} \alpha22\,\beta11 & \alpha22\,\beta12 \\ \alpha22\,\beta21 & \alpha22\,\beta22 \end{pmatrix} & \begin{pmatrix} \alpha23\,\beta11 & \alpha23\,\beta12 \\ \alpha23\,\beta21 & \alpha23\,\beta22 \end{pmatrix} \\
\begin{pmatrix} \alpha31\,\beta11 & \alpha31\,\beta12 \\ \alpha31\,\beta21 & \alpha31\,\beta22 \end{pmatrix} & \begin{pmatrix} \alpha32\,\beta11 & \alpha32\,\beta12 \\ \alpha32\,\beta21 & \alpha32\,\beta22 \end{pmatrix} & \begin{pmatrix} \alpha33\,\beta11 & \alpha33\,\beta12 \\ \alpha33\,\beta21 & \alpha33\,\beta22 \end{pmatrix}
\end{pmatrix}
$$

The product is not commutative:

```
In[24] := β ⊗ α
```

$$
Out[24] = \begin{pmatrix}
\begin{pmatrix} \alpha11\,\beta11 & \alpha12\,\beta11 & \alpha13\,\beta11 \\ \alpha21\,\beta11 & \alpha22\,\beta11 & \alpha23\,\beta11 \\ \alpha31\,\beta11 & \alpha32\,\beta11 & \alpha33\,\beta11 \end{pmatrix} & \begin{pmatrix} \alpha11\,\beta12 & \alpha12\,\beta12 & \alpha13\,\beta12 \\ \alpha21\,\beta12 & \alpha22\,\beta12 & \alpha23\,\beta12 \\ \alpha31\,\beta12 & \alpha32\,\beta12 & \alpha33\,\beta12 \end{pmatrix} \\
\begin{pmatrix} \alpha11\,\beta21 & \alpha12\,\beta21 & \alpha13\,\beta21 \\ \alpha21\,\beta21 & \alpha22\,\beta21 & \alpha23\,\beta21 \\ \alpha31\,\beta21 & \alpha32\,\beta21 & \alpha33\,\beta21 \end{pmatrix} & \begin{pmatrix} \alpha11\,\beta22 & \alpha12\,\beta22 & \alpha13\,\beta22 \\ \alpha21\,\beta22 & \alpha22\,\beta22 & \alpha23\,\beta22 \\ \alpha31\,\beta22 & \alpha32\,\beta22 & \alpha33\,\beta22 \end{pmatrix}
\end{pmatrix}
$$

(For a discussion of tensor products, see [Hal58].)

```
In[25] := Clear[α, β, CircleTimes]
```

2.5.1.4 STRUCTURAL ELEMENTS

Structural elements specify structures. Two examples are

Character	Full Name	Alias
∖	\[Continuation]	:cont:
	\[InvisibleComma]	:,:

In *Mathematica* output, the character ∖ indicates the continuation of an expression onto the next line:

```
In[26] := 80!
```

```
Out[26] = 71569457046263802294811533723186532165584657342365752577109\
          4505822703925548014884266894486728081408000000000000000000
```

Invisible commas are interpreted on input as ordinary commas:

In[27] := ⟨α β δ γ⟩

Out[27] = ⟨α, β, δ, γ⟩

2.5.1.5 SIMILAR-LOOKING CHARACTERS

Beware of characters that look alike but have disparate meanings. Some examples are

Character	Full Name	Alias
μ	\[Mu]	:m:, :mu:
μ	\[Micro]	:mi:
Å	\[Angstrom]	:Ang:
Å	\[CapitalARing]	:Ao:
Σ	\[CapitalSigma]	:S:, :Sigma:
Σ	\[Sum]	:sum:
Π	\[CapitalPi]	:P:, :Pi:
Π	\[Product]	:prod:
U	keyboard U	
∪	\[Union]	:un:
d	keyboard d	
ⅆ	\[DifferentialD]	:dd:
i	keyboard i	
ⅈ	\[ImaginaryI]	:ii:
×	\[Times]	:*:
×	\[Cross]	:cross:
∧	\[And]	:&&:
∧	\[Wedge]	:^:
→	\[Rule]	:->:
→	\[RightArrow]	:_->:
*	keyboard *	
⋆	\[Star]	:star:
\	keyboard \	
\	\[Backslash]	:\:

```
|         keyboard |
|        \[VerticalSeparator]     :|:
'        \[VerticalBar]           :_|:
|        \[LeftBracketingBar]     :l|:
|        \[RightBracketingBar]    :r|:
```

The character ␣ stands for Space entered by pressing the space bar.

2.5.2   Two-Dimensional Forms

In addition to one-dimensional input like that in the earlier versions, *Mathematica* version 3.0 supports two-dimensional input. This section shows several ways to enter two-dimensional forms, describes a number of two-dimensional forms that have built-in meanings, and illustrates the use of two-dimensional forms in solving physics problems.

2.5.2.1   WAYS TO ENTER TWO-DIMENSIONAL FORMS

■ **Palettes**

To enter a two-dimensional form with a palette, highlight or select an expression, choose **BasicInput** or **BasicTypesetting** in the Palettes submenu of the File menu, and click on a two-dimensional form. The two-dimensional form is pasted in the notebook with the current selection already inserted in the primary or selection placeholder "■". Other expressions can be entered into the remaining placeholders "□". To move from one placeholder to another, use the Tab key or the mouse.

Let us create, for example, the two-dimensional input x^a:

1. Enter

    x

2. Highlight or select **x**, choose **BasicInput** in the Palettes submenu of the File menu, and click the button at the upper left-hand corner:

    $x^{\square}$

3. Enter **a** in the placeholder:

    x^a

Except for a few special superscripts such as x^+, $x^\dagger$, and x^*, most superscripts are interpreted as powers by *Mathematica*. However, subscripts have no built-in meanings.

For another example, let us enter the two-dimensional form

$$\sum_{n=1}^{k} x_n + \int_1^4 x^2 \, dx$$

1. Enter

    $x + x^2$

2. Select the first **x** from the left, choose **BasicInput** in the Palettes submenu of the File menu, and click the button at the lower left-hand corner:

$$x_\Box \; + \; x^2$$

3. Enter **n** in the placeholder:

$$x_n \; + \; x^2$$

4. Select x_n, and click the fifth button from the top of the left column in the palette:

$$\sum_{\Box = \Box}^{\Box} x_n \; + \; x^2$$

5. Enter **n**, **1**, and **k** in the appropriate placeholders, using the Tab key to move from one placeholder to another:

$$\sum_{n=1}^{k} x_n \; + \; x^2$$

6. Select x^2, and click the fourth button from the top in the left column of buttons in the palette:

$$\sum_{n=1}^{k} x_n \; + \; \int_{\Box}^{\Box} x^2 \, d\Box$$

7. Type **1**, **4**, and **x** in the respective placeholders, using the Tab key to go from one placeholder to another:

$$\sum_{n=1}^{k} x_n \; + \; \int_{1}^{4} x^2 \, d x$$

■ **Control Keys**

We can use control characters to enter two-dimensional forms:

Control Character	Built-in Meaning
CTRL [^] or CTRL [6]	go to the superscript position
CTRL [_] or CTRL [-]	go to the subscript position
CTRL [@] or CTRL [2]	go into a square root
CTRL [%] or CTRL [5]	go from subscript to superscript or vice versa, from underscript to overscript or vice versa, or to the exponent position in a root
CTRL [/]	go to the denominator for a fraction
CTRL [+] or CTRL [=]	go to the underscript position
CTRL [&] or CTRL [7]	go to the overscript position
CTRL [_]	return from a special position

CTRL [*Key*] stands for Control-*Key*, that is, hold down the Control key and press *Key*. The character ␣ represents Space, that is, the space bar.

Here are several useful characters for entering two-dimensional forms:

Character	Full Name	Alias	Built-in Meaning
Σ	\[Sum]	:sum:	summation sign
Π	\[Product]	:prod:	product sign
$\int$	\[Integral]	:int:	integral sign
d	\[DifferentialD]	:dd:	for use in integrals
∂	\[PartialD]	:pd:	partial derivative operator

Let us create again the two-dimensional input x^a. The key sequence is

x CTRL [^] **a** CTRL [␣]

The key sequences for entering

$$\sum_{n=1}^{k} x_n \; + \; \int_{1}^{4} x^2 \, \mathrm{d}x$$

are

:sum: CTRL [+] n = 1 CTRL [%] k CTRL [␣] x CTRL [_] n CTRL [␣] +
:int: CTRL [_] 1 CTRL [%] 4 CTRL [␣] x CTRL [^] 2 CTRL [␣] :dd: x

■ **Ordinary Characters**

We can use only ordinary printable characters to enter two-dimensional forms:

Characters	Form
\!\(x\^y\)	superscript x^y
\!\(x_y\)	subscript x_y
\!\(x\^y\%z\)	subscript and superscript x_z^y
\!\(\@x\)	square root $\sqrt{x}$
\!\(x\/y\)	built-up fraction $\frac{x}{y}$
\!\(x\+y\)	underscript $\underset{y}{x}$
\!\(x\&y\)	overscript $\overset{y}{x}$
\!\(x\+y\%z\)	underscript and overscript $\underset{z}{\overset{y}{x}}$

Note that each expression is enclosed on the left by the characters "\!\(" and on the right by the characters "\)". We can use the characters "\(" and "\)" within an expression to generate invisible parentheses for grouping elements.

Let us create once more the two-dimensional input x^a. The sequence of characters is

\!\(x\^a\)

To display the two-dimensional form, select the expression and choose **Make 2D** in the Edit menu:

$$x^a$$

The character sequences for entering

$$\sum_{n=1}^{k} x_n \; + \; \int_{1}^{4} x^2 \, dx$$

are

```
\!\(\(\[Sum]\+\(n=1\)\%k\(x\_n\)\)+
    \(\[Integral]\_1\%4\(x\^2\)\[DifferentialD]x\)\)
```

To display the two-dimensional form, select the expression and choose **Make 2D** in the Edit menu:

$$\sum_{n=1}^{k} x_n \; + \; \int_{1}^{4} x^2 \, dx$$

■ Create Table/Matrix/Palette Command

To create matrices, we can use the **BasicInput** or **BasicTypesetting** palettes, or choose **Create Table/Matrix/Palette** in the Input menu and enter the necessary specifications in the dialog box. The Tab key and several control characters are useful in editing matrices:

Key or Control Character	Built-in Meaning
CTRL [,]	add a column
CTRL [↵] (Control-Return)	add a row
TAB	go to the next placeholder
CTRL [␣]	move out of the array

Consider, for example, the matrix

$$\begin{pmatrix} -2 & 1 & 3 \\ 0 & -1 & 1 \\ 1 & 2 & 0 \end{pmatrix}$$

To enter the matrix:

1. Choose **Create Table/Matrix/Palette** in the Input menu, and click the Matrix button at the top of the dialog box:

$$\begin{pmatrix} \square & \square & \square \\ \square & \square & \square \\ \square & \square & \square \end{pmatrix}$$

2. Enter the matrix elements, using the Tab key to move from one placeholder to the next:

$$\begin{pmatrix} -2 & 1 & 3 \\ 0 & -1 & 1 \\ 1 & 2 & 0 \end{pmatrix}$$

We can find the inverse of this matrix:

$$In[28] := \text{Inverse @ } \begin{pmatrix} -2 & 1 & 3 \\ 0 & -1 & 1 \\ 1 & 2 & 0 \end{pmatrix}$$

$$Out[28] = \begin{pmatrix} -\frac{1}{4} & \frac{3}{4} & \frac{1}{2} \\ \frac{1}{8} & -\frac{3}{8} & \frac{1}{4} \\ \frac{1}{8} & \frac{5}{8} & \frac{1}{4} \end{pmatrix}$$

2.5.2.2 SOME TWO-DIMENSIONAL FORMS WITH BUILT-IN MEANINGS

What follow are some two-dimensional forms together with their corresponding one-dimensional forms and built-in meanings:

Two-Dimensional Form	One-Dimensional Form	Built-in Meaning
x^y	`x^y`	power
$\frac{x}{y}$	`x/y`	division
$\sqrt{x}$	`Sqrt[x]`	square root
$\sqrt[n]{x}$	`x^(1/n)`	nth root
$\sum\limits_{i=imin}^{imax} f$	`Sum[f, {i, imin, imax}]`	sum
$\prod\limits_{i=imin}^{imax} f$	`Product[f, {i, imin, imax}]`	product
$\int f \, \mathrm{d}x$	`Integrate[f, x]`	indefinite integral
$\int_{xmin}^{xmax} f \, \mathrm{d}x$	`Integrate[f, {x, xmin, xmax}]`	definite integral
$\partial_x f$	`D[f, x]`	partial derivative
$\partial_{x,y} f$	`D[f, x, y]`	multivariate partial derivative

2.5.2.3 TWO-DIMENSIONAL NOTATION IN PHYSICS

To illustrate the use of two-dimensional forms for input, this section gives several physics examples.

EXAMPLE 2.5.1 Define the gradient of a scalar point function in traditional mathematical notation, and apply it to the electric potential produced by a single point charge placed at the origin of a Cartesian coordinate system in order to obtain the corresponding electric field.

In terms of the Cartesian coordinates, the gradient of f can be defined as

$$\nabla f\_ := \partial_x f\, \hat{\imath} + \partial_y f\, \hat{\jmath} + \partial_z f\, \hat{k}$$

To display this definition in traditional mathematical notation, click the cell bracket and choose **TraditionalForm** in the Convert To submenu of the Cell menu:

$$In[29] := \nabla f\_ := \frac{\partial f}{\partial x} \hat{\imath} + \frac{\partial f}{\partial y} \hat{\jmath} + \frac{\partial f}{\partial z} \hat{k}$$

(Section 2.5.3 will relate the necessary precaution for using TraditionalForm input.)
    The electric potential due to a point charge q located at the origin is

$$In[30] := \phi[x\_, \ y\_, \ z\_] := k\, \frac{q}{\sqrt{x^2 + y^2 + z^2}}$$

In electrostatics, the electric field can be written as

$$\mathbf{E} = -\nabla\phi$$

Thus, the electric field produced by the point charge is

$$In[31] := -\nabla\phi[x, \ y, \ z] \; // \; \text{Factor}$$

$$Out[31] = \frac{k\, q\, (x\, \hat{\imath} + y\, \hat{\jmath} + z\, \hat{k})}{(x^2 + y^2 + z^2)^{3/2}}$$

$$In[32] := \text{Clear}[\phi, \ \text{Del}]$$ ∎

EXAMPLE 2.5.2   For the one-dimensional harmonic oscillator, determine the matrix element of the kinetic energy operator between the first and third excited energy eigenstates.
    In terms of the dimensionless coordinate $\zeta = \sqrt{m\omega/\hbar}\, x$, the kinetic energy operator of the one-dimensional harmonic oscillator is

$$T = \frac{p^2}{2m} = -\frac{\hbar^2}{2m}\frac{d^2}{dx^2} = -\frac{\hbar\omega}{2}\frac{d^2}{d\zeta^2}$$

The energy eigenfunctions are

$$In[33] := u_n\_[\zeta\_] := \sqrt{\frac{1}{\sqrt{\pi}\, n!\, 2^n}}\; H_n[\zeta]\; e^{-\frac{\zeta^2}{2}}$$

where n can assume any nonnegative integer and

$$In[34] := H_n\_[\zeta\_] := \text{HermiteH}[n, \ \zeta]$$

As shown in Section 2.5.1.3, the scalar product of two functions of the variable ζ can be defined as

$$In[35] := \langle\varphi\_, \ \psi\_\rangle := \int_{-\infty}^{\infty} \varphi^*\psi\, d\zeta$$

in which φ^* is the complex conjugate of φ and is given by

$$In[36] := \varphi\_{}^{\ *} := \varphi \; /. \; \text{Complex}[u\_, \ v\_] \to \text{Complex}[u, \ -v]$$

where the character "*****" is entered as the superscript of φ. Thus, the matrix element of the kinetic energy operator between the first and third excited states is

$In[37] := \left\langle u_1 [\xi], -\dfrac{\hbar \omega}{2} \partial_{\xi,\xi} u_3 [\xi] \right\rangle$

$Out[37] = -\dfrac{1}{2}\sqrt{\dfrac{3}{2}}\,\omega\,\hbar$

$In[38] := \text{Clear[Subscript, AngleBracket, SuperStar]}$ ∎

EXAMPLE 2.5.3 The normalized energy eigenfunctions of the hydrogen atom can be written as

$$\psi_{nlm}(r, \theta, \phi) = R_{nl}(r)Y_{lm}(\theta, \phi)$$

where R_{nl} and Y_{lm} are the radial functions and the normalized spherical harmonics, respectively. In terms of the associated Laguerre polynomials L_p^q and the Bohr radius a, the radial functions are given by

$$R_{nl}(r) = a^{-3/2}\dfrac{2}{n^2}\sqrt{\dfrac{(n-l-1)!}{[(n+l)!]^3}}\,F_{nl}\left(\dfrac{2r}{na}\right)$$

with

$$F_{nl}(x) = x^l\,e^{-x/2}\,L_{n-l-1}^{2l+1}(x)$$

The associated Laguerre polynomials $L_p^q(x)$ are related to the generalized Laguerre polynomials `LaguerreL[p, q, x]` of *Mathematica* by

$$L_p^q(x) = (p+q)!\,\text{LaguerreL[p, q, x]}$$

The allowed quantum numbers n, l, and m are given by the rules

$$n = 1, 2, 3, \ldots$$

$$l = 0, 1, 2, \ldots, (n-1)$$

$$m = -l, -l+1, \ldots, 0, 1, 2, \ldots, +l$$

Plot the probability densities for the states specified by the quantum numbers $(n, l, m) = (4, 3, 1)$, $(4, 2, 1)$, and $(4, 2, 0)$. Explain the apparent lack of spherical symmetry in these plots for the hydrogen atom, which has a spherically symmetric Hamiltonian.

We cannot really generate a three-dimensional plot for the probability density $\psi^*\psi$, because it depends on three coordinates (r, θ, ϕ) and, therefore, its plot must be four-dimensional. We can, however, make a density plot effectively using the gray level at each point (more precisely, at each cell) as the fourth coordinate or dimension. That is, the gray-level intensity, varying from 0 (black) to 1 (white), at a point is an increasing function of and, hence, represents the probability density at that point.

As it turns out, the probability densities of the hydrogen atom are independent of the angle ϕ. Only two-dimensional density plots on a plane containing the z or polar axis are necessary, as the three-dimensional plots can readily be obtained by rotating the two-dimensional ones through 360° about the z axis.

The wave functions can be defined by

$$\psi_{n\_l\_m\_}[r\_, \theta\_, \phi\_] := R_{nl}[r] \, Y_{lm}[\theta, \phi]$$

$$In[39] := R_{n\_l\_}[r\_] := a^{-3/2} \frac{2}{n^2} \sqrt{\frac{(n - l - 1)!}{((n + 1)!)^3}} \, F_{nl}\left[\frac{2\,r}{n\,a}\right]$$

$$In[40] := F_{n\_l\_}[x\_] := x^l \, e^{-x/2} (n + 1)! \, LaguerreL[n - l - 1, 2l + 1, x]$$

$$In[41] := Y_{l\_m\_}[\theta\_, \phi\_] := SphericalHarmonicY[l, m, \theta, \phi]$$

where invisible commas $:,:$ have been inserted between all quantum numbers in the foregoing definitions in order to mimic traditional physics notation. While this definition for the wave function is perfectly sound, Section 3.2.3, "Restricting Patterns," will show us how to write a better definition that complies with the rules on the quantum numbers n, l, and m:

$$In[42] := \psi_{n\_Integer?Positive\,l\_Integer?NonNegative\,m\_Integer}[$$
$$r\_, \theta\_, \phi\_] := R_{nl}[r] \, Y_{lm}[\theta, \phi] \; /; \; l \le n - 1 \wedge -l \le m \le l$$

where, as before, invisible commas $:,:$ have been inserted between all quantum numbers in this definition, including between `n_Integer?Positive` and `l_Integer?NonNegative` as well as between `l_Integer?NonNegative` and `m_Integer`. The definitions for R, F, and Y remain the same.

The complex conjugate of the wave function φ can be obtained from

$$In[43] := \varphi\_^* := \varphi \; /. \; Complex[u\_, v\_] \to Complex[u, -v]$$

Let us define a function for generating the density plots:

```
In[44] := hydrogenPlot[n_, l_, m_] :=
            (DensityPlot[
                Evaluate[ψnlm[r, θ, ϕ]* ψnlm[r, θ, ϕ]//.
                  {a -> 1, r -> Sqrt[x^2 + z^2], θ -> ArcCos[z/r]}
                    ],
                {x, -10 (1.5 n + 1) / 2, 10 (1.5 n + 1) / 2},
                {z, -10 (1.5 n + 1) / 2, 10 (1.5 n + 1) / 2},
                Mesh -> False,
                PlotPoints -> 200,
                FrameLabel -> {"x", "z"},
                PlotLabel -> {n, l, m},
                DisplayFunction -> Identity,
                FrameTicks -> None
                    ]
                )
```

`DensityPlot[f, {x, xmin, xmax}, {z, zmin, zmax}]` makes a density plot of f as a function of x and z. In the definition for the function `hydrogenPlot`, the rules for r and θ effect a coordinate transformation from (r, θ) to (x, z) and the rule for a makes the Bohr radius the unit of length. Be reminded again that invisible or ordinary commas must be inserted between the quantum numbers n, l, and m in the preceding definition.

Here are the density plots:

```
In[45] := Show[GraphicsArray[{hydrogenPlot[4, 3, 1],
                              hydrogenPlot[4, 2, 1],
                              hydrogenPlot[4, 2, 0]},
                   GraphicsSpacing -> 0.03]];
```

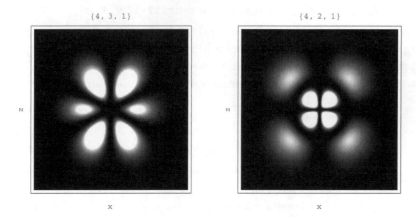

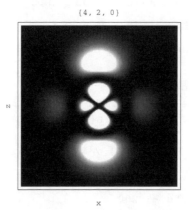

Mathematica returns the graphics in a row. For clarity, they are displayed in two rows here. `GraphicsArray[{g1, g2, ...}]` represents a row of graphics objects. The option `GraphicsSpacing` specifies the spacing between the graphs.

The probability densities exhibit rotational symmetry about the z or polar axis but lack the full spherical symmetry of the central potential. However, we cannot observe these probability densities experimentally. Otherwise, we can establish a preferred or z direction in space in the absence of external fields. All we can actually measure is the average probability density of the n^2 degenerate states in a shell identified by the principal quantum number n. That the average probability density of a shell is spherically symmetric can be verified:

$$In[46] := \text{shell}[n\_] := \frac{1}{n^2} \sum_{l=0}^{n-1} \sum_{m=-l}^{l} \psi_{nlm}[r, \theta, \phi]^* \, \psi_{nlm}[r, \theta, \phi] \; //$$
$$\text{Simplify}$$

$$In[47] := \text{shell}[4]$$

$$Out[47] = \frac{e^{-\frac{r}{2a}}(9216\,a^6 - 13824\,r\,a^5 + 10368\,r^2\,a^4 - 3264\,r^3\,a^3 + 504\,r^4\,a^2 - 36\,r^5\,a + r^6)}{9437184\,a^9\,\pi}$$

The average probability density for the $n = 4$ shell is indeed spherically symmetric, as it depends only on the coordinate r. We can easily verify that the average probabilities for other values of n are also spherically symmetric.

The rotational invariance of the Hamiltonian actually imposes only the m-degeneracy, called essential degeneracy, in a subshell labeled by the quantum numbers n and l. The additional l-degeneracy, called accidental degeneracy, is specific to the Coulomb potential. We can show that the average probability densities of the subshells are spherically symmetric, that is, independent of the angle θ:

$$In[48] := \text{subshell}[n\_, l\_] :=$$
$$\frac{1}{2l + 1} \sum_{m=-l}^{l} \psi_{nlm}[r, \theta, \phi]^* \, \psi_{nlm}[r, \theta, \phi] \; // \; \text{Simplify}$$

$$In[49] := \text{FreeQ[Table[subshell[n, l], \{n, 1, 4\}, \{l, 0, n - 1\}], } \theta]$$

$$Out[49] = \text{True}$$

$$In[50] := \text{Clear[Subscript, SuperStar, hydrogenPlot, shell, subshell]} \quad \blacksquare$$

2.5.3 Input and Output Forms

In addition to InputForm and OutputForm that are used in earlier versions, *Mathematica* version 3.0 also supports StandardForm and TraditionalForm as format types for input and output:

InputForm	a one-dimensional form using only keyboard characters; for both input and output
OutputForm	a form using only keyboard characters; for output only
StandardForm	a form utilizing special characters and two-dimensional forms; the default form for both input and output
TraditionalForm	a form that attempts to imitate traditional mathematical notation; primarily for output

To illustrate these forms or format types, consider the mathematical expression

$$\frac{x^3}{2 + \sin^2 x} + \int f(x)\, dx$$

The current choice of the Default Input FormatType submenu in the Cell menu specifies globally the default form of input. Set Default Input FormatType to **InputForm** and enter

```
x^3/(2 + Sin[x]^2) + Integrate[f[x], x]
```

The four choices of the Default Output FormatType submenu in the Cell menu correspond to the four possible output forms. The current choice specifies globally the form in which output will be displayed:

■ **InputForm**

```
Integrate[f[x], x] + x^3/(2 + Sin[x]^2)
```

■ **OutputForm**

$$\texttt{Integrate[f[x], x]} \; + \; \frac{x^3}{2 + \texttt{Sin[x]}^2}$$

■ **StandardForm**

$$\int \texttt{f[x] dx} + \frac{x^3}{2 + \texttt{Sin[x]}^2}$$

■ **TraditionalForm**

$$\frac{x^3}{\sin^2(x) + 2} + \int f(x)\, dx$$

With Default Input FormatType in the Cell menu set to **StandardForm,** we can enter the same mathematical expression in the form

$$\frac{x^3}{2 + \texttt{Sin[x]}^2} \; + \; \int \texttt{f[x] dx}$$

Mathematica, of course, returns the previous results since this is just another form of the input. Rather than working with the Default Output FormatType submenu, we can use the functions `InputForm`, `OutputForm`, `StandardForm`, and `TraditionalForm` to specify locally the form in which the output will be displayed:

■ **InputForm**

$$In[51] \; := \; \frac{x^3}{2 + \texttt{Sin[x]}^2} \; + \; \int \texttt{f[x] dx} \; // \; \texttt{InputForm}$$

Out[51]//InputForm=
```
Integrate[f[x], x] + x^3/(2 + Sin[x]^2)
```

■ **OutputForm**

$$In[52] \; := \; \frac{x^3}{2 + \texttt{Sin[x]}^2} \; + \; \int \texttt{f[x] dx} \; // \; \texttt{OutputForm}$$

Out[52]//OutputForm=

$$\text{Integrate[f[x], x]} + \frac{x^3}{2 + \text{Sin[x]}^2}$$

■ **StandardForm**

$$In[53] := \frac{x^3}{2 + \text{Sin[x]}^2} + \int f[x]\, dx \; // \; \text{StandardForm}$$

Out[53]//StandardForm=

$$\int f[x]\, dx + \frac{x^3}{2 + \text{Sin[x]}^2}$$

■ **TraditionalForm**

$$In[54] := \frac{x^3}{2 + \text{Sin[x]}^2} + \int f[x]\, dx \; // \; \text{TraditionalForm}$$

Out[54]//TraditionalForm=

$$\frac{x^3}{\sin^2(x) + 2} + \int f(x)\, dx$$

For *Mathematica* input, StandardForm includes InputForm; that is, whatever can be entered in InputForm can also be entered in StandardForm, but the converse is not true. For example, consider again the mathematical expression

$$\frac{x^3}{2 + \sin^2 x} + \int f(x)\, dx$$

With Default Input FormatType in the Cell menu set to **StandardForm**, we can enter

```
x^3/(2 + Sin[x]^2) + Integrate[f[x], x]
```

that was entered in InputForm. On the other hand, with Default Input FormatType set to **InputForm**, we cannot enter

$$\frac{x^3}{2 + \text{Sin[x]}^2} + \int f[x]\, dx$$

which is only appropriate for StandardForm input.

The appeal of StandardForm with the special characters and two-dimensional forms notwithstanding, we must still learn InputForm en route to mastery of *Mathematica*. As will be explained in Section 3.1, the *Mathematica* kernel sees only FullForm, and the one-dimensional InputForm is its close cousin, whereas Standardform bears little resemblance to it. Consequently, it is much easier to debug the code in InputForm than in StandardForm. For interactive *Mathematica* sessions, this advantage may be insignificant. For substantial programs, consider writing the code in InputForm, fix the bugs, and then convert it to StandardForm. Conversion of the code from InputForm to StandardForm can be done with the **Convert To** command in the Cell menu and minor edits of the result may be necessary. Most of the code in the remainder of this book is written in InputForm.

Traditional mathematical notation is discernible, intuitive, natural, and elegant to those who are mathematically proficient, and many mathematicians, scientists, and engineers have long awaited a computer algebra system that would utilize such a notation. The main difficulty lies in its lack of the precision required for computer input and output. For example,

$a(1 + x)$ can mean a times $1 + x$ or the function a with the argument $1 + x$. Also, $\tan^{-1}\theta$ may be interpreted as $(\tan\theta)^{-1}$ or $1/\tan\theta$ rather than the customary inverse tangent or arctangent of θ, as $\tan^2\theta$ is normally taken to be $(\tan\theta)^2$. StandardForm is *Mathematica*'s answer to the call, as it has the required precision for input as well as output and resembles somewhat the familiar mathematical notation. For those who want "the real thing," *Mathematica* offers TraditionalForm, which gives a fair rendition of the traditional mathematical notation, but with the caveat that it is primarily for output only.

TraditionalForm lacks the necessary precision for input as it tries to imitate the traditional mathematical notation. For example,

```
In[55] := {S[1 + α], FresnelS[1 + α], S(1 + α)} // StandardForm

Out[55]//StandardForm=
  {S[1 + α], FresnelS[1 + α], S(1 + α)}


In[56] := {S[1 + α], FresnelS[1 + α], S(1 + α)} // TraditionalForm

Out[56]//TraditionalForm=
```
$$\{S(\alpha + 1), S(\alpha + 1), S\,(\alpha + 1)\}$$

where the first element is a function S with the argument $1 + \alpha$, the second is a Fresnel integral, and the third is S times $(1 + \alpha)$. The three elements appear distinct in StandardForm but look identical in TraditionalForm except for the tiny extra space in the third element. If we input them in TraditionalForm, how can we tell what they actually represent, and how can *Mathematica* distinguish between the first two?

If the lure of TraditionalForm for input becomes irresistible, consider entering the expression in StandardForm, clicking the cell bracket, and then choosing **TraditionalForm** in the Convert To submenu of the Cell menu. *Mathematica* will insert hidden tags in the converted expression, now in TraditionalForm, so as to make it sufficiently precise for input. For example, enter

$$\mathcal{E} := \sum_{k=1}^{f} \dot{q}_k \, \partial_{\dot{q}_k} \, L - L$$

in StandardForm and convert this to

$$\mathcal{E} := \sum_{k=1}^{f} \dot{q}_k \, \frac{\partial L}{\partial \dot{q}_k} - L$$

which is a legitimate expression in TraditionalForm for *Mathematica* input. For another example, enter

```
DSolve[D[x[t], {t, 2}] == -γ D[x[t], t] - ω₀²x[t], x[t], t]
```

and convert it to

$$In[57] := \mathbf{DSolve}\left[\frac{\partial^2 x(t)}{\partial t^2} == -\gamma \frac{\partial x(t)}{\partial t} - \omega_0^2 \, x(t), x(t), t \right]$$

$$Out[57] = \left\{ \left\{ x(t) \to e^{\frac{1}{2}t\left(-\gamma - \sqrt{\gamma^2 - 4\omega_0^2}\right)} c_1 + e^{\frac{1}{2}t\left(\sqrt{\gamma^2 - 4\omega_0^2} - \gamma\right)} c_2 \right\} \right\}$$

where we have evaluated the converted expression to verify its legitimacy as an expression for input.

2.5.4 Evaluation of Subexpressions

Hitherto *Mathematica* always evaluated the whole current cell. Now it can also evaluate only a selected subexpression in the cell. To evaluate a subexpression, select or highlight it and press Command-Return for the Macintosh or Shift+Control+Enter for Windows.

Set Default Output FormatType to **StandardForm** so that we can later use any output, edited or not, as input. Now enter, for example, the integral

$In[58] := \int \sqrt{a + b \, Sin[\theta]^2} \, Cos[\theta] \, d\theta$

$Out[58] = \dfrac{a \, Log\left[2 \sqrt{b} \, Sin[\theta] + 2 \sqrt{a + b \, Sin[\theta]^2}\right]}{2 \sqrt{b}} + \dfrac{1}{2} \, Sin[\theta] \, \sqrt{a + b \, Sin[\theta]^2}$

Differentiating the output and then simplifying the result should recover the integrand:

$In[59] := \partial_\theta \% \, // \, Simplify$

$Out[59] = \dfrac{Cos[\theta] \sqrt{2 a + b - b \, Cos[2 \, \theta]}}{\sqrt{2}}$

Yet this is not exactly the integrand. Applying the trigonometric manipulating functions **TrigExpand, TrigFactor**, or **TrigReduce** to it produces no further changes. Manipulating its subexpressions may be the only way to retrieve the integrand.

We can work directly on the output. As soon as we begin editing the output, *Mathematica* will leave it unaltered but will automatically paste the edited version in a new cell immediately below. We can move the cell to any convenient location and then continue our work.

Let us wrap the function **TrigExpand** around the radicand in the previous output and move it below:

$\dfrac{Cos[\theta] \sqrt{TrigExpand[2 a + b - b \, Cos[2 \, \theta]]}}{\sqrt{2}}$

Select the edited radicand and press Command-Return for the Macintosh or Shift+Control+Enter for Windows. *Mathematica* returns in place

$\dfrac{Cos[\theta] \sqrt{(2 a + b - b \, Cos[\theta]^2 + b \, Sin[\theta]^2)}}{\sqrt{2}}$

Wrap the function **Simplify** around $b - b \, Cos[\theta]^2$:

$\dfrac{Cos[\theta] \sqrt{(2 a + Simplify[b - b \, Cos[\theta]^2] + b \, Sin[\theta]^2)}}{\sqrt{2}}$

Select **Simplify[b - b Cos[θ]²]**, and press Command-Return for the Macintosh or Shift+Control+Enter for Windows. *Mathematica* generates in place

$$\frac{Cos[\theta]\sqrt{(2a+bSin[\theta]^2+bSin[\theta]^2)}}{\sqrt{2}}$$

Finally, apply the function `Simplify` to the entire expression:

$$In[60] \; := \; \frac{Cos[\theta]\sqrt{(2a+bSin[\theta]^2+bSin[\theta]^2)}}{\sqrt{2}} \; // \; Simplify$$

$$Out[60] \; = \; Cos[\theta]\sqrt{a+bSin[\theta]^2}$$

This is the integrand!

For manipulating subexpressions and expressions, the AlgebraicManipulation palette of the Palettes submenu in the File menu provides an alternative to entering transforming functions such as `Simplify` and `TrigExpand` directly from the keyboard. Select the radicand in the numerator of the output

$$\frac{Cos[\theta]\sqrt{2a+b-bCos[2\theta]}}{\sqrt{2}}$$

and click `TrigExpand[■]` in the palette. *Mathematica* gives

$$\frac{Cos[\theta]\sqrt{(2a+b-bCos[\theta]^2+bSin[\theta]^2)}}{\sqrt{2}}$$

In this expression, highlight `b - b Cos[θ]²` and click `Simplify[■]` in the palette. *Mathematica* generates in place

$$\frac{Cos[\theta]\sqrt{(2a+bSin[\theta]^2+bSin[\theta]^2)}}{\sqrt{2}}$$

Finally, select the entire expression and click `Simplify[■]`. *Mathematica* returns in place

$$Cos[\theta]\sqrt{a+bSin[\theta]^2}$$

which is the integrand.

2.5.5   Exercises

For these exercises, set Default Input FormatType and Default Output FormatType in the Cell menu to **StandardForm**.

1. Utilizing special characters and two-dimensional forms, redo Examples (a) 2.1.1, (b) 2.1.5, (c) 2.2.3, (d) 2.2.7, (e) 2.3.12, (f) 2.4.7, and (g) 2.4.11.

2. Using special characters and two-dimensional forms, enter the following mathematical expressions in StandardForm, then convert them to TraditionalForm with the Convert To submenu in the Cell menu, and finally evaluate the converted expressions:
   (a)

$$\sum_{n=0}^{\infty} n\frac{1}{n!}\left(\frac{t}{\tau_0}\right)^n e^{-t/\tau_0}$$

(b)

$$\frac{\sum_{n=0}^{\infty} n \exp\left(-\frac{n\varepsilon}{kT}\right)}{\sum_{n=0}^{\infty} \exp\left(-\frac{n\varepsilon}{kT}\right)}$$

(c)

$$\gamma \frac{d^2 n}{dz^2} + \frac{z}{2} \frac{dn}{dz} = 0$$

3. Reduce the integral

$$-\frac{1}{\omega_1} \int_{\omega_1 t}^{0} \exp\left(-\gamma t\right) \exp\left(\left[(\gamma - \beta)/\omega_1\right] z\right) \sin z \, dz$$

to

$$\frac{-e^{-t\beta} (\beta - \gamma) \sin(t\omega_1) + e^{-t\gamma} \omega_1 - e^{-t\beta} \cos(t\omega_1) \omega_1}{(\beta - \gamma)^2 + \omega_1^2}$$

2.6 Problems

1. Determine the sum of the infinite series

$$1 + \frac{1}{3^4} + \frac{1}{5^4} + \frac{1}{7^4} + \cdots$$

Answer: `1.01468`

2. Invert the 5×5 Hilbert matrix, the matrix whose (i, j)th element is $1/(i + j - 1)$ for $1 \le i, j \le 5$.

Answer:

```
25      -300     1050     -1400    630
-300    4800     -18900   26880    -12600
1050    -18900   79380    -117600  56700
-1400   26880    -117600  179200   -88200
630     -12600   56700    -88200   44100
```

3. Solve the system of equations and verify the solution(s).

$$2x + y = 3z$$
$$6x + 24 + 4y = 0$$
$$20 - 5x + 2z = 0$$

Answer:

$$\{\{x \to \frac{24}{7},\ y \to -(\frac{78}{7}),\ z \to -(\frac{10}{7})\}\}$$

4. Solve the system of equations and verify the solution(s).

$$x^2 + y^2 = 1$$
$$x + 3y = 0$$

Answers:

$$\{\{x \rightarrow \frac{-3}{\text{Sqrt}[10]}, \ y \rightarrow \frac{1}{\text{Sqrt}[10]}\},$$
$$\{x \rightarrow \frac{3}{\text{Sqrt}[10]}, \ y \rightarrow -(\frac{1}{\text{Sqrt}[10]})\}\}$$

5. Plot the elliptic paraboloid given by the equation

$$\frac{x^2}{a^2} + \frac{y^2}{b^2} = 2cz$$

where $a = b = 1$ and $c = 1/2$. With the **3D ViewPoint Selector** show the picture from several viewpoints.

Answer: (Only one of the four elliptic paraboloids is included here.)

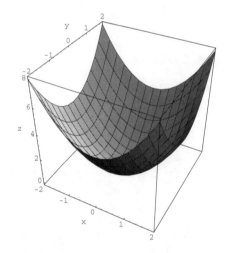

6. The graphs in Problem 5 do not look like elliptic paraboloids, which have rotational symmetry about the z axis. Use the **ParametricPlot3D** function to replot the graphs.

Answer: (Only one of the four elliptic paraboloids is included here.)

7. Consider again Problem 5 on the elliptic paraboloid. Let a plane be defined by the equation

$$ax + by + cz + d = 0$$

with $a = 0.25$, $b = 0.25$, $c = -1$, and $d = 2.5$. Superpose this plane on the elliptic paraboloid and produce the graph

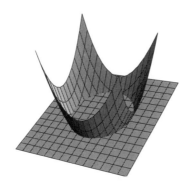

8. Find the solution for $0 < x < 3$ of the following equation to five significant figures:

$$e^{-x} \sin x = x^2 - \frac{1}{2}$$

Answer:

```
{x -> 0.90453}
```

9. Plot a sequence of seven graphs corresponding to different values of λ from 3π to 9π. Each graph has three curves representing the following equations:

$$y(x) = 1$$
$$y(x) = -1$$
$$y(x) = \cos x + \frac{\lambda}{2} \frac{\sin x}{x}$$

Then animate them to demonstrate the effects of varying λ.

Answer: (Only one of the seven graphs is included here.)

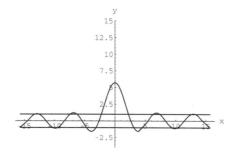

10. Consider two waves with the same amplitude and frequency but with opposite directions of propagation. The resultant wave is a standing wave described by the equation

$$y = 2A \cos kx \cos \omega t$$

Animate the standing wave. (*Hint:* Plot a number of graphs for different values of time *t*.)

Answer: (Only one of the sixteen graphs is included here.)

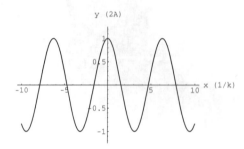

11. Create a spinning sphere.

Answer: (Only one of the six spheres is included here.)

12. Determine

$$\sum_{k=1}^{n} k^4$$

Answer:

$$\frac{(-1 - n)\ n\ (1 + 2\ n)\ (1 - 3\ n - 3\ n^2)}{30}$$

13. The current *i* in an electric circuit is given by

$$\frac{di}{dt} + 2i = \sin t$$

with $i(0) = 0$. (a) Determine $i(t)$ numerically for t in the range 0 to 1.1. (b) Determine $i(t)$ analytically. (c) Compare the numerical and analytical solutions for t from 0 to 1 in steps of 0.1.

Answers:

$$\{\{i \rightarrow \text{InterpolatingFunction}[\{0., 1.1\}, <>]\}\}$$

$$\{\{i[t] \rightarrow \frac{1}{5\ E^2\ t} - \frac{Cos[t]}{5} + \frac{2\ Sin[t]}{5}\}\}$$

t	numerical	analytical
0	0	0
0.1	0.00468042	0.00467868
0.2	0.0175201	0.0175184
0.3	0.0369043	0.0369031
0.4	0.0614216	0.0614209
0.5	0.0898301	0.0898296
0.6	0.121029	0.121029
0.7	0.154038	0.154038
0.8	0.187982	0.18798
0.9	0.22207	0.222069
1.	0.255596	0.255595

14. Consider the differential equation

$$\frac{dx}{dt} + x + e^t x^2 = 0$$

with $x(0) = 1$. (a) Determine $x(t)$ numerically for t in the range 0 to 0.4. (b) Determine $x(t)$ analytically. (c) Compare the numerical and analytical solutions for t from 0 to 0.4 in steps of 0.1.

Answers:

$$\{\{x \rightarrow \text{InterpolatingFunction}[\{0., 0.4\}, <>]\}\}$$

$$\{\{x[t] \rightarrow \frac{1}{E^t\ (1\ +\ t)}\}\}$$

t	numerical	analytical
0	1.	1
0.1	0.822582	0.822579
0.2	0.682276	0.682276
0.3	0.56986	0.56986
0.4	0.478801	0.4788

15. (a) Find a power series approximation, up to terms of order x^7, for the function $\sin x \cos x$ about $x = 0$. (b) Draw a graph of the function $\sin x \cos x$ together with its power series approximation. (c) Examine the graph to determine the region in which the power series is a good approximation of the function.

Answers:

$$x - \frac{2\ x^3}{3} + \frac{2\ x^5}{15} - \frac{4\ x^7}{315} + O[x]^8$$

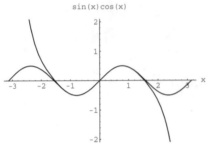

The power series is a good approximation for $x \approx -1.6$ to 1.6.

16. Consider the differential equation

$$-\frac{d^2\Psi}{dx^2} + V(x)\Psi = E\Psi$$

with $V(x) = -6\,\mathrm{sech}^2(x)$

Also, $\Psi(0) = 1$, $\Psi'(0) = 0$, and $E = -4$. Determine numerically and plot $\Psi(x)$ for x in the range -5 to 5.

Answers:

$$\{\{psi \rightarrow InterpolatingFunction[\{-5., 5.\}, <>]\}\}$$

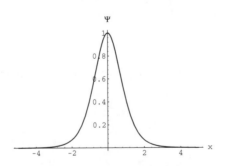

17. Three equal point charges are at the corners of an equilateral triangle. For the plane of the triangle, locate the point(s) of equilibrium where the electric field vanishes.

18. Two one-dimensional harmonic oscillators are attached, as in Figure 2.6.1. While m_2 is held in place, m_1 is moved to the left a distance a. At time $t = 0$, both masses are released. Determine their subsequent motion. What are the normal or characteristic frequencies of the system? Let x_1 and x_2 be the displacements of m_1 and m_2, respectively.

Each displacement is measured from the equilibrium position, and the direction to the right is positive. Consider only the case of equal mass and identical spring constant, that is, $m_1 = m_2 = m$ and $k_1 = k_2 = k$.

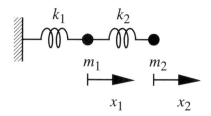

Figure 2.6.1 Two attached one-dimensional harmonic oscillators

19. Let $a = 1$ in Equation 2.2.1 and

$$f(t) = \begin{cases} \sin \omega_0 t & |t| < T \\ 0 & |t| > T \end{cases}$$

Show that the Fourier transform of $f(t)$ is

$$g(\omega) = -\frac{i}{\sqrt{2\pi}} \left\{ \frac{\sin(\omega - \omega_0)T}{(\omega - \omega_0)} - \frac{\sin(\omega + \omega_0)T}{(\omega + \omega_0)} \right\}$$

Verify the uncertainty relation

$$\Delta\omega\,\Delta t \approx 4\pi$$

20. If the wave function in configuration space is the Gaussian wave packet

$$\psi(x) = A \exp\left[\frac{i}{\hbar}p_0 x - \frac{(x - x_0)^2}{2b^2}\right]$$

where A and b are positive real constants, show that the wave function in momentum space is

$$\varphi(p) = A\frac{b}{\sqrt{\hbar}} \exp\left[\frac{i}{\hbar}x_0(p_0 - p) - \frac{b^2(p - p_0)^2}{2\hbar^2}\right]$$

Verify that the product of the uncertainty in position Δx and the uncertainty in momentum Δp is of order h, which is Planck's constant. *Hint: Mathematica* can use a little bit of help in evaluating the Fourier transform.

21. A one-dimensional relativistic particle of rest mass m under the action of a constant force F starts from rest at the origin at time $t = 0$. The equation of motion is

$$F = \frac{dp}{dt}$$

where p is the relativistic momentum.(a) Show that the velocity of the particle is

$$v = \frac{(F/m)t}{\sqrt{1 + (Ft/mc)^2}}$$

and its position is

$$x = \frac{mc^2}{F}\left[\sqrt{1 + \left(\frac{Ft}{mc}\right)^2} - 1\right]$$

(b) Plot the relativistic position together with the classical one as a function of time. *Hint:* Adopt a convenient system of units.

CHAPTER 3

Programming
in *Mathematica*

We have examined the dazzling numerical, symbolic, and graphical capabilities of *Mathematica*. Yet *Mathematica*'s real power rests on its programming capabilities. Many programming concepts such as assignments, transformation rules, and lists were introduced in Chapter 2. This chapter covers five programming topics: expressions, patterns, functions, procedures, and graphics. It also discusses three programming styles and the writing of packages.

3.1 Expressions

Chapter 2 showed that *Mathematica* works with many kinds of objects such as algebraic expressions, mathematical functions, equations, functions for specifying operations, assignments, transformation rules, graphics, and lists. They are all expressions. This section discusses the building blocks and internal representation of expressions. It also considers their manipulation and evaluation.

3.1.1 Atoms

Most *Mathematica* objects are composed of parts. For example, the list {a, b, c} has four parts. We can extract these parts with the function **Part** introduced in Section 2.4.3:

```
In[1] := Table[Part[{a, b, c}, i], {i, 0, 3}]

Out[1] = {List, a, b, c}
```

There are several types of special *Mathematica* objects, called atoms, that do not have parts or are indivisible. The following table lists the types, descriptions, and examples of these objects:

Type	Description	Example
Symbol	symbol	x, E, Sin
String	character string "*ccc*"	"I Love Lucy"
Integer	integer	50
Real	approximate real number $n.m$ where n and m are integers	1.2345
Rational	rational number n/m where n and m are integers and $m \neq 0$	2/3
Complex	complex number $a + b$ I where a and b are integers, approximate real numbers, or rational numbers	2.12 + 3/5 I

The function AtomQ tests whether an object is one of the atomic types. AtomQ [*object*] yields **True** if *object* is an atom, and yields **False** otherwise:

In[2] := **AtomQ[E]**

Out[2] = True

In[3] := **AtomQ[2.12 + 3/5 I]**

Out[3] = True

In[4] := **AtomQ[1 + x^2 == 0]**

Out[4] = False

In[5] := **AtomQ[{a, b, c}]**

Out[5] = False

For an atom, the function Head gives its type:

In[6] := **Head[x]**

Out[6] = Symbol

In[7] := **Head["I Love Lucy"]**

Out[7] = String

In[8] := **Head[50]**

Out[8] = Integer

In[9] := **Head[1.2345]**

Out[9] = Real

In[10] := **Head[2/3]**

Out[10] = Rational

3.1.2 *Internal Representation*

Mathematica handles many kinds of objects; the following table lists several of them together with examples:

Algebraic expressions	`x + y`
Mathematical functions	`Sin[x]`
Equations	`y''[x] + y[x] + 2 x == 0`
Functions for specifying operations	`D[1/3 x^3, x]`
Assignments	`a = b`
Transformation rules	`a -> b`
Graphics	`-Graphics-`
Lists	`{a, b, c}`

These objects are all expressions.

Mathematica represents expressions internally in a uniform way. The building blocks of expressions are atoms, which are the simplest expressions. There is a rule for obtaining more complicated expressions from simpler ones: If *h, e1, e2, . . .* are expressions, then *h*[*e1, e2, e3, . . .*] is also an expression, where *h* is called the head of the expression and the *ei* are called the elements of the expression. Expressions may contain no elements and take the form *h*[]. *Mathematica* expressions are the atoms and the objects represented internally in the form *h*[*e1, e2, e3, . . .*], which are called normal expressions.

The input forms of expressions may differ from their internal forms. **FullForm**[*expr*] shows the internal representation of *expr*:

In[11] := **FullForm[x + y]**

Out[11]//FullForm=
      Plus[x, y]

In[12] := **FullForm[Sin[x]]**

Out[12]//FullForm=
      Sin[x]

In[13] := **FullForm[y''[x] + y[x] + 2 x == 0]**

Out[13]//FullForm=
          Equal[Plus[Times[2, x], y[x], Derivative[2][y][x]], 0]

In[14] := **FullForm[a -> b]**

Out[14]//FullForm=
          Rule[a, b]

The output forms of expressions may also be different from their internal forms. We can use **FullForm** to show their internal representations. For example, let us plot *x* for *x* from 0 to 1:

In[15] := **Plot[x, {x, 0, 1}]**

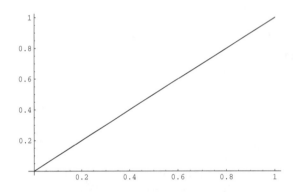

Out[15] = -Graphics-

Here is the internal representation of the graphics object:

In[16] := **Short[FullForm[%], 4]**

Out[16]//Short=
          Graphics[List[List[Line[<<1>>]]],
             List[Rule[PlotRange, Automatic],
                Rule[AspectRatio, Power[GoldenRatio, -1]],
                RuleDelayed[DisplayFunction, $DisplayFunction], <<21>>,
                RuleDelayed[FormatType, $FormatType]]]

FullForm returned many lines. The printing of the output has been shortened with **Short**[*expr, n*], which reduces the printing of *expr* to about *n* lines. The <<*n*>> indicates that *n* elements have been omitted.

To see the full form of an expression, sometimes it is necessary to wrap the function **Hold** around the expression to prevent its evaluation before the internal representation is revealed.

For example, consider

In[17] := **FullForm[D[1/3 x^3, x]]**

Out[17]//FullForm=
          Power[x, 2]

In this case, *Mathematica* returns the full form of the result of differentiation. To see the internal representation of **D[1/3 + x^3, x]**, enter

In[18] := **FullForm[Hold[D[1/3 x^3, x]]]**

Out[18]//FullForm=
          Hold[D[Times[1, Power[3, −1], Power[x, 3]], x]]

Hold[*expr*] maintains *expr* in an unevaluated form.

   FullForm[*expr*] shows the full form of *expr* even if it contains special symbols such as **->**, **/.**, and **/@**:

In[19] := **FullForm[a -> b]**

Out[19]//FullForm=
          Rule[a, b]

In[20] := **FullForm[Hold[(f /@ expr) /. f -> g]]**

Out[20]//FullForm=
          Hold[ReplaceAll[Map[f, expr], Rule[f, g]]]

   Knowing the internal representation of expressions is essential to programming. For example, let us determine the complex conjugate of $(a + ib)(c + id)$ where $a, b, c,$ and d are real. To find the complex conjugate, we simply replace i by $-i$, since $a, b, c,$ and d are real:

In[21] := **(a + I b)(c - I d) /. I -> -I**

Out[21] = (a − I b) (c − I d)

Yet *Mathematica* returns an incorrect result. Let us use **FullForm** to reveal the source of error:

In[22] := **FullForm[(a + I b)(c - I d)]**

Out[22]//FullForm=
          Times[Plus[a, Times[Complex[0, 1], b]],
          Plus[c, Times[Complex[0, −1], d]]]

We see that *Mathematica* represents **I** as **Complex[0, 1]** and **-I** as **Complex[0, -1]** internally. To determine the complex conjugate, enter

In[23] := **(a + I b)(c - I d) /. Complex[x_, y_] -> Complex[x, -y]**

Out[23] = (a − I b) (c + I d)

This time *Mathematica* yields the correct answer. The notion of "blanks" was introduced in Section 2.2.13, "Defining Functions." Here the symbol *arg_* means any expression to be named *arg* on the right-hand side of the transformation rule.

The function `Head` introduced in Section 3.1.1 can also be used to extract the head of an expression:

In[24] := `Head[1 + x + x^2]`

Out[24] = Plus

`FullForm` shows that `Plus` is indeed the head of the expression:

In[25] := `FullForm[1 + x + x^2]`

Out[25]//FullForm=
             Plus[1, x, Power[x, 2]]

`Length[`*expr*`]` gives the index n in the internal form h`[`*e1*`,` *e2*`,` *e3*`,` ...`,` *en*`]` of *expr*:

In[26] := `Length[1 + x + x^2]`

Out[26] = 3

For expressions containing no elements, `Length` returns the number 0:

In[27] := `Length[h[]]`

Out[27] = 0

3.1.3 Manipulation

3.1.3.1 OBTAINING PARTS OF EXPRESSIONS

The positions of the parts of an expression are determined according to the internal form h`[`*e1*`,` *e2*`,` ...`,` *en*`]` of the expression: h is part 0 and *ei* is the ith part. If the ith part is itself an expression, the position of its jth part is $\{i, j\}$. This indexing scheme continues if a subexpression of a subexpression is also an expression.

The parts of an expression that require exactly n indices for specifying their positions are said to be at level n. Level 0 refers to the entire expression. `Level[`*expr*`,` *levelspec*`]` gives a list of all subexpressions of *expr* on levels specified by *levelspec*. The second argument, *levelspec*, of the function is entered according to the level-specification scheme

n	levels 1 through n
`Infinity`	all levels
`{`n`}`	level n only
`{`$n1$`,` $n2$`}`	levels $n1$ through $n2$

If the option `Heads -> True` is included as its third argument, `Level` returns a list that includes also the heads of subexpressions on levels specified by *levelspec*. Consider the expression `(1 + x + 2 (x + 3)^m)`. `FullForm` reveals its internal representation:

In[28] := `FullForm[(1 + x + 2 (x + 3)^m)]`

Out[28]//FullForm=
             Plus[1, x, Times[2, Power[Plus[3, x], m]]]

Here is a list of subexpressions at the level $n = 4$:

```
In[29] := Level[(1 + x + 2 (x + 3)^m), {4}]

Out[29] = {3, x}
```

The function `TreeForm` is helpful in seeing the levels of expressions. `TreeForm[expr]` prints out *expr* in a form that shows its level structure:

```
In[30] := TreeForm[(1 + x + 2 (x + 3)^m)]

Out[30]//TreeForm=
  Plus[1, x, |                                      ]
             Times[2, |                     ]
                      Power[|          , m]
                            Plus[3, x]
```

We see that the fourth level indeed consists of the parts 3 and x. Here is a list of subexpressions including heads on levels 3 through 4:

```
In[31] := Level[(1 + x + 2 (x + 3)^m), {3, 4}, Heads -> True]

Out[31] = {Power, Plus, 3, x, 3 + x, m}
```

Lists are examples of expressions. The functions introduced in Section 2.4.3 for obtaining list and sublist elements can be generalized to functions for obtaining pieces of expressions:

`First[`*expr*`]`	the first element
`Last[`*expr*`]`	the last element
`Part[`*expr, n*`]` or *expr*`[[`*n*`]]`	the *n*th part
`Part[`*expr, i, j, ...*`]` or *expr*`[[`*i, j, ...*`]]`	the part at the position $\{i, j, \ldots\}$ of *expr*
`Part[`*expr,* `{`*n1, n2, ...*`}]` or *expr*`[[{`*n1, n2, ...*`}]]`	*expr* with the *n1*th, *n2*th, ... elements
`Take[`*expr, n*`]`	*expr* with the first *n* elements
`Take[`*expr,* `{`*m, n*`}]`	*expr* with the *m*th through *n*th elements
`Rest[`*expr*`]`	*expr* without the first element
`Drop[`*expr, n*`]`	*expr* without the first *n* elements
`Drop[`*expr,* `{`*m, n*`}]`	*expr* without the *m*th through *n*th elements

Let us use the expression `(1 + x + 2 x^2 + 3 x^3 + Sin[x])` to illustrate the use of these functions. Seeing the full form of this expression can be helpful in understanding the results of function calls.

```
In[32] := FullForm[(1 + x + 2 x^2 + 3 x^3 + Sin[x])]

Out[32]//FullForm=
        Plus[1, x, Times[2, Power[x, 2]], Times[3, Power[x, 3]], Sin[x]]
```

We can pick out one part at a time. Here is the last element:

In[33] := **Last[(1 + x + 2 x^2 + 3 x^3 + Sin[x])]**

Out[33] = Sin[x]

Here is the fourth element:

In[34] := **Part[(1 + x + 2 x^2 + 3 x^3 + Sin[x]), 4]**

Out[34] = 3 x^3

Here is the part at the position {4, 2, 1}:

In[35] := **Part[(1 + x + 2 x^2 + 3 x^3 + Sin[x]), 4, 2, 1]**

Out[35] = x

We can obtain expressions with some of the elements. Here is the expression with the first, third, and fourth elements:

In[36] := **Part[(1 + x + 2 x^2 + 3 x^3 + Sin[x]), {1, 3, 4}]**

Out[36] = 1 + 2 x^2 + 3 x^3

Here is the expression with the first two elements:

In[37] := **Take[(1 + x + 2 x^2 + 3 x^3 + Sin[x]), 2]**

Out[37] = 1 + x

Here is the expression without the second through fourth elements:

In[38] := **Drop[(1 + x + 2 x^2 + 3 x^3 + Sin[x]), {2, 4}]**

Out[38] = 1 + Sin[x]

The functions in this section pick out parts according to their positions in expressions. The function **Position** allows us to determine the positions of parts in expressions:

Position[*expr, part***]**	give a list of the positions at which *part* appears in *expr*
Position[*expr, part, levelspec***]**	give only positions on levels specified by *levelspec*

Let us determine the positions of the symbol **x**:

In[39] := **Position[(1 + x + 2 x^2 + 3 x^3 + Sin[x]), x]**

Out[39] = {{2}, {3, 2, 1}, {4, 2, 1}, {5, 1}}

We can determine the positions at the third level only:

In[40] := **Position[(1 + x + 2 x^2 + 3 x^3 + Sin[x]), x, {3}]**

Out[40] = {{3, 2, 1}, {4, 2, 1}}

The function **Select** focuses on the properties rather than the positions of elements in expressions:

Select [*expr*, *criterion*]	*expr* with the elements for which *criterion* gives **True**
Select [*expr*, *criterion*, *n*]	*expr* with the first *n* elements for which *criterion* gives **True**

AtomQ and the built-in predicates described in Section 2.4.3 can be used as criteria for the **Select** function. Let us employ again the expression (1 + x + 2 x^2 + 3 x^3 + Sin[x]) to demonstrate the use of **Select**. We can obtain an expression with the elements that are of the atomic types:

```
In[41] := Select[(1 + x + 2 x^2 + 3 x^3 + Sin[x]), AtomQ]

Out[41] = 1 + x
```

Here is the expression with only the first atom:

```
In[42] := Select[(1 + x + 2 x^2 + 3 x^3 + Sin[x]), AtomQ, 1]

Out[42] = 1
```

We can also generate an expression with the elements that are odd numbers:

```
In[43] := Select[(1 + x + 2 x^2 + 3 x^3 + Sin[x]), OddQ]

Out[43] = 1
```

Note that only the name, that is, without an argument, of a function or predicate is entered as *criterion* for the **Select** function.

Several other built-in functions that test the properties of expressions can also be used to define criteria for the **Select** function:

Function	Yields True if
PolynomialQ [*expr*, *var*]	*expr* is a polynomial in *var*
PolynomialQ [*expr*, {*var1*, ...}]	*expr* is a polynomial in the *vari*
VectorQ [*expr*, *test*]	*test* gives **True** when applied to each of the elements in *expr*
MatrixQ [*expr*, *test*]	*test* gives **True** when applied to each of the matrix elements in *expr*
SameQ [*lhs*, *rhs*] or *lhs* === *rhs*	the expression *lhs* is identical to *rhs*
UnsameQ [*lhs*, *rhs*] or *lhs* =!= *rhs*	the expressions *lhs* and *rhs* are not identical
OrderedQ [*h*[*e1*, *e2*, ...]]	the *ei* are in canonical order
MemberQ [*expr*, *form*]	an element of *expr* matches *form*
FreeQ [*expr*, *form*]	no subexpression in *expr* matches *form*
MatchQ [*expr*, *form*]	*form* matches *expr*

Let us demonstrate how to construct criteria with these functions. We define a function that, when applied to an expression, returns `True` if the expression is of the form $a_0 + a_1x + a_2x^2 + \cdots + a_nx^n$, for any integer $n \geq 0$ and any a_i:

```
In[44] := test1[expr_] := PolynomialQ[expr, x]
```

With this function, we can obtain an expression with the elements that are polynomials in the variable x:

```
In[45] := Select[(1 + x + 2 x^2 + 3 x^3 + Sin[x]), test1]
```
```
Out[45] = 1 + x + 2 x^2 + 3 x^3
```

Here is a function for testing whether an expression is identical to `Sin[x]`:

```
In[46] := newtest[expr_] := (expr === Sin[x])
```

With the function `newtest`, we can pick out the term `Sin[x]`:

```
In[47] := Select[(1 + x + 2 x^2 + 3 x^3 + Sin[x]), newtest]
```
```
Out[47] = Sin[x]
```

The following function tests whether an expression is free of integers:

```
In[48] := mytest[expr_] := FreeQ[expr, _Integer]
```

(Section 3.2.3.1 will introduce the pattern _*h*, which stands for any single expression with the head *h*.) This function allows us to generate an expression with the elements that contain no integers:

```
In[49] := Select[(1 + x + 2 x^2 + 3 x^3 + Sin[x]), mytest]
```
```
Out[49] = x + Sin[x]
```

We can, of course, define criteria without using these built-in functions. Here is a function for testing whether an expression is of length greater than or equal to two:

```
In[50] := ourtest[expr_] := (Length[expr] >= 2)
```

Let us generate an expression with the elements of appropriate lengths:

```
In[51] := Select[(1 + x + 2 x^2 + 3 x^3 + Sin[x]), ourtest]
```
```
Out[51] = 2 x^2 + 3 x^3
```

It is always a good practice to clear symbols that are no longer needed:

```
In[52] := Clear[test1, newtest, mytest, ourtest]
```

3.1.3.2   CHANGING PARTS OF EXPRESSIONS

The functions introduced in Section 2.4.4 for adding, deleting, and replacing list and sublist elements can be generalized to ones for changing parts of expressions:

`Prepend[`*expr, part*`]`	add *part* at the first position of *expr*
`Append[`*expr, part*`]`	add *part* at the last position of *expr*

`Insert[`*expr, part, i*`]`	insert *part* at position *i* in *expr*
`Insert[`*expr, part,* `{`*i, j,* `...}]`	insert *part* at position $\{i, j, \ldots\}$ in *expr*
`Insert[`*expr, part,* `{{`*i1, j1,* `...},` `{`*i2,* `...},` `...}]`	insert *part* at positions $\{i1, j1, \ldots\}$, $\{i2, \ldots\}, \ldots$ in *expr*
`Delete[`*expr, i*`]`	delete the part at position *i* in *expr*
`Delete[`*expr,* `{`*i, j,* `...}]`	delete the part at position $\{i, j, \ldots\}$ in *expr*
`Delete[`*expr,* `{{`*i1, j1,* `...},` `{`*i2,* `...},` `...}]`	delete parts at positions $\{i1, j1, \ldots\}$, $\{i2, \ldots\}, \ldots$ in *expr*
`ReplacePart[`*expr, part, i*`]`	replace the part at position *i* in *expr* with *part*
`ReplacePart[`*expr, part,* `{`*i, j,* `...}]`	replace the *part* at position $\{i, j, \ldots\}$ with *part*
`ReplacePart[`*expr, part,* `{{`*i1,* *j1,* `...},` `{`*i2,* `...},` `...}]`	replace parts at positions $\{i1, j1, \ldots\}, \{i2, \ldots\}, \ldots$ with *part*

Let us illustrate the use of these functions with the generic expression `f[a, g[b], c, d]`:

In[53] `:= Prepend[f[a, g[b], c, d], e]`

Out[53] `= f[e, a, g[b], c, d]`

In[54] `:= Insert[f[a, g[b], c, d], e, 3]`

Out[54] `= f[a, g[b], e, c, d]`

Now consider the expression

In[55] `:= x/(1 + x)^2`

Out[55] $= \dfrac{x}{(1 + x)^2}$

The function `Insert` allows us to add `x^2` after `x` in the denominator:

In[56] `:= Insert[%, x^2, {2, 1, 3}]`

Out[56] $= \dfrac{x}{(1 + x + x^2)^2}$

We can remove `x^2` with the function `Delete`:

In[57] `:= Delete[%, {2, 1, 3}]`

Out[57] $= \dfrac{x}{(1 + x)^2}$

Let us replace `y` in the expression `x/y` by `(1 + x^2)`:

In[58] `:= ReplacePart[x/y, (1 + x^2), {2, 1}]`

Out[58] $= \dfrac{x}{1 + x^2}$

To see that **y** is at the position {2, 1}, use the function **FullForm** or **Position**:

In[59] := **FullForm[x/y]**

Out[59]//FullForm=
        Times[x, Power[y, -1]]

In using these functions, we must remember that *Mathematica* tries to evaluate the expressions whenever it can before their parts are changed. Consider replacing the second element of **D[x + y^2, x]** by **y**:

In[60] := **ReplacePart[D[x + y^2, x], y, 2]**

        ReplacePart::part: Part {2} of 1 does not exist.

Out[60] = ReplacePart[1, y, 2]

Mathematica returns an error message because the derivative was evaluated to **1** before its second argument could be replaced by **y**. To obtain the desired result, enter

In[61] := **ReleaseHold[**
                ReplacePart[Hold[D[x + y^2, x]], y, {1, 2}]
                ]

Out[61] = 2 y

ReleaseHold[*expr*] removes **Hold** in *expr*.

If the heads of the expressions possess special properties such as associativity and commutativity, the results of function calls would reflect these properties. For example, consider the function call

In[62] := **Prepend[Plus[1, x, x^2], x^3]**

Out[62] = $1 + x + x^2 + x^3$

Contrary to expectation, the element x^3 appears at the end rather than the beginning. This is because *Mathematica* sorts the elements of output into a standard order since **Plus** is both associative and commutative. Section 3.2.5 will discuss assigning attributes to the heads of expressions.

3.1.3.3 REARRANGING EXPRESSIONS

The functions introduced in Section 2.4.5 for rearranging lists can be generalized to ones for rearranging expressions. Several of them are

Sort[*expr*]	sort the elements of *expr* into canonical order
Union[*expr*]	give a sorted version of *expr* in which all duplicated elements have been dropped
Reverse[*expr*]	reverse the order of the elements in *expr*
Permutations[*expr*]	generate a list of *expr* with all possible permutations of the elements

Let us use the expression `f[d, c, b, a, a]` to illustrate the use of these functions. `Sort` arranges the elements into a standard order:

```
In[63] := Sort[f[d, c, b, a, a]]

Out[63] = f[a, a, b, c, d]
```

`Union` sorts the elements and discards the duplicated ones:

```
In[64] := Union[f[d, c, b, a, a]]

Out[64] = f[a, b, c, d]
```

`Permutations` gives a list of the expression with all possible permutations of its elements:

```
In[65] := Short[Permutations[f[d, c, b, a, a]], 4]

Out[65]//Short=
          {f[d, c, b, a, a], f[d, c, a, b, a], f[d, c, a, a, b],
           f[d, b, c, a, a], f[d, b, a, c, a], f[d, b, a, a, c],
           f[d, a, c, b, a], f[d, a, c, a, b], <<48>>, f[a, a, c, d, b],
           f[a, a, c, b, d], f[a, a, b, d, c], f[a, a, b, c, d]}
```

The printing of the output has been shortened with the function `Short` introduced in Section 3.1.2.

Again, if the heads of expressions possess special properties, the results of function calls would reflect these properties:

```
In[66] := Permutations[Plus[1, x, x^2]]

Out[66] = {1 + x + x², 1 + x + x², 1 + x + x², 1 + x + x², 1 + x + x²,
           1 + x + x²}
```

The elements of each permutation have been arranged into the same standard order since `Plus` is both associative and commutative.

3.1.3.4 RESTRUCTURING EXPRESSIONS

Most functions introduced in Section 2.4.6 for restructuring lists can be generalized to ones for restructuring expressions:

`Partition[`*expr, n*`]`	partition without overlap the elements of *expr* into subexpressions with the same head as *expr* and *n* elements
`Partition[`*expr, n, d*`]`	partition with offset *d* the elements of *expr* into subexpressions with the same head as *expr* and *n* elements
`Flatten[`*expr*`]`	flatten out all subexpressions with the same head as *expr*
`Flatten[`*expr, n*`]`	flatten to level *n*

`Flatten[expr, n, h]`	flatten subexpressions with head _h_ to level _n_
`FlattenAt[expr, i]`	flatten only the _i_th element of _expr_
`FlattenAt[expr, {i, j, ...}]`	flatten only the part of _expr_ at position _{i, j, ...}_
`FlattenAt[expr,` `   {{i1, j1, ...}, {i2, ...}}]`	flatten parts of _expr_ at several positions

We can group without overlap and in pairs the elements of the expression `f[a, b, c, d, e]`:

In[67] := `Partition[f[a, b, c, d, e], 2]`

Out[67] = `f[f[a, b], f[c, d]]`

Note that the leftover element at the end is dropped! We can also group the elements in triples with an offset of one element between successive triples:

In[68] := `Partition[f[a, b, c, d, e], 3, 1]`

Out[68] = `f[f[a, b, c], f[b, c, d], f[c, d, e]]`

To demonstrate several properties of the functions `Flatten` and `FlattenAt`, we assign an expression to the variable `myexpr`:

In[69] := `myexpr = f[a, h[b, h[c, d]], f[a, h[d]], h[e, f[g]],`
`                f[a, f[b, f[c, d]]]]`

Out[69] = `f[a, h[b, h[c, d]], f[a, h[d]], h[e, f[g]], f[a, f[b, f[c, d]]]]`

`Flatten` removes the heads of subexpressions with the same head as the expression:

In[70] := `Flatten[myexpr]`

Out[70] = `f[a, h[b, h[c, d]], a, h[d], h[e, f[g]], a, b, c, d]`

Observe that none of the elements with the head `h` are affected, including the element `h[e, f[g]]`. We can flatten subexpressions with the head `h` to the second level:

In[71] := `Flatten[myexpr, 2, h]`

Out[71] = `f[a, b, c, d, f[a, h[d]], e, f[g], f[a, f[b, f[c, d]]]]`

This time none of the elements with the head `f` are affected. Now apply the function `FlattenAt` to the fifth element of `myexpr`:

In[72] := `FlattenAt[myexpr, 5]`

Out[72] = `f[a, h[b, h[c, d]], f[a, h[d]], h[e, f[g]], a, f[b, f[c, d]]]`

Note that the subexpressions of the fifth element remain unchanged. We can flatten only the part at the position {2, 2}:

In[73] := `FlattenAt[myexpr, {2, 2}]`

Out[73] = `f[a, h[b, c, d], f[a, h[d]], h[e, f[g]], f[a, f[b, f[c, d]]]]`

For a more concrete example of restructuring expressions, we define the function

$In[74] := $ `q[x_] := 1/ (1 + x)`

With this function, we can generate a nested expression:

$In[75] := $ `Nest[q, x, 4]`

$$Out[75] = \cfrac{1}{1 + \cfrac{1}{1 + \cfrac{1}{1 + \cfrac{1}{1 + x}}}}$$

`Nest[`*f, expr, n*`]` gives an expression with *f* applied *n* times to *expr*. The function `FullForm` shows the internal representation of the nested expression:

$In[76] := $ `FullForm[%]`

$Out[76]//FullForm=$
```
          Power[Plus[1, Power[Plus[1,
             Power[Plus[1, Power[Plus[1, x], -1]], -1]], -1]], -1]
```

We can reduce this structure by flattening its subexpressions that have `Power` as their heads:

$In[77] := $ `FlattenAt[%, {{1, 2}, {1, 2, 1, 2}, {1, 2, 1, 2, 1, 2}}]`

$$Out[77] = \frac{1}{1 + x}$$

For another example, consider the polynomial

$In[78] := $ `x^2/3 + x^3 + (5 x^5)/3 + x^7 + x^10 + (4 x^17)/3`

$$Out[78] = \frac{x^2}{3} + x^3 + \frac{5 x^5}{3} + x^7 + x^{10} + \frac{4 x^{17}}{3}$$

We would like to generate a list of the exponents. First, we flatten those subexpressions with the head `Times` to the first level:

$In[79] := $ `Flatten[%, 1, Times]`

$$Out[79] = \frac{10}{3} + x^2 + x^3 + x^5 + x^7 + x^{10} + x^{17}$$

The function `Table` gives the desired list:

$In[80] := $ `Table[%[[i, 2]], {i, 2, Length[%]}]`

$Out[80] = $ `{2, 3, 5, 7, 10, 17}`

There are other functions for restructuring expressions. Several of them are

`Thread[`*f[args]*`]`	thread *f* over any lists that appear in *args*
`Thread[`*f[args]*`, h]`	thread *f* over any objects with head *h* that appear in *args*
`Outer[`*f, list1, list2, ...*`]`	give the generalized outer product of the *listi*
`Inner[`*f, list1, list2, g*`]`	give the generalized inner product of the *listi*

Thread applies a function to the corresponding elements of several lists:

```
In[81] := Thread[f[{a1, a2, a3}, {b1, b2, b3}]]

Out[81] = {f[a1, b1], f[a2, b2], f[a3, b3]}
```

Here is a way to make a list of rules:

```
In[82] := Thread[Rule[{a1, a2, a3}, {b1, b2, b3}]]

Out[82] = {a1 -> b1, a2 -> b2, a3 -> b3}
```

With a head specified as an additional argument, **Thread** threads a function over the function's arguments that have the specified head:

```
In[83] := Thread[f[g[a1, a2, a3], g[b1, b2, b3], h[c1, c2, c3]], g]

Out[83] = g[f[a1, b1, h[c1, c2, c3]], f[a2, b2, h[c1, c2, c3]],
             f[a3, b3, h[c1, c2, c3]]]
```

Note that the argument that does not have the head g appears in all subexpressions.

Outer applies a function to all possible combinations of the elements from several lists, each combination consisting of one element from each list:

```
In[84] := Outer[f, {a1, a2, a3}, {b1, b2, b3}]

Out[84] = {{f[a1, b1], f[a1, b2], f[a1, b3]},
           {f[a2, b1], f[a2, b2], f[a2, b3]},
           {f[a3, b1], f[a3, b2], f[a3, b3]}}
```

Inner generates an expression with the specified head and with elements from threading a function over two lists:

```
In[85] := Inner[f, {a1, a2, a3}, {b1, b2, b3}, g]

Out[85] = g[f[a1, b1], f[a2, b2], f[a3, b3]]
```

3.1.3.5 OPERATING ON EXPRESSIONS

The functions **Map**, **MapAt**, and **Apply** discussed in Section 2.4.8 for operating on lists can be generalized to functions for operating on expressions. This section also introduces the function **MapAll**.

Map[*f, expr*] or *f* /@ *expr*	apply *f* to each element on the first level in *expr*
Map[*f, expr, levelspec*]	apply *f* to parts of *expr* specified by *levelspec*
MapAll[*f, expr*] or *f* //@ *expr*	apply *f* to every subexpression in *expr*

Let us illustrate the use of these functions with the expression h[p[a, b, c], q[r, s, t]]. **Map**[*f, expr*] or *f* /@ *expr* wraps *f* around each element of *expr*:

```
In[86] := f /@ h[p[a, b, c], q[r, s, t]]

Out[86] = h[f[p[a, b, c]], f[q[r, s, t]]]
```

We can apply *f* to the parts of *expr* at specified levels:

```
In[87] := Map[f, h[p[a, b, c], q[r, s, t]], {2}]

Out[87] = h[p[f[a], f[b], f[c]], q[f[r], f[s], f[t]]]
```

Here *f* is applied at the second level. If the option **Heads -> True** is specified, the function *f* is also applied to the heads at the specified levels:

```
In[88] := Map[f, h[p[a, b, c], q[r, s, t]], {2}, Heads -> True]

Out[88] = h[f[p][f[a], f[b], f[c]], f[q][f[r], f[s], f[t]]]
```

MapAll[*f, expr*] applies *f* to every subexpression of *expr*, including *expr* itself:

```
In[89] := MapAll[f, h[p[a, b, c], q[r, s, t]]]

Out[89] = f[h[f[p[f[a], f[b], f[c]]], f[q[f[r], f[s], f[t]]]]]
```

Sometimes, it is desirable to apply separately a function to only some parts of an expression.

MapAt [*f, expr, n*]	apply *f* to the element at position *n* in *expr*
MapAt [*f, expr,* {*i, j,* ...}]	apply *f* to the part of *expr* at position {*i, j,* ...}
MapAt [*f, expr,* {{*i1, j1,* ...}, {*i2, j1,* ...}, ...}]	apply *f* to parts of *expr* at several positions

Consider the expression

```
In[90] := myexpr = a + b/c + (d + e/g)/(r + s/t)
```

$$Out[90] = a + \frac{b}{c} + \frac{d + \frac{e}{g}}{r + \frac{s}{t}}$$

How do we apply the function *f* to the denominators? We first use **FullForm** to reveal the expression's internal representation upon which the positions of its parts are based:

```
In[91] := FullForm[myexpr]

Out[91]//FullForm=
  Plus[a, Times[b, Power[c, -1]],
    Times[Plus[d, Times[e, Power[g, -1]]],
      Power[Plus[r, Times[s, Power[t, -1]]], -1]]]
```

Note that denominators are represented in the form **Power[*denominator*, -1]**. We can obtain the positions of **Power** at specified levels of the expression with the function **Position**:

```
In[92] := Position[myexpr, Power, {3}]

Out[92] = {{2, 2, 0}, {3, 2, 0}}
```

Replacing the 0 in each sublist here by 1 gives the positions of the denominators since in each subexpression of the form `Power` [*denominator*, -1], part 0 is `Power` and part 1 is *denominator*:

In[93] := **% /. 0 -> 1**

Out[93] = **{{2, 2, 1}, {3, 2, 1}}**

Let **MapAt** apply *f* to the denominators at the third level:

In[94] := **MapAt[f, myexpr, %]**

$$Out[94] = a + \frac{b}{f[c]} + \frac{d + \dfrac{e}{g}}{f[r + \dfrac{s}{t}]}$$

MapAt can also apply *f* to denominators at other levels:

In[95] := **MapAt[f, myexpr, Position[myexpr, Power, {5,6}] /. 0 -> 1]**

$$Out[95] = a + \frac{b}{c} + \frac{d + \dfrac{e}{f[g]}}{r + \dfrac{s}{f[t]}}$$

In[96] := **Clear[myexpr]**

The method presented here for mapping at the denominators works well for the given expression. For other algebraic expressions, there is a more general and complex algorithm for applying a function to their numerators or denominators (see [Cop91]).

The function **Apply** changes heads of expressions:

> **Apply** [*f*, *expr*] or *f* **@@** *expr*   replace the head of *expr* by *f*
>
> **Apply** [*f*, *expr*, *levelspec*]   replace heads in parts of *expr* specified by *levelspec*

Let us illustrate the use of **Apply** with the expression, **h[p[a, b, c], q[r, s, t]]**, introduced earlier in this section:

In[97] := **Apply[f, h[p[a, b, c], q[r, s, t]]]**

Out[97] = **f[p[a, b, c], q[r, s, t]]**

Apply replaces **h** with **f** as the head of the expression. We can also replace the heads of subexpressions by giving a level specification:

In[98] := **Apply[f, h[p[a, b, c], q[r, s, t]], {1}]**

Out[98] = **h[f[a, b, c], f[r, s, t]]**

3.1.3.6 MANIPULATING EQUATIONS

This section presents two examples on manipulating equations: one taken from introductory mechanics and the other extracted from Section 5.3, "Charged Particle in Crossed Electric and Magnetic Fields."

EXAMPLE 3.1.1 A projectile moves in the xy plane where the y axis is directed vertically upward. It is initially at the origin, and its initial velocity $\mathbf{v}_0$ makes an angle θ_0 with the positive x direction. Derive an equation relating the x and y coordinates of the projectile, and determine the horizontal range. Neglect air resistance.

The x and y components of the projectile acceleration can be expressed as

$$\frac{d^2x}{dt^2} = 0 \tag{3.1.1}$$

$$\frac{d^2y}{dt^2} = -g \tag{3.1.2}$$

where g is the magnitude of free-fall acceleration. The components of the initial velocity are

$$v_{x0} = v_0 \cos\theta_0 \tag{3.1.3}$$

$$v_{y0} = v_0 \sin\theta_0 \tag{3.1.4}$$

and the initial coordinates are

$$x_0 = 0 \tag{3.1.5}$$

$$y_0 = 0 \tag{3.1.6}$$

Let `theta0` be the alias of θ_0. `DSolve` solves Equations 3.1.1 and 3.1.2 subject to the initial conditions (Equations 3.1.3 to 3.1.6):

```
In[99] := Clear[g, v0, theta0, x, y, t, R]
```

```
In[100] := DSolve[{x''[t] == 0, y''[t] == - g,
               x'[0] == v0 Cos[theta0], y'[0] == v0 Sin[theta0],
               x[0] == 0, y[0] == 0},
              {x[t], y[t]}, t]
```

$$Out[100] = \{\{x[t] \to t\ v0\ Cos[theta0],$$
$$y[t] \to \frac{-(g\ t^2) + 2\ t\ v0\ Sin[theta0]}{2}\}\}$$

These rules provide the equations relating the x and y coordinates to time t:

```
In[101] := {x, y} == {x[t], y[t]} /. %[[1]]
```

$$Out[101] = \{x, y\} == \{t\ v0\ Cos[theta0],\ \frac{-(g\ t^2) + 2\ t\ v0\ Sin[theta0]}{2}\}$$

Solve these equations for y in terms of x by eliminating t:

```
In[102] := Solve[%, y, t] // ExpandAll // Simplify
```

$$Out[102] = \{\{y \to \frac{-(g\ x^2\ Sec[theta0]^2)}{2\ v0^2} + x\ Tan[theta0]\}\}$$

Solve[*eqns, vars, elims*] attempts to solve the equations for *vars,* eliminating the variables *elims.* Let us turn the preceding rule for y into an equation:

$In[103] := $ **Apply[Equal, %[[1,1]]]**

$Out[103] = y == \dfrac{-(g\ x^2\ Sec[theta0]^2)}{2\ v0^2} + x\ Tan[theta0]$

This equation gives the relation between the x and y coordinates; this is the equation of the trajectory of the projectile. To determine the horizontal range, solve this equation for x with $y = 0$:

$In[104] := $ **Solve[% /. y -> 0, x] // Simplify**

$Out[104] = \{\{x \rightarrow 0\}, \{x \rightarrow \dfrac{v0^2\ Sin[2\ theta0]}{g}\}\}$

The first rule specifies the initial x coordinate, whereas the second gives the range:

$In[105] := $ **R == x /. %[[2]]**

$Out[105] = R == \dfrac{v0^2\ Sin[2\ theta0]}{g}$

Note that R is a mathematical variable without an assigned value:

$In[106] := $ **R**

$Out[106] = R$

To turn the last equation into an assignment, replace **Equal** (==) with **Set** (=) or **SetDelayed** (:=),

$In[107] := $ **Set @@ %%**

$Out[107] = \dfrac{v0^2\ Sin[2\ theta0]}{g}$

where we have used the special input form for **Apply**. Now R has an assigned value:

$In[108] := $ **R**

$Out[108] = \dfrac{v0^2\ Sin[2\ theta0]}{g}$

$In[109] := $ **Clear[R]**   ■

❋ **3.0 EXAMPLE 3.1.2**   Utilizing the typesetting capabilities of *Mathematica* version 3.0, redo Example 3.1.1 in TraditionalForm.

$In[110] := $ **Clear[g, Subscript, x, y, t, R]**

(To see why we have applied the function `Clear` to `Subscript` rather than v_0 and θ_0, refer to Exercise 15 in Section 3.3.7.) Heeding the admonition given in Section 2.5.3 about the lack of necessary precision of TraditionalForm for input, we enter in StandardForm

```
DSolve[{D[x[t], {t, 2}] == 0,
       D[y[t], {t, 2}] == -g,
       (D[x[t], t] /. t -> 0) == v₀ Cos[θ₀],
       (D[y[t], t] /. t -> 0) == v₀ Sin[θ₀],
       x[0] == 0, y[0] == 0},
      {x[t], y[t]}, t]
```

To convert this input to TraditionalForm, paste a copy of it below, click the bracket of the new cell, and choose **TraditionalForm** in the Convert To submenu of the Cell menu:

$In[111] := \mathbf{DSolve}\!\left[\left\{\frac{\partial^2 x(t)}{\partial t^2} == 0, \frac{\partial^2 y(t)}{\partial t^2} == -g, \left(\frac{\partial x(t)}{\partial t} /. t \to 0\right) == v_0 \cos(\theta_0),\right.\right.$
$\left.\left.\left(\frac{\partial y(t)}{\partial t} /. t \to 0\right) == v_0 \sin(\theta_0), x(0) == 0, y(0) == 0\right\}, \{x(t), y(t)\}, t\right]$

$Out[111] = \left\{\left\{x(t) \to t \cos(\theta_0)\, v_0, y(t) \to \frac{1}{2}\,(2\,t\,\sin(\theta_0)\, v_0 - g\,t^2)\right\}\right\}$

where we have set Default Output FormatType in the Cell menu to **TraditionalForm** before the evaluation of the input. These rules provide the equations relating the x and y coordinates to time t:

$In[112] := \{\mathbf{x, y}\} == \{\mathbf{x(t), y(t)}\} \,/.\, (\%[\![1]\!])$

$Out[112] = \{x, y\} == \{t \cos(\theta_0)\, v_0, \frac{1}{2}\,(2\,t\,\sin(\theta_0)\, v_0 - g\,t^2)\}$

Solve these equations for y in terms of x by eliminating t:

$In[113] := \mathbf{Simplify[ExpandAll[Solve[\%, y, t]]]}$

$Out[113] = \left\{\left\{y \to x\,\tan(\theta_0) - \frac{g\,x^2\,\sec^2(\theta_0)}{2\,v_0^2}\right\}\right\}$

Let us turn the preceding rule for y into an equation:

$In[114] := \mathbf{Equal@@(\%[\![1, 1]\!])}$

$Out[114] = y == x\tan(\theta_0) - \frac{g\,x^2\,\sec^2(\theta_0)}{2\,v_0^2}$

This equation gives the relation between the x and y coordinates; this is the equation of the trajectory of the projectile. To determine the horizontal range, solve this equation for x with $y = 0$:

$In[115] := \mathbf{Simplify[Solve[\% \,/.\, y \to 0, x]]}$

$Out[115] = \left\{\{x \to 0\}, \left\{x \to \frac{\sin(2\,\theta_0)\, v_0^2}{g}\right\}\right\}$

Whereas the first rule specifies the initial x coordinate, the second gives the range:

In[116] := **R = x /. (%[[2]])**

Out[116] = $\dfrac{\sin(2\,\theta_0)\,v_0^2}{g}$

In[117] := **Clear[R]**

Before leaving this example, let us restore Default Output FormatType in the Cell menu to **OutputForm**.   ∎

EXAMPLE 3.1.3 Without using **DSolve**, obtain the solution to the differential equations

$$\frac{dx}{dt} = by \tag{3.1.7}$$

$$\frac{dy}{dt} = -bx + e \tag{3.1.8}$$

where b and e are constants.

Instead of using **DSolve**, we solve these equations here with the "paper and pencil" approach. Though the procedure is rather tedious, it can be argued that this approach may be more appropriate for undergraduate education since it elucidates the fundamentals of mathematics, which normal use of *Mathematica* often obscures. (For a computer algebra program based on this approach, see [WMS94].)

In[118] := **Clear[t, b, e]**

Let Equations 3.1.7 and 3.1.8 be assigned to a variable:

In[119] := **eqn = {x'[t] == b y[t], y'[t] == -b x[t] + e}**

Out[119] = {x'[t] == b y[t], y'[t] == e - b x[t]}

Now differentiate both sides of the first equation,

In[120] := **D[eqn[[1]], t]**

Out[120] = x''[t] == b y'[t]

solve the resulting equation and the second equation for **x''[t]** by eliminating **y'[t]**,

In[121] := **Solve[{%, eqn[[2]]}, x''[t], y'[t]] // ExpandAll**

Out[121] = {{x''[t] -> b e - b^2 x[t]}}

and turn this rule into an equation,

In[122] := **Equal @@ %[[1, 1]]**

Out[122] = x''[t] == b e - b^2 x[t]

This equation has the same form as the equation of motion of a harmonic oscillator acted on by a constant force. The solution to this equation is well known:

In[123] := **xsol = C1 E^(-I b t) + C2 E^(I b t) + e/b**

$Out[123] = \dfrac{e}{b} + \dfrac{C1}{E^{I\ b\ t}} + C2\ E^{I\ b\ t}$

Similarly, differentiate both sides of Equation 3.1.8, solve the resulting equation and Equation 3.1.7 for **y''[t]** with the elimination of **x'[t]**,

In[124] := **Solve[{D[eqn[[2]], t], eqn[[1]]}, y''[t], x'[t]]//ExpandAll**

$Out[124] = \{\{y''[t]\ \texttt{->}\ -(b^2\ y[t])\}\}$

and turn the rule into an equation,

In[125] := **Equal @@ %[[1, 1]]**

$Out[125] = y''[t]\ \texttt{==}\ -(b^2\ y[t])$

Again, this equation has the same form as the equation of motion of a harmonic oscillator, and its solution is known:

In[126] := **ysol = C3 E^(-I b t) + C4 E^(I b t)**

$Out[126] = \dfrac{C3}{E^{I\ b\ t}} + C4\ E^{I\ b\ t}$

The difficulty is that there are four arbitrary constants **c1**, **c2**, **c3**, and **c4** for two first-order differential equations (Equations 3.1.7 and 3.1.8) that permit only two initial conditions. The problem arises because new solutions that may not satisfy the original equations were introduced when these equations were differentiated. To eliminate these extraneous solutions, substitute **xsol** and **ysol** and their derivatives into Equations 3.1.7 and 3.1.8,

In[127] := **eqn /. {x[t] -> xsol, x'[t] -> D[xsol, t],**
                   y[t] -> ysol, y'[t] -> D[ysol, t]}

$Out[127] = \{\dfrac{-I\ b\ C1}{E^{I\ b\ t}} + I\ b\ C2\ E^{I\ b\ t}\ \texttt{==}\ b\ (\dfrac{C3}{E^{I\ b\ t}} + C4\ E^{I\ b\ t}),$

$\dfrac{-I\ b\ C3}{E^{I\ b\ t}} + I\ b\ C4\ E^{I\ b\ t}\ \texttt{==}\ e - b\ (\dfrac{e}{b} + \dfrac{C1}{E^{I\ b\ t}} + C2\ E^{I\ b\ t})\}$

and solve for **c3** and **c4**,

In[128] := **Solve[%, {C3, C4}]//Simplify//Flatten**

$Out[128] = \{C3\ \texttt{->}\ -I\ C1,\ C4\ \texttt{->}\ I\ C2\}$

Thus, replacing **c3** and **c4** in **ysol** according to these rules gives the solution to Equations 3.1.7 and 3.1.8:

In[129] := **{ysol /. %, xsol}**

$Out[129] = \{\dfrac{-I\ C1}{E^{I\ b\ t}} + I\ C2\ E^{I\ b\ t},\ \dfrac{e}{b} + \dfrac{C1}{E^{I\ b\ t}} + C2\ E^{I\ b\ t}\}$

Of course, DSolve can solve Equations 3.1.7 and 3.1.8 with almost no effort from us:

❋ 3.0

In[130] := (DSolve[{x'[t]/b == y[t], y'[t] == -b x[t] + e},
                     {y[t], x[t]}, t]//Simplify//TrigToExp)/.
                     {C[1] -> I(C[2] - C[1]),
                      C[2] ->   (C[2] + C[1])}//ExpandAll

Out[130] = {{y[t] -> $\dfrac{-\text{I C}[1]}{\text{E}^{\text{I b t}}}$ + I E$^{\text{I b t}}$ C[2],

                x[t] -> $\dfrac{e}{b}$ + $\dfrac{\text{C}[1]}{\text{E}^{\text{I b t}}}$ + E$^{\text{I b t}}$ C[2]}}

❋ 2.2

DSolve[{x'[t]/b == y[t], y'[t] == -b x[t] + e},
        {y[t], x[t]}, t]

{{y[t] -> -I E$^{-\text{I b t}}$ C[1] + I E$^{\text{I b t}}$ C[2],

   x[t] -> $\dfrac{e}{b}$ + E$^{-\text{I b t}}$ C[1] + E$^{\text{I b t}}$ C[2]}}

In[131] := **Clear[eqn, xsol, ysol]**   ∎

3.1.4 Exercises

1. Determine the internal representations of the following:
   (a) $\sqrt{x^2 + 8x + 16}$
   (b) $\dfrac{3}{x-1} + \dfrac{2}{(x-1)^2}$
   (c) $\sin(x + y)$
   (d) $\sin(x + y) = \sin x \cos y + \cos x \sin y$
   (e) $\displaystyle\int_0^{\pi/2} \sin x \, dx$
   (f) $\dfrac{a + ib}{c - id}$

2. Determine the internal form of each expression.
   (a) {a, {b, {c, d}}, {e, f}}
   (b) Integrate[x^2, x]
   (c) Factor[x^10 - y^10]
   (d) f[a] + g[c, d] /. f[x_] -> x^3
   (e) Sin /@ (a + b + c)
   (f) x^2/y^2

3. Consider the expression

    ```
    x^2/(x - 1) == 1/(x - 1)
    ```

    Obtain a list of subexpressions from levels 1 through 3.

4. Consider the expression

    ```
    a + b/c + (d + e)/(1 + f/(1 - g/h))
    ```

    Pick out the subexpression g/h.

5. Determine the positions of a^3 at the sixth level of the expression

    ```
    Sqrt[(24 a^2 b)/(1 + 3 a^3)] Sqrt[a^3 b^2]
    ```

6. In the expression

    ```
    1 + 3 x + 3 x^2 + x^3 + 3 y + 6 x y + 3 x^2 y + 3 y^2 +
        3 x y^2 + y^3
    ```

    remove all elements containing y.

7. Insert b after every element in the expression

    ```
    h[e1, e2, e3, e4, e5, e6, e7]
    ```

8. Replace x^2 by y^3 in the expression

    ```
    (x + y)/(1 + 3 x + 3 x^2 + x^3 + 3 y + 6 x y +
        3 x^2 y + 3 y^2 + 3 x y^2 + y^3)
    ```

9. From the expression

    ```
    a + b/c + (d + e)/(f + g)
    ```

    obtain the expression

    $$a + b + \frac{1}{c} + d + e + \frac{1}{f + g}$$

10. With Map, square both sides of the equation

    ```
    Sqrt[1 + x] == x^2 + 2
    ```

11. Apply *f* to the numerators of the expression

    ```
    a + b/c + (d + e/f)/(r + s/t)
    ```

12. Turn the rule

    ```
    x^2 + x -> 1
    ```

    into an equation.

13. Write a function that returns the complex conjugate of an expression the parts of which, with the exception of the constant **I**, are all real.

3.2 *Patterns*

The power of *Mathematica* comes from its pattern-matching capabilities. In Section 2.2.13, we introduced the pattern *arg_* in order to define functions. This section shows that _, called a blank, is the basic pattern that stands for any expression. We can construct other patterns with blanks and also name or restrict patterns. Pattern-matching capabilities of *Mathematica* allow us to define functions that accept only arguments matching specific patterns. In pattern-matching, *Mathematica* examines structural rather than mathematical equivalence and considers attributes of functions. Patterns can have default values and can contain alternative or repeated patterns. This section concludes with a discussion of multiple blanks. Whereas single blanks stand for single expressions, multiple blanks represent sequences of expressions.

3.2.1 *Blanks*

The basic pattern is the "blank," which stands for any single expression. Its special input form and internal form are _ and **Blank[]**, respectively. For example, **h[_]** represents the class of expressions of the form **h**[*element*], where *element* can be any single expression.

```
In[1] := {h[], h[x^2], h[Sin[x]], h[3 + 4I],
            g[x], h[b, c], h[a, b, c, d]} /. h[_] -> "MATCH"
```

```
Out[1] = {h[], MATCH, MATCH, MATCH, g[x], h[b, c], h[a, b, c, d]}
```

The expressions **h[x^2]**, **h[Sin[x]]**, **h[3 + 4I]** match the pattern **h[_]** because each of them has the head **h** and a single element.

The pattern **{_, _}** stands for a list with any two elements. The function **Position** introduced in Section 3.1.3.1 gives a list of the positions at which parts matching a pattern appear in an expression.

```
In[2] := Position[{{}, {a}, {b, c}, {c, {b, d}}}, {_, _}]
```

```
Out[2] = {{3}, {4, 2}, {4}}
```

Position returns a list of the positions of **{b, c}**, **{b, d}**, and **{c, {b, d}}**, which match the pattern.

From a list of expressions, we can pick out with the function **Cases** those matching the pattern "**_^_**", which represents any expression to any power. **Cases**[*expr, pattern, levelspec*] gives a list of parts that match *pattern* and that are on levels of *expr* specified by *levelspec*. If the third argument is omitted, *levelspec* takes the default value **{1}**.

```
In[3] := Cases[{3, x, a^4, Sqrt[1 + x]}, _^_]
```

$$Out[3] = \{a^4, \text{Sqrt}[1 + x]\}$$

Note that 3 and **x**, which are mathematically equivalent to Sqrt[3]^2 and x^1, respectively, do not match the pattern "_^_". Section 3.2.4 will show that pattern-matching is based on internal forms of expressions and structural equivalence.

   From the list of expressions, we can also delete with the function DeleteCases those matching the pattern "_^_". DeleteCases[*expr*, *pattern*, *levelspec*] removes all parts that match *pattern* and that are on levels of *expr* specified by *levelspec*. If the third argument is omitted, *levelspec* takes the default value {1}.

   In[4] := DeleteCases[{3, x, a^4, Sqrt[1 + x]}, _^_]

   Out[4] = {3, x}

3.2.2 *Naming Patterns*

It is often necessary to assign names to patterns. Names of patterns must be symbols; that is, Head[*name*] must return Symbol.

❀ **3.0**   Two-dimensional forms, for example, cannot be names of patterns, as they are not symbols:

   In[5] := Head /@ {x^y, x$_y$, x$_z^y$, x$_y$, x†, x$^+$, x*, x̂, x̄}

   Out[5] = {Power, Subscript, Power, Underscript, SuperDagger, SuperPlus, SuperStar, OverHat, OverBar}

   The pattern x_ stands for any single expression to be named *x* on the right-hand side of a definition or transformation rule. Its full form is Pattern[x, Blank[]].

   In[6] := (h[a] + h[b, c] + h[a, a]) h[d, e, f] /. h[x_, y_] -> x^y

   Out[6] = (a^a + b^c + h[a]) h[d, e, f]

Both h[b, c] and h[a, a] match h[x_, y_]. Yet only h[a, a] matches h[x_, x_], because a pattern containing pieces with the same name (**x**, here) can only match expressions the corresponding pieces of which are identical:

   In[7] := (h[a] + h[b, c] + h[a, a]) h[d, e, f] /. h[x_, x_] -> x^x

   Out[7] = (a^a + h[a] + h[b, c]) h[d, e, f]

   Let myfunc be a function that requires an argument matching the pattern "x_^n_".

   In[8] := myfunc[x_^n_] := {x, n}

Apply myfunc to the elements of the list introduced at the end of Section 3.2.1:

   In[9] := Map[myfunc, {3, x, a^4, Sqrt[1 + x]}]

   Out[9] = {myfunc[3], myfunc[x], {a, 4}, {1 + x, $\frac{1}{2}$}}

Again, only **a^4** and **Sqrt[1 + x]** match the pattern "**x_^n_**".

> *In[10]* := **Clear[myfunc]**

For a pattern more complex than a single blank, *x:pattern* represents any expression matching *pattern*, and this expression is named *x* on the right-hand side of a definition or transformation rule:

> *In[11]* := **Sin[1 + a^2] /. h:Sin[x_ + y_] -> {h, x, y}**
>
> *Out[11]* = {Sin[1 + a^2], 1, a^2}

Here the expression matching the pattern **Sin[x_ + y_]** is named **h** on the right-hand side of the transformation rule.

As mentioned earlier, a pattern with pieces of the same name can only match expressions in which the corresponding pieces are identical. Whereas **{Sin[a], Sin[a]}** matches **{h_, h:Sin[x_]}**,

> *In[12]* := **{Sin[a], Sin[a]} /. {h_, h:Sin[x_]} -> h**
>
> *Out[12]* = Sin[a]

{Cos[a], Sin[a]} does not match the pattern because the elements are not the same,

> *In[13]* := **{Cos[a], Sin[a]} /. {h_, h:Sin[x_]} -> h**
>
> *Out[13]* = {Cos[a], Sin[a]}

3.2.3   Restricting Patterns

Patterns stand for classes of expressions. How can we construct patterns that represent subclasses of a class of expressions? We can delineate subclasses of expressions with head specifications, conditions, and tests.

3.2.3.1   TYPES

The pattern *_h* stands for any single expression with the head *h*. The pattern *x_h* represents any single expression with the head *h*, and this expression is named *x* on the right-hand side of a definition or transformation rule. Section 3.1.1 gave the heads of atomic expressions: **Symbol**, **String**, **Integer**, **Real**, **Rational**, and **Complex**. For any list, the head is **List**.

With the function **Union** introduced in Section 2.4.5 and the function **Cases** introduced in Section 3.2.1, we can pick out the variables of an algebraic expression:

> *In[14]* := **Expand[(x + y + z)^3]**
>
> *Out[14]* = x^3 + 3 x^2 y + 3 x y^2 + y^3 + 3 x^2 z + 6 x y z + 3 y^2 z +
> 3 x z^2 + 3 y z^2 + z^3

> *In[15]* := **Cases[%, _Symbol, Infinity]//Union**
>
> *Out[15]* = {x, y, z}

With **Infinity** as the level specification, **Cases[***expr, pattern, levelspec***]** returns a list of parts that match *pattern* and appear at any level of *expr*.

We can square the rational numbers in a list:

```
In[16] := {x, "good", 3, 2/3, 5/7} /. x_Rational -> x^2
```

$$Out[16] = \{x, \text{good}, 3, \frac{4}{9}, \frac{25}{49}\}$$

The geometric mean of n quantities is defined to be the product of these quantities to the $1/n$ power. Here is a function for computing the geometric mean of the elements of a list:

```
In[17] := geomean[x_List] := Apply[Times, x]^(1/ Length[x])
```

We can determine the geometric mean of a list of four quantities:

```
In[18] := geomean[{a1, a2, a3, a4}]
```

$$Out[18] = (\text{a1 a2 a3 a4})^{1/4}$$

The argument of the function must be a list:

```
In[19] := geomean[2, 3, 4]
```

```
Out[19] = geomean[2, 3, 4]
```

Mathematica returns the function unevaluated since the argument is not a list.

```
In[20] := Clear[geomean]
```

Consider defining a function that returns the Legendre polynomial of order n with the formula of Rodrigues:

$$P_n(x) = \frac{1}{2^n n!} \frac{d^n (x^2 - 1)^n}{dx^n}$$

where $n = 0, 1, 2, 3, \ldots$ and $-1 \leq x \leq 1$. Let **legdPoly** be the alias of P.

```
In[21] := legdPoly[n_Integer, x_] :=
                    ((1/(2^n n!)) D[(x^2 - 1)^n, {x, n}])//Simplify
```

We can tabulate the first few Legendre polynomials:

```
In[22] := Do[Print["n = ", i, "    ", legdPoly[i, x]], {i, 0, 3}]
```

$$n = 0 \quad 1$$
$$n = 1 \quad x$$
$$n = 2 \quad \frac{-1 + 3 x^2}{2}$$
$$n = 3 \quad \frac{x (-3 + 5 x^2)}{2}$$

Note that the first argument must match the pattern **n_Integer**, that is, it must be an integer:

```
In[23] := legdPoly[2.5, x]
```

```
Out[23] = legdPoly[2.5, x]
```

Mathematica returns the function unevaluated since the first argument is not an integer.

3.2.3.2 TESTS

We can restrict a pattern with a test. The pattern *pattern?test* stands for any expression that matches *pattern,* and on which the application of *test* gives `True`. The built-in functions introduced in Sections 2.4.3 and 3.1.3.1 for testing properties of expressions can be used with this pattern.

Let us count the number of even integers in a list of ten random integers from 0 to 100.

```
In[24] := Table[Random[Integer, {1, 100}], {10}]
```

```
Out[24] = {10, 8, 10, 1, 11, 24, 86, 33, 1, 26}
```

`Count[`*expr, pattern, levelspec*`]` gives the total number of subexpressions that match *pattern* and that are on levels of *expr* specified by *levelspec.* If the third argument is omitted, *levelspec* takes the default value `{1}`.

```
In[25] := Count[%, _?EvenQ]
```

```
Out[25] = 6
```

In Section 3.2.3.1, there is a flaw in the definition of a function that returns the Legendre polynomial of order *n.* The first argument must be a nonnegative integer! The flaw can be corrected by introducing the pattern `n_Integer?NonNegative` for the argument.

```
In[26] := Clear[legdPoly]
```

```
In[27] := legdPoly[n_Integer?NonNegative, x_] :=
                ((1/(2^n n!)) D[(x^2 - 1)^n, {x, n}])//Simplify
```

```
In[28] := legdPoly[2, x]
```

$$Out[28] = \frac{-1 + 3\ x^2}{2}$$

If the first argument is a negative integer, *Mathematica* returns the function unevaluated:

```
In[29] := legdPoly[-2, x]
```

```
Out[29] = legdPoly[-2, x]
```

3.2.3.3 CONDITIONS

The pattern object *pattern* `/;` *condition* stands for any expression matching *pattern* provided that the evaluation of *condition* yields `True`. The operator `/;` can be construed as "such that," "provided that," "only if," or "whenever."

From a nested list, we can pick out sublists that are vectors of numbers:

```
In[30] := Cases[{a, {a, b}, {a, {b, c}}, 3, {3, 4}, {5, {6, 7}}},
            x_List /; VectorQ[x, NumberQ], Infinity]
```

```
Out[30] = {{3, 4}, {6, 7}}
```

The logical operator `&&`, which has the meaning "and," allows imposition of multiple conditions on a pattern. Upon evaluation, *condition1* `&&` *condition2* returns `True` when both *condition1* and *condition2* yield `True`. Consider selecting the integers between 10 to 80 from a list of ten random integers in the range 1 to 100:

```
In[31] := Table[Random[Integer, {1, 100}], {10}]

Out[31] = {30, 1, 36, 29, 91, 52, 71, 69, 56, 97}
```

```
In[32] := Cases[%, x_Integer/;(x > 10 && x < 80)]

Out[32] = {30, 36, 29, 52, 71, 69, 56}
```

In Section 3.2.3.1, there is another flaw in the definition of a function that returns the Legendre polynomial of order *n*. The second argument must be restricted to the interval $[-1, 1]$. We can correct this defect by introducing the pattern `x_ /; Abs[x] <= 1` for the second argument and by modifying the body of the function to prevent the substitution of a numerical value for **x** before the derivative is taken.

```
In[33] := Clear[legdPoly]
```

```
In[34] := legdPoly[n_Integer?NonNegative, x_ /; Abs[x] <= 1] :=
              ((1/(2^n n!)) D[(y^2 - 1)^n, {y, n}] /. y -> x)//Simplify
```

Let us plot `legdPoly[2, x]` as a function of **x** from –1 to 1:

```
In[35] := Plot[legdPoly[2, x], {x, -1, 1},
              AxesLabel -> {"x", "legdPoly[2, x]"}];
```

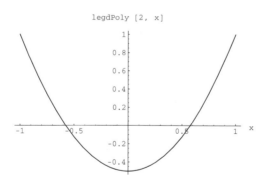

If we try to plot `legdPoly[2, x]` outside the interval $[-1, 1]$, *Mathematica* gives an error message:

```
In[36] := Plot[legdPoly[2, x], {x, -2, 2},
              AxesLabel -> {"x", "legdPoly[2, x]"}];
```

```
        Plot::plnr: legdPoly[2, x] is not a machine-size real number at
            x = -2..
```

```
Plot::plnr: legdPoly[2, x] is not a machine-size real number at
   x = -1.83773.

Plot::plnr: legdPoly[2, x] is not a machine-size real number at
   x = -1.66076.

General::stop:
   Further output of Plot::plnr
      will be suppressed during this calculation.
```

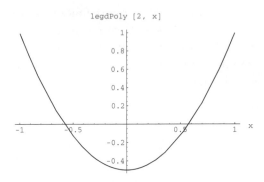

Mathematica plots the function only in the range for which it is defined. Note that `Abs[x] <= 1` is not evaluated if **x** does not have a numerical value:

In[37] := **Abs[x] <= 1**

Out[37] = Abs[x] <= 1

Consequently, the function no longer works if **x** is a mathematical variable:

In[38] := **legdPoly[1, x]**

Out[38] = legdPoly[1, x]

We fixed one flaw but introduced another. The function should work regardless of whether the second argument is a mathematical symbol or a number. We can solve this problem by defining separate functions for arguments matching different patterns: one for the pattern **x_Symbol** and another for the pattern **x_Real /; Abs[x] <= 1**:

In[39] := **legdPoly[n_Integer?NonNegative, x_Symbol] :=**
                    ((1/(2^n n!)) D[(x^2 - 1)^n, {x, n}])//Simplify

In[40] := **legdPoly[n_Integer?NonNegative, x_Real /; Abs[x] <= 1] :=**
                    ((1/(2^n n!)) D[(y^2 - 1)^n, {y, n}] /. y -> x)//Simplify

Now the function works for both types of second arguments:

In[41] := **legdPoly[2, x]**

Out[41] = $\dfrac{-1 + 3 x^2}{2}$

In[42] := **legdPoly[2, 0.5]**

Out[42] = -0.125

Of course, we can define a function that returns the Legendre polynomial of order *n* in terms of the built-in function **LegendreP**:

In[43] := **newLegdP[n_Integer?NonNegative, x_Symbol] := LegendreP[n, x]**

In[44] := **newLegdP[n_Integer?NonNegative, x_Real /; Abs[x] <= 1] :=**
LegendreP[n, x]

Here are **newLegdP[2, x]** and **newLegdP[2, 0.5]**:

In[45] := **newLegdP[2, x]**

$$Out[45] = -(\frac{1}{2}) + \frac{3\ x^2}{2}$$

In[46] := **newLegdP[2, 0.5]**

Out[46] = -0.125

In[47] := **Clear[legdPoly, newLegdP]**

Consider plotting the potential well

$$V(x) = \begin{cases} 0 & |x| > a \\ -V_0 & |x| < a \end{cases}$$

Let *V* be measured in units of V_0 and *x* in units of *a*, that is, $V_0 = 1$ and $a = 1$. With **v** as the alias of *V*, we give two separate definitions for two regions of **x** values:

In[48] := **v[x_ /; Abs[x] > 1] := 0**
v[x_ /; Abs[x] < 1] := -1

Here is a plot for the potential well:

In[50] := **Plot[v[x], {x, -2.5, 2.5},**
AxesLabel -> {" x (a)", " V (Vo)"},
PlotStyle -> Thickness[0.01]];

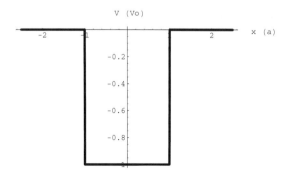

Names in a condition constraining a pattern must appear in the pattern. Consider the following definitions for the potential well:

```
In[51] := Clear[v]
```

```
In[52] := v[x_ /; Abs[x] > a] := 0
          v[x_ /; Abs[x] < a] := -1
```

The symbol **a** does not appear on the left-hand side of the **/;** operator.

```
In[54] := v[0.5]
```

```
Out[54] = v[0.5]
```

Mathematica returns the function unevaluated unless there is a global value for **a**.

```
In[55] := Clear[v]
```

In addition to putting conditions on patterns, we can also impose conditions on entire transformation rules and definitions. The transformation rule *lhs* **:>** *rhs* **/;** *condition* and the definition *lhs* **:=** *rhs* **/;** *condition* apply only if the evaluation of *condition* yields **True**.

Let us specify a condition for the application of a transformation rule:

```
In[56] := {f[2], f[a], g[6], f[5]} /. f[x_] :> x /; NumberQ[x]
```

```
Out[56] = {2, f[a], g[6], 5}
```

Note that the **:>** operator is used. Using the operator **->** causes the right-hand side of the rule, that is, *rhs* **/;** *condition*, to be evaluated immediately:

❀ **3.0**

```
In[57] := {f[2], f[a], g[6], f[5]} /. f[x_] -> x /; NumberQ[x]
```

```
Out[57] = {2 /; NumberQ[2], a /; NumberQ[a], g[6], 5 /; NumberQ[5]}
```

❀ **2.2**

```
{f[2], f[a], g[6], f[5]} /. f[x_] -> x /; NumberQ[x]
```

```
Condition::obscf:
        Warning: Obsolete use of Condition (x /; NumberQ[x]
              ) evaluated to Fail.
```

```
{Fail, Fail, g[6], Fail}
```

We can put mathematical constraints on the arguments of a function.

```
In[58] := func[x_, y_] := Sqrt[x - y] /; x > y
```

In[59] := **func[5, 4]**

Out[59] = 1

The definition is not used if the first argument is not larger than the second:

In[60] := **func[5, 5]**

Out[60] = func[5, 5]

Note the operator := rather than the = operator is used in the definition. Using the = operator causes immediate evaluation of the right-hand side of the definition, that is, *rhs* /; *condition*:

In[61] := **Clear[func]**

❀ **3.0**

In[62] := **func[x_, y_] = Sqrt[x - y] /; x > y**

Out[62] = Sqrt[x - y] /; x > y

❀ **2.2**

func[x_, y_] = Sqrt[x - y] /; x > y

Condition::obscf:
        Warning: Obsolete use of Condition (Sqrt[x - y] /; x > y
) evaluated to Fail.

Fail

3.2.4 *Structural Equivalence*

In matching patterns, *Mathematica* looks for structural equivalences between expressions and patterns. Mathematically equivalent expressions match different patterns if their structures are different. For example, a^2 - b^2 and (a + b)(a - b) are mathematically equivalent. The expression a^2 - b^2 matches the pattern x_^2 - y_^2 because they have the same structure:

In[63] := **MatchQ[a^2 - b^2, x_^2 - y_^2]**

Out[63] = True

The expression Sin[a]^2 - Cos[b]^2 also matches the pattern x_^2 - y_^2 because they are structurally equivalent:

In[64] := **MatchQ[Sin[a]^2 - Cos[b]^2, x_^2 - y_^2]**

Out[64] = True

Yet `(a + b)(a - b)` does not match `x_^2 - y_^2` because their structures are different:

> `In[65] := MatchQ[(a + b)(a - b), x_^2 - y_^2]`
>
> `Out[65] = False`

Though `a^2 - b^2` and `(a + b)(a - b)` are mathematically equivalent, they do not match the same pattern because they have different structures. A pattern that `(a + b)(a - b)` matches is `(x_ + y_)(x_ - y_)`:

> `In[66] := MatchQ[(a + b)(a - b), (x_ + y_)(x_ - y_)]`
>
> `Out[66] = True`

   The structures of expressions that *Mathematica* considers are those of the internal forms of the expressions. For example, `(a - b)` and `(a - 2b)` appear to have the same structure as the pattern `(x_ - y_)`. Whereas `(a - b)` matches the pattern, `(a - 2b)` does not:

> `In[67] := MatchQ[a - b, x_ - y_]`
>
> `Out[67] = True`

> `In[68] := MatchQ[a - 2b, x_ - y_]`
>
> `Out[68] = False`

An examination of the internal forms reveals that `a - 2b` and `x_ + y_` have different structures:

> `In[69] := FullForm[a - 2b]`
>
> `Out[69]//FullForm=`
>   `Plus [a, Times[-2, b]]`

> `In[70] := FullForm[x_ - y_]`
>
> `Out[70]//FullForm=`
>   `Plus [Pattern[x, Blank[]], Times[-1, Pattern[y, Blank[]]]]`

Though `a` can fill in for `Pattern[x, Blank[]]` and `b` can fill in for `Pattern[y, Blank[]]`, the first argument of `Times` is `-2` in the internal form of `a - 2b`, whereas it is `-1` in the internal form of "`x_ - y_`".

3.2.5   Attributes

We can assign general properties, called attributes, to functions. In evaluation and pattern-matching, *Mathematica* considers these attributes. Section 2.4.8 discussed the attribute `Listable`. A function with the attribute `Listable` automatically acts on each element of a list that appears as its argument. This section examines two attributes that are important to pattern-matching: `Orderless` and `Flat`. Section 3.3.5 will describe the attribute `Protected`. With on-line help, we can access the definitions of the remaining *Mathematica* attributes that are not used as often: `OneIdentity`, `Constant`, `Locked`, `ReadProtected`, `HoldFirst`,

HoldRest, HoldAll, Temporary, and Stub. (*Mathematica* version 3.0 provides several additional attributes: NumericFunction, HoldAllComplete, NHoldFirst, NHoldRest, NHoldAll, and SequenceHold.)

The function Attributes lists the attributes of functions. Attributes[*f*] gives the attributes assigned to *f*. The operator ?? also shows the attributes of functions.

There are two ways to assign attributes to functions:

> Attributes[*f*] = {*attr1*, *attr2*, ...}   set the attributes of *f*
>
> SetAttributes[*f*, *attr*]   add *attr* to the attributes of *f*

If we are to specify for a function any of the attributes that affect pattern-matching, namely, Orderless, Flat, and OneIdentity, we must make the assignment before giving definitions or transformation rules to the function. (See Exercise 5 of Section 3.2.9.)

There are two ways to remove the attributes of functions:

> Attributes[*f*] = {}   set *f* to have no attributes
>
> ClearAttributes[*f*, *attr*]   remove *attr* from attributes of *f*

We can also use ClearAll to remove the attributes of functions. However, in addition to removing the attributes, ClearAll[*symbol1*, *symbol2*, ...] clears all values, definitions, messages, and defaults associated with the symbols. (Be aware that Clear[*symbol1*, *symbol2*, ...], which we have been using, clears only values and definitions for the specified symbols.)

The arguments of a function with the attribute Orderless are automatically sorted into standard order. Consequently, if the function f has the attribute Orderless, f[a, c, b], f[b, a, c], and so forth are equivalent to f[a, b, c]. In other words, f is a commutative function. *Mathematica* recognizes this property in matching patterns.

Addition and multiplication are commutative; the functions Plus and Times have the attribute Orderless. Let us illustrate how this attribute affects pattern-matching:

In[71] := {{a + b, a}, {a + b, b}} /. {x_ + y_, x_} :> {x, y}

Out[71] = {{a, b}, {b, a}}

The list {a + b, a} matches the pattern with x -> a and y -> b. The list {a + b, b} also matches the pattern because *Mathematica* takes account of the attribute Orderless for the function Plus and rewrites the list as {b + a, b}. In this case, the list matches the pattern with x -> b and y -> a.

The arguments of a function with the attribute Flat are automatically flattened out. Therefore, if the function f has the attribute Flat, f[f[a], f[b, c]], f[f[a], b, f[f[c]]], and so forth are equivalent to f[a, b, c]. In other words, f is an associative function. *Mathematica* takes this property into consideration in matching patterns.

In addition to being commutative, addition and multiplication are also associative; the functions Plus and Times have the attribute Flat. Let us demonstrate with an example the combined effects of the attributes Orderless and Flat.

In[72] := **(a + b + c) /.(x_ + y_) :> {x, y}**

Out[72] = {a, b + c}

The expression matches the pattern with **x -> a** and **y -> b + c**. Because the function **Plus** has the attributes **Orderless** and **Flat**, there are many other possible matches. *Mathematica* returns the first match it finds. To see the other matches, we attach a condition to the rule. The function **Print**, which is never evaluated to **True**, can be used as a condition to show the possible matches in the order attempted by *Mathematica*.

In[73] := **(a + b + c) /.(x_ + y_) :> {x, y} /; Print[{{x}, {y}}];**

```
{{a}, {b + c}}
{{b}, {a + c}}
{{c}, {a + b}}
{{a + b}, {c}}
{{a + c}, {b}}
{{b + c}, {a}}
{{a}, {b}}
{{b}, {a}}
{{a}, {c}}
{{c}, {a}}
{{b}, {c}}
{{c}, {b}}
```

❋ **3.0**   **ReplaceList** also reveals all ways that the entire expression can match the pattern. **ReplaceList** [*expr, rules*] attempts to transform the entire expression *expr* by applying a rule or list of rules in all possible ways, and it returns a list of the results obtained.

In[74] := **ReplaceList[(a + b + c), (x_ + y_) -> {x, y}]**

Out[74] = {{a, b + c}, {b, a + c}, {c, a + b}, {a + b, c}, {a + c, b}, {b + c, a}}

(Why did **ReplaceList** not return the matches: {a, b}, {b, a}, {a, c}, {c, a}, {b, c}, {c, b}?)

3.2.6 *Defaults*

The pattern *x_:v* stands for any single expression that, if omitted, has default value *v*. The expression is to be named *x* on the right-hand side of a definition or transformation rule. For example, the list **{a, b}** matches the pattern **{x_, y_:d}**,

In[75] := **{a, b} /. {x_, y_:d} :> {x^2, y^2}**

Out[75] = {a^2, b^2}

and the rule applies. The rule also applies even when the second element of the list is omitted:

In[76] := **{a} /. {x_, y_:d} :> {x^2, y^2}**

Out[76] = $\{a^2, d^2\}$

In this case, the second element is given the default value **d**.

The pattern *x_h:v* stands for an expression that, if omitted, has default value *v*. The expression has the head *h* and is to be named *x* on the right-hand side of a definition or transformation rule. For example, the list **{a, 2}** matches the pattern **{x_, y_Integer:10}**,

In[77] := **{a, 2} /. {x_, y_Integer:10} :> {x^2, y^2}**

Out[77] = $\{a^2, 4\}$

and the rule applies. What happens if the second element is omitted?

In[78] := **{a} /. {x_, y_Integer:10} :> {x^2, y^2}**

Out[78] = $\{a^2, 100\}$

In this case, the second element assumes the default value **10**. For a list to match the pattern **{x_, y_Integer:10}**, it must have one or two elements and, if present, the second element must be an integer.

In[79] := **{a, 2, 3} /. {x_, y_Integer:10} :> {x^2, y^2}**

Out[79] = $\{a, 2, 3\}$

The list fails to match the pattern because it has three elements.

In[80] := **{a, b} /. {x_, y_Integer:10} :> {x^2, y^2}**

Out[80] = $\{a, b\}$

The list does not match the pattern because its second element is not an integer.

The pattern *x_.* represents an expression with a built-in default value. The expression is to be named *x* on the right-hand side of a definition or transformation rule. We can use this pattern to give a built-in default value to a piece in each of three patterns. In the pattern "*x_ + y_.*", the default value for *y* is 0; in "*x_ y_.*", the default for *y* is 1; in "*x_^y_.*", the default for *y* is 1. For example, **a + b** matches the pattern "**x_ + y_.**",

In[81] := **a + b /. x_ + y_. :> x^2 + y^2**

Out[81] = $a^2 + b^2$

and if the second term of the sum is omitted, the expression still matches the pattern,

In[82] := **a /. x_ + y_. :> x^2 + y^2**

Out[82] = a^2

where the default value 0 is used for the second term that is missing. The expression **a b** matches the pattern "**x_ y_.**",

In[83] := **a b /. x_ y_. :> x^2 + y^2**

Out[83] = $a^2 + b^2$

and if the second term of the product is omitted, it still matches the pattern,

In[84] := a /. x_ y_. :> x^2 + y^2

Out[84] = 1 + a^2

where the default value of 1 is used for the missing term in the product.

EXAMPLE 3.2.1 Construct a pattern that represents any difference of two terms.

   In Section 3.2.4, we found that a - 2b does not match the pattern "x_ - y_". Neither does a - 5 match the pattern "x_ - y_":

In[85] := MatchQ[a - 5, x_ - y_]

Out[85] = False

FullForm reveals the reason for the mismatch.

In[86] := FullForm[x_ - y_]

Out[86]//FullForm=
         Plus [Pattern[x, Blank[]], Times[-1, Pattern[y, Blank[]]]]

In[87] := FullForm[a - 5]

Out[87]//FullForm=
         Plus [-5, a]

The problem is that -5 is an atom and therefore does not match **Times[-1, Pattern[y, Blank[]]]**. Is there a pattern that matches the difference of any two expressions? A pattern is "x_ + n_?Negative y_.". The built-in default value for y is 1.

In[88] := Cases[{a - Sqrt[b], a - 2b, a - b/3, a^2 - 5b^3, a - 5,
                Tan[x] - Sin[y], a + b, Tan[x] + Sin[x]},
             x_ + n_?Negative y_.]

Out[88] = {a - Sqrt[b], a - 2 b, a - $\dfrac{b}{3}$, a^2 - 5 b^3, -5 + a,
          -Sin[y] + Tan[x]}

Note that a + b and **Tan[x]** + **Sin[x]** do not match the pattern.   ■

EXAMPLE 3.2.2 Construct a pattern that matches any expression of the form $a_n x^n$ where a_n, x, and n are a number, a variable, and a positive integer, respectively.

   The pattern a_ x_^n_ matches the expression 3 x^2:

In[89] := MatchQ[3 x^2, a_ x_^n_]

Out[89] = True

Yet it does not match the expression x^2:

In[90] := MatchQ[x^2, a_ x_^n_]

Out[90] = False

Neither does it match the expression 2 **x**:

In[91] :_ **MatchQ[2 x, a_ x_^μ_]**

Out[91] = False

The pattern **a_. x_Symbol^n_. /; NumberQ[a]&&Positive[n]&&IntegerQ[n]** matches any expression of the form $a_n x^n$ where a_n, x, and n are a number, a variable, and a positive integer, respectively:

In[92] := **Cases[{3 x^2, x^2, 2 x, x^1.5, 1/y, 3, a + b,**
                    2 y^(-2), y^2 + 1, x/y^2, Sqrt[x], Tan[x]},
                 a_. x_Symbol^n_./;NumberQ[a]&&Positive[n]&&IntegerQ[n]]

Out[92] = {3 x^2, x^2, 2 x}   ■

3.2.7 Alternative or Repeated Patterns

The pattern *patt1* | *patt2* | ... is a pattern that can have one of several forms. For example, the pattern **_Integer** | **_Symbol** | **_Rational** matches any integer, variable, or rational number:

In[93] := **Cases[{3, x, 2/3, 2 + 3 I, Tan[x], Sqrt[x], 1.5},**
                 _Integer | _Symbol | _Rational]

Out[93] = {3, x, $\frac{2}{3}$}

Cases picks out the integer, symbol, and rational number from the list.

EXAMPLE 3.2.3 Construct a pattern that stands for any monomial, which is a number or a number times a variable to some positive integer power.

A monomial is a number or an expression of the form $a_n x^n$ where a_n, x, and n are a number, a variable, and a positive integer, respectively. The pattern constructed in Example 3.2.2 matches the latter. We only need to include in the pattern an alternative that represents a number:

In[94] := **Cases[{3 x^2, x^2, 2 x, x^1.5, 1/y, 3, a + b, 1.5,**
                    2 y^(-2), y^2 + 1, x/y^2, Sqrt[x], Tan[x]},
                 (a_. x_Symbol^n_./;Positive[n]&&IntegerQ[n]&&NumberQ[a])
                 |a_ /; NumberQ[a]]

Out[94] = {3 x^2, x^2, 2 x, 3, 1.5}

Cases picks out all the monomials in the list.   ■

The pattern *expr*.. is a pattern or other expression repeated one or more times. The pattern *expr*... is a pattern or other expression repeated zero or more times. For example, {a, a, a, a, b, b} matches the pattern {x_ .., y_ ..}:

In[95] := **{a, a, a, a, b, b} /. {x_ .., y_ ..} :> {x, y}**

Out[95] = {a, b}

3.2.8 *Multiple Blanks*

The "double blank" stands for any sequence of one or more expressions. Its special input form and internal form are `__` and `BlankSequence[]`, respectively. The "triple blank" represents any sequence of zero or more expressions. Its special input form and internal form are `___` and `BlankNullSequence[]`, respectively. We can set up other patterns with single, double, and triple blanks and also name or restrict these patterns.

Let us define a function that lists its argument:

```
In[96] := myfunc[x_] := {x}
```

```
In[97] := myfunc[a]
```

```
Out[97] = {a}
```

This function does not accept more than one argument:

```
In[98] := myfunc[a, b, c]
```

```
Out[98] = myfunc[a, b, c]
```

With a double blank, we can define a function that lists one or more arguments:

```
In[99] := ourfunc[x__] := {x}
```

```
In[100] := ourfunc[a]
```

```
Out[100] = {a}
```

```
In[101] := ourfunc[a, b, c, d]
```

```
Out[101] = {a, b, c, d}
```

The pattern `__h` stands for any sequence of one or more expressions, each with a head h. Thus, `__List` represents any sequence of one or more lists, and `{__List}` matches any list of lists.

```
In[102] := Cases[{{}, {a}, {a, b}, {{a}, {b, c}}}, {__List}]
```

```
Out[102] = {{{a}, {b, c}}}
```

There is only one nested list.

There are often several possible matches of an expression to a pattern containing multiple blanks:

```
In[103] := {a, b, c, d} /.{x__, y__} :> "SMILE" /; Print[{{x}, {y}}];

          {{a}, {b, c, d}}
          {{a, b}, {c, d}}
          {{a, b, c}, {d}}
```

To reveal all the possible matches in the order attempted by *Mathematica*, we attached, as a condition to the transformation rule, `Print[{{x}, {y}}]`; this prints each match to the pattern but is never evaluated to `True` for the rule to actually apply.

EXAMPLE 3.2.4 Evaluate the constant **E** to 60 digits, determine the number of digits between the decimal point and a repeated sequence of numbers, and identify the sequence.

To compute **E** to 60 digits, use the function **N**:

In[104] := **N[E, 60]**

Out[104] = 2.71828182845904523536028747135266249775724709369995957496700

ToString[*expr*] gives a string corresponding to the printed form of *expr*.

In[105] := **ToString[%]**

Out[105] = 2.71828182845904523536028747135266249775724709369995957496700

The function **Characters** gives a list of the characters in a string:

In[106] := **charactersOfE = Characters[%]**

Out[106] = {2, ., 7, 1, 8, 2, 8, 1, 8, 2, 8, 4, 5, 9, 0, 4, 5, 2, 3, 5,
              3, 6, 0, 2, 8, 7, 4, 7, 1, 3, 5, 2, 6, 6, 2, 4, 9, 7, 7, 5,
              7, 2, 4, 7, 0, 9, 3, 6, 9, 9, 9, 5, 9, 5, 7, 4, 9, 6, 7, 0}

Finally, determine how many digits fall between the decimal point and a repeated sequence of numbers:

In[107] := **Print["digits   Sequence\n"];**
             Timing[
               charactersOfE /.
                 {"2", ".", a___, x__/; Length[{x}] > 1, x__, ___} :>
                 "Ha!" /; Print[Length[{a}], "   ", x];
                   ]

```
digits    Sequence

1         1828
48        95
```

{197.383 Second, Null}

To search for all the possible matches to the pattern {"2", ".", a___, x__/; Length[{x}] > 1, x__, ___}, we attached, as a condition to the rule, the function **Print**, which is never evaluated to **True**. In addition to preventing *Mathematica* from identifying only the first match to the pattern, **Print[Length[{a}], "   ", x]** prints for each match the number of digits between the decimal point and the repeated sequence of numbers and also prints the value of **x**, which is the repeated sequence. There are two repeated sequences: 1828 appearing after 1 digit, and 95 appearing after 48 digits. The evaluation took 197 seconds on a Macintosh IIci and 27 seconds on a Gateway 2000 4DX2-66V. (In this book, all timings have been performed with *Mathematica* version 2.2.) ■

In[110] := **Clear[myfunc, ourfunc, charactersOfE]**

3.2.9 Exercises

1. Write a function that computes the mean of a list of one or more quantities. Determine the mean of a list of n random integers in the range 0 to 1000, where n itself is a random integer in the range 1 to 100.

2. Write a function to expand only the operand of the `Log` function in an expression. Transform `(1 - x + Log[(a - b)^2])^2` to `(1 - x + Log[a^2 - 2 a b + b^2])^2`.

3. Consider the associated Legendre polynomials $P_n^m(x)$, where n must be a nonnegative integer, m can take on only the values $-n, -n+1, \ldots, n-1, n$, and x is confined to the interval $[-1, 1]$. P_n^0 equals P_n, which is the Legendre polynomial of order n. For positive m, P_n^m can be determined from the equation

$$P_n^m(x) = (-1)^m (1 - x^2)^{m/2} \frac{d^m P_n(x)}{dx^m}$$

For negative m, P_n^m can be obtained from the formula

$$P_n^{-m}(x) = (-1)^m \frac{(n-m)!}{(n+m)!} P_n^m(x)$$

Define a function for generating the associated Legendre polynomials. Plot P_2^1. Determine P_n^m for $n = 0, 1, 2$ and all possible m. Verify the results with the built-in function `LegendreP`.

4. Produce the decimal expansion of π out to 770 places, and determine the number of digits past the decimal point before six 9's appear in a row.

5. Define a function that takes three arguments: a symbol, an integer, and a complex number. However, the order in which arguments are specified is not important. In other words, the function should accept a symbol, an integer, or a complex number as an argument at any position, but no two arguments can be of the same type. The function returns an ordered list of the arguments: the symbol followed by the integer and then the complex number. For example, `f[2 + I, g, 5]` gives `{g, 5, 2 + I}`. *Hint:* For the function, assignment of attributes must come before the definition.

6. Generate a list of ordered triplets (x, y, z), where x, y, z, and the length of the list are nonnegative random integers from 0 to 10. From this list, obtain a list of those triplets satisfying the condition that $x + y + z$ is an even number. Also, obtain the list of those triplets with $x + y + z$ an odd number. For example, from the generated list

```
{{6, 6, 1}, {5, 8, 1}, {1, 2, 10}, {1, 8, 0}, {3, 0, 5}, {0, 8, 4},
   {6, 4, 8}}
```

obtain the list

```
{{5, 8, 1}, {3, 0, 5}, {0, 8, 4}, {6, 4, 8}}
```

and the list

```
{{6, 6, 1}, {1, 2, 10}, {1, 8, 0}}
```

(7.) Construct a pattern that matches any quotient.

8. Define a function that converts only a product of integers into a list of the integers. For example,

```
prod2list[2]
```

```
prod2list[2]
```

```
prod2list[2.0 3 4]
```

```
prod2list[2. 3 4]
```

```
prod2list[20 3 40 5 60 7 80]
```

```
{20, 3, 40, 5, 60, 7, 80}
```

```
prod2list[22
          3
          4
          55
          6
          9]
```

```
{22, 3, 4, 55, 6, 9}
```

Hint: With on-line help, obtain information about the *Mathematica* attribute **HoldAll**. Also, examine the internal forms of products.

3.3 Functions

3.3.1 Pure Functions

Sections 2.2.4 and 2.2.13 stressed the hazards of named functions and programming variables with global values and promoted the practice of clearing them as soon as they are no longer needed. Elegant *Mathematica* programs minimize the number of temporary programming variables and named functions. To reduce the number of temporary variables, these programs prefer replacements with transformation rules over assignments. To reduce the number of temporary named functions, they make frequent use of pure or anonymous functions, which are functions without names. Unless a function is to be invoked repeatedly or its arguments are to be constrained, consider using a pure function. **Function**[*var*, *body*] is a pure function with a single formal parameter *var*. **Function**[*var*, *body*][*arg*] returns the result from evaluating *body* in which the formal variable *var* is replaced everywhere by the argument *arg*.

To specify an operation such as multiplying an expression by 3, we can define a named function

In[1] := **f[var_] := 3 var**

Let us apply this function to an argument **y**:

In[2] := **f[y]**

Out[2] = 3 y

We can also use the pure function **Function[var, 3 var]**:

In[3] := **Function[var, 3 var][y]**

Out[3] = 3 y

The pure function returns the same result.

Since the name of the formal parameter is irrelevant, *Mathematica* allows the use of two short forms: **Function[***body***]** or *body***&**. In these forms the formal parameter is **#**. Let us go back to the example of multiplying an expression by 3. The pure function can now take the forms

In[4] := **Function[3#][y]**

Out[4] = 3 y

In[5] := **3#&[y]**

Out[5] = 3 y

The function **Select** together with a pure function allows us to select numbers greater than 4 from a list:

In[6] := **Select[{1, a, x^2, 3, 5, 1 + x, 7}, # > 4 &]**

Out[6] = {5, 7}

In Section 3.1.3.1, we defined the function **test1** for testing whether an expression is a polynomial in the variable *x*:

In[7] := **test1[expr_] := PolynomialQ[expr, x]**

With this function, we can obtain an expression with the elements that are polynomials in the variable *x*:

In[8] := **Select[(1 + x + 2 x^2 + 3 x^3 + Sin[x]), test1]**

Out[8] = $1 + x + 2 x^2 + 3 x^3$

Using a pure function makes **test1** unnecessary:

In[9] := **Select[(1 + x + 2 x^2 + 3 x^3 + Sin[x]),**
          Function[var, PolynomialQ[var, x]]]

Out[9] = $1 + x + 2 x^2 + 3 x^3$

We can also use the short form for the pure function:

In[10] := `Select[(1 + x + 2 x^2 + 3 x^3 + Sin[x]), PolynomialQ[#, x]&]`

Out[10] = $1 + x + 2 x^2 + 3 x^3$

In Section 3.1.3.4, we defined a function to illustrate the restructuring of expressions:

In[11] := `q[x_] := 1/ (1 + x)`

With this function, we can generate a nested expression:

In[12] := `Nest[q, x, 4]`

Out[12] = $\cfrac{1}{1 + \cfrac{1}{1 + \cfrac{1}{1 + \cfrac{1}{1 + x}}}}$

Using a pure function, we no longer need the function q:

In[13] := `Nest[Function[1/(1 + #)], x, 4]`

Out[13] = $\cfrac{1}{1 + \cfrac{1}{1 + \cfrac{1}{1 + \cfrac{1}{1 + x}}}}$

We can also use the ampersand notation for the pure function:

In[14] := `Nest[(1/(1 + #))&, x, 4]`

Out[14] = $\cfrac{1}{1 + \cfrac{1}{1 + \cfrac{1}{1 + \cfrac{1}{1 + x}}}}$

A pure function can have more than one parameter. `Function[{x1, x2, ...}, body]` is a pure function with a list of formal parameters $x1, x2, \ldots$. In the short forms, the formal parameters are **#** (or **#1**), **#2**, ..., and **#**n for the first, second, ..., and nth variables in the pure function, respectively. The expression **##** stands for the sequence of all variables in a pure function, and **##**n represents the sequence of variables starting with the nth one.

Here is a function for computing the sum of the squares of its two arguments:

In[15] := `myfunc[x_, y_] := x^2 + y^2`

In[16] := `myfunc[a, b]`

Out[16] = $a^2 + b^2$

The same operation can be done with a pure function:

In[17] := `#1^2 + #2^2 & [a, b]`

Out[17] = $a^2 + b^2$

In Section 3.2.8, we defined a function that lists one or more arguments:

In[18] := `ourfunc[x__] := {x}`

```
In[19] := ourfunc[a]

Out[19] = {a}


In[20] := ourfunc[a, b, c, d]

Out[20] = {a, b, c, d}
```

The same operations can be performed with the pure function `{##}&`:

```
In[21] := {##}&[a]

Out[21] = {a}


In[22] := {##}&[a, b, c, d]

Out[22] = {a, b, c, d}


In[23] := ClearAll[f, test1, q, myfunc, ourfunc]
```

By including an additional argument, we can assign attributes to a pure function. `Function[{x1, x2, ...}, body, {attributes}]` is a pure function that has the specified attributes. Consider a pure function that is equivalent to the function `p`:

```
In[24] := Function[x, p[x]][{a, b, c}]

Out[24] = p[{a, b, c}]
```

With the attribute `Listable`, the pure function is automatically applied to each element of the list that appears as its argument:

```
In[25] := Function[x, p[x], Listable][{a, b, c}]

Out[25] = {p[a], p[b], p[c]}
```

We conclude this section on pure functions with an example from vector analysis.

EXAMPLE 3.3.1 Show that

$$\nabla(\mathbf{A} \cdot \mathbf{B}) = \mathbf{A} \times (\nabla \times \mathbf{B}) + \mathbf{B} \times (\nabla \times \mathbf{A}) + (\mathbf{A} \cdot \nabla)\mathbf{B} + (\mathbf{B} \cdot \nabla)\mathbf{A}$$

where $\mathbf{A}$ and $\mathbf{B}$ are any two vector point functions. In terms of the basis vectors $\mathbf{i}$, $\mathbf{j}$, and $\mathbf{k}$ of the Cartesian coordinate system and the components of the vector functions, the third term on the *rhs* of the vector identity can be written as

$$(\mathbf{A} \cdot \nabla)\mathbf{B} = +\left[A_x\frac{\partial B_x}{\partial x} + A_y\frac{\partial B_x}{\partial y} + A_z\frac{\partial B_x}{\partial z}\right]\mathbf{i}$$

$$+\left[A_x\frac{\partial B_y}{\partial x} + A_y\frac{\partial B_y}{\partial y} + A_z\frac{\partial B_y}{\partial z}\right]\mathbf{j}$$

$$+\left[A_x\frac{\partial B_z}{\partial x} + A_y\frac{\partial B_z}{\partial y} + A_z\frac{\partial B_z}{\partial z}\right]\mathbf{k}$$

and a similar expression can be given for the fourth term.

The package `Calculus`VectorAnalysis`` contains many functions for vector analysis, and those relevant to this example were defined in Example 2.4.10 in Section 2.4.9.

■ *Mathematica* **Version 3.0 Specifics**

In[26] := **Needs["Calculus`VectorAnalysis`"]**

CoordinateSystem gives the name of the default coordinate system:

In[27] := **CoordinateSystem**

Out[27] = Cartesian

Coordinates[] returns the default names of coordinate variables in the default coordinate system:

In[28] := **Coordinates[]**

Out[28] = {Xx, Yy, Zz}

To conform with traditional notation, let us change the names of the coordinate variables to **x**, **y**, and **z**:

In[29] := **SetCoordinates[Cartesian[x, y, z]]**

Out[29] = Cartesian[x, y, z]

We can verify the new names of the coordinate variables:

In[30] := **Coordinates[]**

Out[30] = {x, y, z}

■ *Mathematica* **Version 2.2 Specifics**

Needs["Calculus`VectorAnalysis`"]

x::shdw: Warning: Symbol x appears in multiple contexts
     {Calculus`VectorAnalysis`, Global`}; definitions in context
     Calculus`VectorAnalysis`
       may shadow or be shadowed by other definitions.

y::shdw: Warning: Symbol y appears in multiple contexts
     {Calculus`VectorAnalysis`, Global`}; definitions in context
     Calculus`VectorAnalysis`
       may shadow or be shadowed by other definitions.

Mathematica warns that we have invoked the names **x** and **y** before loading the package. We should remove these symbols with the function **Remove**:

Remove[x, y]

CoordinateSystem gives the name of the default coordinate system:

CoordinateSystem

Cartesian

Coordinates[] returns the default names of coordinate variables in the default coordinate system:

Coordinates[]

{x, y, z}

■ **Vector Identity**

Mathematica represents three-dimensional vectors by three-element lists. In keeping with the convention of reserving capitalized names for built-in *Mathematica* objects, we let **a** and **b** be aliases of vector functions **A** and **B**, respectively, and

```
In[31] := a = {ax[x, y, z], ay[x, y, z], az[x, y, z]};
          b = {bx[x, y, z], by[x, y, z], bz[x, y, z]};
```

To prove the vector identity, show that *lhs − rhs* evaluates to the null vector $\{0, 0, 0\}$:

```
In[33] := Grad[DotProduct[a, b]] -
          CrossProduct[a, Curl[b]] -
          CrossProduct[b, Curl[a]] -
          (DotProduct[a, {D[#, x], D[#, y], D[#, z]}]&) /@ b -
          (DotProduct[b, {D[#, x], D[#, y], D[#, z]}]&) /@ a
```

```
Out[33] = {0, 0, 0}
```

We have used pure functions in the last two terms.

```
In[34] := ClearAll[a, b]
```

■

3.3.2　*Selecting a Definition*

In *Mathematica*, we can give a function more than one definition. When the function is called, that is, invoked, how does *Mathematica* select which definition to use? It starts with the most specific definition, tries one at a time the more specific definitions before the more general ones, and without going any further uses the first definition that matches the call. For example, consider the definitions

```
In[35] := ClearAll[f]
          f[y_] := 3 y
          f[x_Integer] := x^2
          f[5] := 1/25
```

and the function call **f[7]**. *Mathematica* first tries the definition for **f[5]**, which is the most specific because **5** is a subclass of the class of expressions represented by **x_Integer**, which in turn stands for a subclass of the class of expressions represented by **y_**. Since **f[7]** does not match **f[5]**, *Mathematica* proceeds to the definition for **f[x_Integer]**, where it finds a match. The second definition is then used, and the result is **49**:

```
In[39] := f[7]
```

```
Out[39] = 49
```

The **?** operator shows how *Mathematica* orders these definitions:

```
In[40] := ?f

          Global`f
          f[5] := 1/25

          f[x_Integer] := x^2

          f[y_] := 3*y
```

Definitions often lack definite order in generality. In such cases, *Mathematica* gives higher priority to definitions that are entered earlier. For instance, consider the definitions

```
In[41] := ClearAll[f]
          f[{x_, y_}] := x^3 - y^3
          f[x_Real?Positive] := Sqrt[x]
          f[x_Integer] := x^2
```

Since none of these definitions can be considered to be more specific, *Mathematica* simply stores them in the order they are given:

```
In[45] := ?f

          Global`f
          f[{x_, y_}] := x^3 - y^3

          f[(x_Real)?Positive] := Sqrt[x]

          f[x_Integer] := x^2
```

Mathematica's notion of generality is at times inscrutable. For example, consider the definitions

```
In[46] := ClearAll[f]
          f[x_Integer] := x^2
          f[x_Integer?Positive] := 1/x
          f[x_Integer /; x > 50] := -x
```

Obviously, `f[x_Integer /; x > 50]` is a special case of `f[x_Integer?Positive]`, which is in turn a special case of `f[x_Integer]`. Instead of ordering these definitions as expected, *Mathematica* stores them according to the order in which they are entered:

```
In[50] := ?f

          Global`f
          f[x_Integer] := x^2

          f[(x_Integer)?Positive] := 1/x

          f[x_Integer /; x > 50] := -x
```

For `f[60]`, *Mathematica* returns

```
In[51] := f[60]

Out[51] = 3600
```

It is important to realize that the value of `f[60]` depends in this case on the order in which the definitions are given.

```
In[52] := ClearAll[f]
          f[x_Integer?Positive] := 1/x
          f[x_Integer /; x > 50] := -x
          f[x_Integer] := x^2
```

```
In[56] := ?f

       Global`f
       f[(x_Integer)?Positive] := 1/x

       f[x_Integer /; x > 50] := -x

       f[x_Integer] := x^2
```

Now, the result is

```
In[57] := f[60]
```

$$Out[57] = \frac{1}{60}$$

Therefore, it is often a good idea to check how *Mathematica* orders the definitions with the ? operator to ensure that the appropriate definition is used.

```
In[58] := ClearAll[f]
```

3.3.3   *Recursive Functions and Dynamic Programming*

Consider, for example, a recursion relation

$$H_n(z) = 2zH_{n-1}(z) - 2(n-1)H_{n-2}(z)$$

for the Hermite polynomials, where n can be any integer greater than 1 and the initial conditions are

$$H_0(z) = 1$$

$$H_1(z) = 2z$$

(The recursion relation [Gos92, Mes62] for the Hermite polynomials is also known as a recurrence formula [MF53] or a recurrence relation [Has91]. Some authors use the terms recursion and recurrence interchangeably [Lib92, Vve93]. See [Tho92] for an interesting discussion of recursion versus recurrence. For a contrasting view, see [MF53].)

In programming, a function is recursive if it is defined in terms of itself. We can determine the Hermite polynomials with the recursive function

```
In[59] := hermite[n_Integer /; n > 1, z_] := Expand @
             (2 z hermite[n - 1, z] - 2 (n - 1) hermite[n - 2, z])
```

and the initial conditions

```
In[60] := hermite[0, z_] := 1
          hermite[1, z_] := 2 z
```

where **hermite** is an alias for *H*. Unless n is small, generating Hermite polynomials with the recursive function is computation-intensive because a call to **hermite[n, z]** invokes two calls, one to **hermite[$n-1$, z]** and another to **hermite[$n-2$, z]**, each of these calls in turn triggers two additional ones, and so on until each branch eventually terminates with a call to **hermite[1, z]** or **hermite[0, z]**. Evaluating **hermite[20, z]**, for instance, requires a total of 21,891 calls to the function **hermite** (see Exercise 7a in Section 3.3.7) and takes

305 seconds on a Macintosh IIci and 42 seconds on a Gateway 2000 4DX2-66V. (In this book, all timings have been performed with *Mathematica* version 2.2.)

```
In[62] := Timing[hermite[20, z];]
```

> {304.8 Second, Null}

The following table shows the number of calls to each **hermite[k, z]**, for $0 \le k \le 20$, in the evaluation of **hermite[20, z]** (see Exercise 8 in Section 3.3.7):

k	Number of Function Calls
0	4181
1	6765
2	4181
3	2584
4	1597
5	987
6	610
7	377
8	233
9	144
10	89
11	55
12	34
13	21
14	13
15	8
16	5
17	3
18	2
19	1
20	1

Most of these function calls, however, are unnecessary if the function is so written that *Mathematica* remembers, in the form of rules or definitions, the values that it has computed for the function with specific arguments. The method of creating and storing new rules for a function during an evaluation is called dynamic programming. The syntax is rather simple:

$$func[arg1\_, arg2\_, \ldots] := func[arg1, arg2, \ldots] = rhs$$

For the Hermite polynomials, the definition becomes

```
In[63] := ClearAll[hermite]
          hermite[n_Integer /; n > 1, z_] := hermite[n, z] = Expand @
                   (2 z hermite[n - 1, z] - 2 (n - 1) hermite[n - 2, z])
```

while the initial conditions remain the same:

```
In[65] := hermite[0, z_] := 1
          hermite[1, z_] := 2 z
```

Mathematica stores the value of the function as soon as it is computed, and consequently there is now only one computation for each **hermite[k, z]**, for $k \leq n$, in the evaluation of **hermite[n, z]** (see Exercise 6b in Section 3.3.7). When *Mathematica* needs a value for a specific **hermite[k, z]**, it simply looks up the appropriate definition stored in the global rule base in a manner described in Section 3.3.2. Evaluating **hermite[20, z]** with dynamic programming, for example, only requires a total of 39 instead of 21,891 calls (see Exercise 7c in Section 3.3.7) to the function **hermite** and takes 1.5 instead of 305 seconds on a Macintosh IIci and 0.17 instead of 42 seconds on a Gateway 2000 4DX2-66V.

```
In[67] := Timing[hermite[20, z];]

          {1.46667 Second, Null}
```

The following table shows the number of calls to each **hermite[k, z]**, for $0 \leq k \leq 20$, in the evaluation of **hermite[20, z]** (see Exercise 7c in Section 3.3.7):

k	Number of Function Calls
0	1
1	2
2	2
3	2
4	2
5	2
6	2
7	2
8	2
9	2
10	2
11	2
12	2
13	2
14	2
15	2
16	2
17	2
18	2
19	1
20	1

The ? operator reveals why there is such a reduction of function calls.

```
In[68] := ClearAll[hermite]
          hermite[n_Integer /; n > 1, z_] := hermite[n, z] = Expand @
                  (2 z hermite[n - 1, z] - 2 (n - 1) hermite[n - 2, z])
          hermite[0, z_] := 1
          hermite[1, z_] := 2 z
```

```
In[72] := ?hermite

          Global`hermite
          hermite[n_Integer /; n > 1, z_] :=
            hermite[n, z] =
             Expand[2*z*hermite[n - 1, z] - 2*(n - 1)*hermite[n - 2, z]]

          hermite[0, z_] := 1

          hermite[1, z_] := 2*z
```

Before an evaluation of **hermite**[n, z], there are just three definitions for **hermite** in the global rule base that *Mathematica* searches during computations.

```
In[73] := hermite[8, z]
```

$$Out[73] = 1680 - 13440\ z^2 + 13440\ z^4 - 3584\ z^6 + 256\ z^8$$

```
In[74] := ?hermite

          Global`hermite
          hermite[2, z] = -2 + 4*z^2

          hermite[3, z] = -12*z + 8*z^3

          hermite[4, z] = 12 - 48*z^2 + 16*z^4

          hermite[5, z] = 120*z - 160*z^3 + 32*z^5

          hermite[6, z] = -120 + 720*z^2 - 480*z^4 + 64*z^6

          hermite[7, z] = -1680*z + 3360*z^3 - 1344*z^5 + 128*z^7

          hermite[8, z] = 1680 - 13440*z^2 + 13440*z^4 - 3584*z^6 +
                          256*z^8

          hermite[n_Integer /; n > 1, z_] :=
            hermite[n, z] =
             Expand[2*z*hermite[n - 1, z] - 2*(n - 1)*hermite[n - 2, z]]

          hermite[0, z_] := 1

          hermite[1, z_] := 2*z
```

After an evaluation of, for instance, **hermite**[8, z], there are seven additional rules in the global rule base, one for each **hermite**[k, z], with k ranging from 2 to 8. Evaluating **hermite**[9, z] now requires only 3 (instead of 109) function calls: **hermite**[9, z], `hermite[8, z]`, and `hermite[7, z]`, since the rules for the latter two are already stored in the global rule base and no recomputations are necessary. (For further discussion of dynamic programming, see [GKW93, Gra94].)

In[75] := **ClearAll[hermite]**

3.3.4 Functional Iterations

We sometimes wish to apply a function repeatedly to an expression a specified number of times or until a criterion is satisfied. Several built-in functions are available for this purpose:

Nest[*f*, *expr*, *n*]	give the result of applying *f* nested *n* times to *expr*
NestList[*f*, *expr*, *n*]	return a list of the results of applying *f* nested 0 through *n* times to *expr*; that is, generate the list {*expr*, *f*[*expr*], *f*[*f*[*expr*]], ...} with *n* + 1 elements
FixedPoint[*f*, *expr*]	start with *expr*, then apply *f* repeatedly until the result no longer changes
FixedPoint[*f*, *expr*, *n*]	stop after at most *n* steps; the default value for *n* is 65,536
FixedPoint[*f*, *expr*, **SameTest** -> *comp*]	stop when the function *comp* applied to two successive results yields **True**; the default setting is **SameTest -> SameQ**
FixedPointList[*f*, *expr*]	generate a list giving the results of applying *f* repeatedly, starting with *expr*, until the results no longer change; that is, give the list {*expr*, *f*[*expr*], *f*[*f*[*expr*]], ...}, stopping when the elements no longer change
FixedPointList[*f*, *expr*, *n*]	stop after at most *n* steps, that is, the list having at most *n* + 1 elements
FixedPointList[*f*, *expr*, **SameTest** -> *comp*]	terminate the list when the function *comp* applied to two successive elements yields **True**; the default setting is **SameTest -> SameQ**

For an example, we generate a geometric progression to nine terms:

In[76] := **Plus @@ NestList[r #&, a, 8]**

Out[76] = $a + a\,r + a\,r^2 + a\,r^3 + a\,r^4 + a\,r^5 + a\,r^6 + a\,r^7 + a\,r^8$

With **FixedPoint**, we can implement the Newton-Raphson algorithm for evaluating square roots. To determine the square root of a number *r*, start with an initial guess x_0 and obtain successively better approximations with the formula

$$x_{n+1} = (x_n + r/x_n)/2$$

Starting with $x_0 = 3.0$, we compute, for example, the square root of 11:

```
In[77] := r = 11;
```

```
In[78] := FixedPoint[(# + r/#)/2&, 3.0]
```

```
Out[78] = 3.31662
```

Let us verify that the computed value is indeed correct:

```
In[79] := % == Sqrt[11.0]
```

```
Out[79] = True
```

```
In[80] := ClearAll[r]
```

To further illustrate the use of these built-in functions, we introduce the logistic map, a simple population growth model exhibiting chaotic behavior.

EXAMPLE 3.3.2   Consider the logistic map given by the one-dimensional difference equation

$$x_{n+1} = f(x_n)$$

where $n = 0, 1, \ldots$ and

$$f(x) = \mu x(1 - x)$$

with $0 \le x \le 1$ and $0 \le \mu \le 4$. (a) Show that the points $x = 0.6$ and $x = 0$ are stable fixed points when the map parameter μ assumes the values 2.5 and 0.24, respectively. (b) Plot on the same graph x_n versus n with $\mu = 3.64$ and for two slightly different initial points $x_0 = 0.500$ and 0.501.

(a) A point x satisfying the condition

$$x = f(x)$$

is called a fixed point. A fixed point is stable, or attracting, if upon repeated application of the map f all points converge to it.

Consider the case $\mu = 2.5$ and let mu be the alias of μ:

```
In[81] := mu = 2.5;
```

The point $x = 0.6$ is a fixed point of the map

```
In[82] := f[x_] := mu x (1 - x)
```

because

```
In[83] := 0.6 == f[0.6]
```

```
Out[83] = True
```

To show that $x = 0.6$ is a stable or attracting fixed point, we apply f repeatedly to a number of randomly selected initial points:

```
In[84] := Table[FixedPoint[f, Random[]], {10}]
```

```
Out[84] = {0.6, 0.6, 0.6, 0.6, 0.6, 0.6, 0.6, 0.6, 0.6, 0.6}
```

The arbitrary starting points all converge to $x = 0.6$. Note that the points $x = 0$ and 1 are exceptions; they do not converge to $x = 0.6$ because $f(0) = f(1) = 0$.

Now consider the case $\mu = 0.24$:

```
In[85] := mu = 0.24;
```

The point x = 0 is a fixed point since

```
In[86] := 0 == f[0]
```

```
Out[86] = True
```

Let us apply f repeatedly to an arbitrary initial point, say, $x_0 = 0.612$:

```
In[87] := FixedPoint[f, 0.612]
```

```
Out[87] = $Aborted
```

The computation was manually aborted because *Mathematica* kept running for a long time without returning any result. **FixedPointList** reveals the problem.

```
In[88] := FixedPointList[f, 0.612, 39]
```

```
Out[88] = {0.612,   0.0569894,   0.012898,   0.00305559,   0.000731101,
           0.000175336,   0.0000420733,   0.0000100972,   2.42329 10^{-6},
           5.81589 10^{-7},   1.39581 10^{-7},   3.34995 10^{-8},   8.03988 10^{-9},
           1.92957 10^{-9},   4.63097 10^{-10},   1.11143 10^{-10},   2.66744 10^{-11},
           6.40186 10^{-12},   1.53645 10^{-12},   3.68747 10^{-13},   8.84993 10^{-14},
           2.12398 10^{-14},   5.09756 10^{-15},   1.22341 10^{-15},   2.93619 10^{-16},
           7.04686 10^{-17},   1.69125 10^{-17},   4.05899 10^{-18},   9.74158 10^{-19},
           2.33798 10^{-19},   5.61115 10^{-20},   1.34668 10^{-20},   3.23202 10^{-21},
           7.75686 10^{-22},   1.86165 10^{-22},   4.46795 10^{-23},   1.07231 10^{-23},
           2.57354 10^{-24},   6.17649 10^{-25},   1.48236 10^{-25}}
```

FixedPoint and **FixedPointList** stop when the comparison function *comp* in the option specification **SameTest -> *comp*** yields **True** upon applying to two successive results; the default option setting is **SameTest -> SameQ**.

```
In[89] := SameQ[%[[-1]], %[[-2]]]
```

```
Out[89] = False
```

where the negative indices -1 and -2 refer, respectively, to the first and second elements from the end of the list. The problem is that **SameQ** returns **False** even though the two numbers being compared are practically zero. We can specify a comparison function less stringent than **SameQ**.

```
In[90] := Table[Chop[
            FixedPoint[f, Random[],
                SameTest -> (Abs[#1 - #2] < 10^-11 &)
                ], 10^-10
            ], {10}
        ]
```

Out[90] = {0, 0, 0, 0, 0, 0, 0, 0, 0, 0}

The pure function `Abs[#1 - #2] < 10^-11 &` returns `True` when the absolute value of the difference between two successive iterates is less than 10^{-11}. The point $x = 0$ is indeed a stable fixed point, as randomly selected initial points all converge to it upon repeated application of the map f. With the function `Chop`, we have replaced small real numbers by zeros. `Chop[`*expr*, *tol*`]` replaces in *expr* the approximate real numbers that differ from zero by less than *tol* with the exact integer 0.

(b) `MultipleListPlot[`*list1*, *list2*, `...]` is defined in the package with the name `Graphics`MultipleListPlot``, which plots the lists on the same graph. To use the function, we must first load the package:

```
In[91] := Needs["Graphics`MultipleListPlot`"]
```

Consider the case with $\mu = 3.64$ and two slightly different initial points $x_0 = 0.500$ and 0.501:

```
In[92] := mu = 3.64;
```

```
In[93] := MultipleListPlot[
            Transpose[
                {Range[0, 44], NestList[f, 0.500, 44]}
                ],
            Transpose[
                {Range[0, 44], NestList[f, 0.501, 44]}
                ],
            PlotJoined -> True, PlotRange -> {0, 1},
            AxesLabel -> {"n", "xn"}
                ];
```

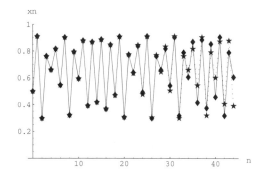

While `NestList[f, x0, n]` returns the list {$x0$, $f[x0]$, $f[f[x0]]$, ...} with $n + 1$ elements, `Transpose[{Range[0, n], NestList[f, x0, n]}]` gives the list {{0, $x0$}, {1, $f[x0]$}, {2, $f[f[x0]]$}, ...} with the same number of elements. Let us zoom in on the portion of the graph for n ranging from 30 to 44:

In[94] := `Show[%, PlotRange -> {{30, 44}, {0, 1}}];`

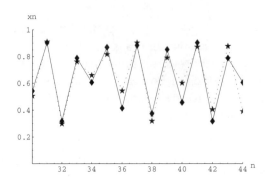

There is significant divergence between the two curves. The values of x_{44} for $x_0 = 0.500$ and 0.501 are

In[95] := `Nest[f, 0.500, 44]`

Out[95] = 0.605808

In[96] := `Nest[f, 0.501, 44]`

Out[96] = 0.390978

A 0.2% separation between the initial points leads to a 35.5% spread of the iterates after only 44 iterations! Sensitive dependence of the iterates x_n on the initial condition is a characteristic of chaotic behavior.

In[97] := `ClearAll[f, mu]`   ■

3.3.5 *Protection*

As discussed in Section 3.2.5, we can assign general properties, called attributes, to functions. `SetAttributes[f, attr]` adds *attr* to the attributes of function *f* and `ClearAttributes[f, attr]` removes *attr* from the attributes of *f*.

Many built-in *Mathematica* functions have the attribute `Protected`, which prevents us from inadvertently giving any definitions for these functions. For example, consider the function `D` for computing partial derivatives. `Attributes[D]` returns a list of its attributes:

In[98] := `Attributes[D]`

Out[98] = {Protected, ReadProtected}

As D has the attribute `Protected`, giving it any definition, be it consistent or inconsistent with built-in rules, triggers an error message:

In[99] := **D[u_ + v_, x_] := D[u, x] + D[v, x]**

       SetDelayed::write: Tag D in D[(u_) + (v_), x_] is Protected.

Out[99] = $Failed

Mathematica, however, allows us to specify rules for built-in functions provided that we first remove the protection. To clear `Protected`, use either `ClearAttributes` or `Unprotect`. `Unprotect[s1, s2, ...]` removes the attribute `Protected` for the symbols *si*.

Section 3.3.6 will show that there are occasions when altering the default rule base for built-in functions is appropriate. We, however, modify built-in functions at our peril because we may add new rules that are mathematically invalid. Furthermore, we may override built-in rules with erroneous ones as *Mathematica* applies user-defined rules before built-in ones. Let us, for example, evaluate

In[100] := **D[x Sin[x], x]**

Out[100] = x Cos[x] + Sin[x]

Now unprotect D,

In[101] := **Unprotect[D];**

and enter the definition

In[102] := **D[u_ v_, x_] := D[v, x] - v D[u, x]**

After giving the rule, we should restore the protection from accidental definitions with either `SetAttributes` or `Protect`. `Protect[s1, s2, ...]` sets the attribute `Protected` for the symbols *si*.

In[103] := **Protect[D];**

The erroneous rule that we just gave for D overrides the built-in product rule and can lead to disastrous results. Let us reevaluate the preceding derivative:

In[104] := **D[x Sin[x], x]**

Out[104] = Cos[x] - Sin[x]

The answer is no longer $x \cos x + \sin x$; it is now mathematically incorrect!

We can assign the attribute `Protected` to user-defined functions. For example,

In[105] := **myfunc[x_] := x^2 + 1**

In[106] := **Protect[myfunc];**

In[107] := **Attributes[myfunc]**

Out[107] = {Protected}

Giving another definition for `myfunc` triggers an error message:

In[108] := `myfunc[x_, y_] := x^2 + y^2`

>   SetDelayed::write: Tag myfunc in myfunc[x_, y_] is Protected.

Out[108] = `$Failed`

After unprotecting `myfunc`, we may give the definition that was rejected:

In[109] := `Unprotect[myfunc];`

In[110] := `myfunc[x_, y_] := x^2 + y^2`

To verify that _Mathematica_ recognizes this definition, let us call `myfunc` with two arguments:

In[111] := `myfunc[a, b]`

Out[111] = $a^2 + b^2$

In[112] := `ClearAll[myfunc]`

3.3.6   *Upvalues and Downvalues*

Familiar to us are definitions of the forms

> *f*[*args*] = *rhs*
>
> *f*[*args*] := *rhs*

These definitions are associated with the symbol *f* and are called downvalues for *f*. As examples,

In[113] := `f[0] = 1;`
            `f[x_, y_] := x^2 + y^3`
            `f[g[x_], h[y_]] := p[x, y]`

define downvalues for `f`.

Definitions of the forms

> *f/*: *g*[..., *f*, ...] = *rhs*
>
> *f/*: *g*[..., *f*, ...] := *rhs*
>
> *f/*: *g*[..., *f*[*args*], ...] = *rhs*
>
> *f/*: *g*[..., *f*[*args*], ...] := *rhs*

are also associated with the symbol *f*, but they are called upvalues for *f*. As examples,

In[116] := `f/: Re[f] = 0;`
            `f/: Log[f[x_]] := q[x]`
            `f/: g[f[x_], h[y_]] := w[x + y]`

define upvalues for `f`.

Upvalues for *f* can be given in compact forms

g [*f*] ^= *rhs*
g [*f*] ^:= *rhs*
g [*f*[*args*]] ^= *rhs*
g [*f*[*args*]] ^:= *rhs*

The first two definitions in the previous group of examples can be expressed as

```
Re[f]  ^= 0;
Log[f[x_]]  ^:= q[x]
```

For functions with more than one argument,

g [*arg1*, *arg2*, ...] ^= *rhs*
g [*arg1*, *arg2*, ...] ^:= *rhs*

define upvalues for the heads of all *argi*. The third definition in the earlier examples can be given as

```
g[f[x_], h[y_]]  ^:= w[x + y]
```

Whereas the earlier definition is associated only with **f**, this definition is associated with both **f** and **h**.

To illustrate the use of downvalues and upvalues, consider adding new rules for the built-in function **Abs**. **Abs**[*z*] gives the absolute value of the real or complex number *z*. If *expr* is not a number, *Mathematica* does not always simplify **Abs**[*expr*]. For instance,

```
In[119] := ClearAll[a, b]
```

```
In[120] := Abs[a + I b]
```

```
Out[120] = Abs[a + I b]
```

where **a** and **b** are mathematical variables, that is, variables without assigned values. *Mathematica* does not transform $|a + ib|$ to $\sqrt{a^2 + b^2}$. We can add new rules for simplifying **Abs**[*expr*] when *expr* consists of numbers and variables some of which are assumed real and either positive or negative. To declare variable **a** to be real and positive, use the definitions

```
In[121] := Im[a]  ^= 0;
           Positive[a]  ^= True;
```

which define upvalues for **a**. Similarly, to declare variable **b** to be real and negative, use the definitions

```
In[123] := Im[b]  ^= 0;
           Negative[b]  ^= True;
```

which define upvalues for **b**. Before specifying new rules for **Abs**, we must first unprotect the symbol since it has the attribute **Protected**:

In[125] := **Unprotect[Abs];**

For simplifying **Abs**[*expr*] where *expr* is a product or quotient of numbers, variables, and expressions of the forms u^x and $x + yi$ in which u may be imaginary and x and y are real, add the definitions

```
In[126] := Abs[x_ /; (Im[x] == 0 && Positive[x])] := x
           Abs[x_ /; (Im[x] == 0 && Negative[x])] := -x
           Abs[u_^x_ /; Im[x] == 0] := Abs[u]^x
           Abs[u_ v_] := Abs[u] Abs[v]
           Abs[(x_ + y_. Complex[0, n_]) /;
             (MatchQ[Im /@ Level[{x, y}, {-1}], {(0)..}] &&
              FreeQ[{x, y},
                    w_^z_ /; (!IntegerQ[z] && !(Positive[w] === True))
                   ]
             )
            ] := Sqrt[x^2 + (n y)^2]
```

which define downvalues for **Abs**. The last rule is valid only if x and y are real, as **Complex[0, n_]** stands for any pure imaginary number *ni*. The two conditions restricting the pattern **(x_ + y_. Complex[0, n_])** ensure that x and y are indeed real. **Level[{x, y}, {-1}]** gives a list of the atoms in **{x, y}**, that is, in x and y. The condition **MatchQ[Im /@ Level[{x, y}, {-1}], {(0)..}]** requires that these atoms are all real, because the pattern **{(0)..}** represents a list of one or more zeros. The second condition with **FreeQ** prohibits combinations of these real atoms from becoming imaginary. **FreeQ**[*expr*, *form*] yields **True** if no subexpression in *expr* matches *form*, and it yields **False** otherwise. The condition **FreeQ[{x, y}, w_^z_ /; (!IntegerQ[z] && !(Positive[w] === True))]** insists that, for an expression to match the pattern **(x_ + y_. Complex[0, n_])**, x and y must not contain any subexpression of the form w^z unless z is an integer or w is a positive. To avoid unintended definitions, we should restore the protection for **Abs**:

In[131] := **Protect[Abs];**

Let us apply these definitions to several examples:

In[132] := **Abs[b]**

Out[132] = -b

In[133] := **Abs[(3^(3/2))/(a^(2/3)) + (a^2/b^3) I]**

$$Out[133] = \text{Sqrt}\left[\frac{27}{a^{4/3}} + \frac{a^4}{b^6}\right]$$

In[134] := **Abs[((2/3) a - 3 b^2 a^(-7) I) b u^2/v^(2/3)]**

$$Out[134] = -\left(\frac{b \ \text{Sqrt}\left[\frac{4\,a^2}{9} + \frac{9\,b^4}{a^{14}}\right] \ \text{Abs}[u]^2}{\text{Abs}[v]^{2/3}}\right)$$

In[135] := `Abs[(a^3 - (b^2/a^4) I)(u + a)^2/((2 + 3 a I)^5 v^3)]`

$$Out[135] = \frac{\text{Sqrt}[a^6 + \frac{b^4}{a^8}] \; \text{Abs}[a + u]^2}{(4 + 9 \; a^2)^{5/2} \; \text{Abs}[v]^3}$$

Despite the negative signs in two results, they are positive since **b** was declared negative. (See Exercise 14 of Section 3.3.7 for simplifying **Abs** *[expr]* with the built-in function **ComplexExpand**.)

For a class of objects of a particular type, we often have to decide whether to define special arithmetic operators for them or augment the definitions of built-in operators in order to extend their domains to include these objects. Furthermore, we need to choose between giving the definitions as upvalues or downvalues. Consider, for example, three-dimensional vectors. *Mathematica* usually represents them as three-element lists. We can represent them as the objects **vector** *[Ax, Ay, Az]* where the arguments are the Cartesian components of a vector. Rather than defining a special function, say **vectorPlus**, for adding vectors, we may find it more convenient to give a new definition to **Plus** for vector addition. The rule can be specified as a downvalue for **Plus** as it is the head of the expression **vector[Ax_, Ay_, Az_] + vector[Bx_, By_, Bz_]**:

In[136] := `Head[vector[Ax_, Ay_, Az_] + vector[Bx_, By_, Bz_]]`

Out[136] = `Plus`

Since *Mathematica* tries user-defined rules before built-in ones, the new rule will be tried in every invocation of **Plus**, even when the computations do not involve vectors, and consequently will slow *Mathematica* down. For efficiency, we should give a rule for addition as an upvalue for **vector**:

In[137] := `vector[Ax_, Ay_, Az_] + vector[Bx_, By_, Bz_] ^:=`
                      `vector[Ax + Bx, Ay + By, Az + Bz]`

This rule is associated with **vector** rather than **Plus**. Let us add, for example, three vectors:

In[138] := `vector[a1, a2, a3] + vector[b1, b2, b3] + vector[c1, c2, c3]`

Out[138] = `vector[a1 + b1 + c1, a2 + b2 + c2, a3 + b3 + c3]`

In[139] := `ClearAll[a, b, f, h, vector]`

3.3.7 Exercises

1. Write named functions that perform the operations specified by the pure functions:

   (a) `(1 + #^3)&`

   (b) `(1/#1^#2)&`

   (c) `{#1, #2^#3}&`

   (d) `(-I D[#, x])&`

   (e) `((#^3)& /@ #)&`

   (f) `Function[x, Apply[And, Map[OddQ, x]]]`

   (g) `Function[x, Delete[x, Random[Integer, {1, Length[x]}]]]`

2. Define pure functions corresponding to the named functions:
   (a) `f[x_] := 1/(1 + x)`
   (b) `g[x_, y_] := (x + y)^2`
   (c) `mytest[expr_] := FreeQ[expr, Integer]`
   (d) `ourtest[expr_] := Length[expr] >= 2`
   (e) `test1[x_, y_] := x > y`

3. Use `Cases` and a pure function to obtain from a list those elements that are integers greater than 3.

4. Use `Select` and a pure function to pick out from a list of pairs those pairs in which the sum of the elements is smaller than 5. *Hint:* Use also `Plus` and `Apply`.

*5. Write a function for tabulating the number of times each distinct element appears in a list. For example, `tabulate[{b, b, 1, a, a, a, Sin[x], Sin[x]}]` should return

1	1
a	3
b	2
Sin[x]	2

Hint: Use `Union`, `Count`, `Map`, `Transpose`, and `TableForm`.

6. `Trace[expr]` generates a list of all expressions used in the evaluation of *expr*. With the function `Trace`, describe how *Mathematica* evaluates the Hermite polynomials `hermite[n, z]` discussed in Section 3.3.3 when (a) dynamic programming is not used and (b) dynamic programming is used. Consider only the cases with $n = 2, 3, 4,$ and 5.

7. Whereas `Trace[expr]`, introduced in Problem 6, generates a list of all expressions used in the evaluation of *expr*, `Trace[expr, form]` includes only expressions that match the pattern *form*. (a) Write a function `totalCalls[n_, z_]` that returns the total number of function calls to `hermite` in the evaluation of `hermite[n, z]` when dynamic programming is not used. Evaluate `totalCalls[20, z]`. (b) Write a function `partialCalls[n_, k_, z_]` that returns the number of function calls to `hermite[k, z]` in the evaluation of `hermite[n, z]`, with $k \leq n$, when dynamic programming is not used. Evaluate `partialCalls[20, 3, z]`. (c) Repeat parts (a) and (b) concerning the evaluation of `hermite[n, z]` but for the cases when dynamic programming is used.

8. Using the function `partialCalls` defined in Problem 7b in the forms `Table[partialCalls[20, k, z], {k, 0, 20}]` or `Do[Print[partialCalls[20, k, z]], {k, 0, 20}]` for the task of listing the number of function calls to `hermite[k, z]` in the evaluation of `hermite[20, z]` with k ranging from 0 to 20 is rather expensive

because there is a fresh evaluation of `Trace[hermite[20, z], hermite[k, z]]` for each k. Find a frugal method for accomplishing this task. *Hint:* With `FullForm`, examine the internal representation of the result of evaluating `Trace[hermite[20, z], hermite[k, z]]` when k, say, is 10. Instead of `FullForm`, `InputForm` may also be used.

9. `Nest`, `NestList`, `FixedPoint`, and `FixedPointList` apply repeatedly a function of one argument to an expression. `Fold` and `FoldList` work with functions of two arguments. `FoldList[f, x, {a, b, ...}]` gives `{x, f[x, a], f[f[x, a], b], ...}` and `Fold[f, x, list]` yields the last element of `FoldList[f, x, list]`. Without using *Mathematica*, determine (a) `FoldList[Plus, 0, list]`, (b) `FoldList[Times, 1, list]`, and (c) `Fold[10 #1 + #2 &, 0, digits]`, where *list* and *digits* stand for an arbitrary list and any list of integers from 0 to 9, respectively. Verify your conclusions with *Mathematica*.

10. Implement the Newton-Raphson method for finding roots (more precisely, zeros) of functions. To determine a root of a function $f(x)$, start with an approximate value x_0 and obtain successively better approximations with the formula

$$x_{i+1} = x_i - f(x_i)/f'(x_i)$$

Determine the first five positive roots of the Bessel function of the first kind $J_0(x)$.

11. Consider the tent map given by the one-dimensional difference equation

$$x_{n+1} = f(x_n)$$

where $n = 0, 1, \ldots$ and

$$f(x) = \mu\left(1 - 2\left|x - \frac{1}{2}\right|\right) = 2\mu\begin{cases} x, & \text{if } 0 \leq x \leq \frac{1}{2} \\ 1 - x, & \text{if } \frac{1}{2} \leq x \leq 1 \end{cases}$$

with $0 \leq \mu \leq 1$. (a) Verify that the point $x = 0$ is a stable fixed point when the map parameter μ is less than $1/2$. (b) Show that there are two fixed points when the map parameter μ is greater than $1/2$ and that neither point is stable or attracting.

12. Using only built-in functions and pure functions, write a function that takes as its arguments a list

```
{a1, a2, ..., an}
```

as well as a nested list

```
{{{b11, c11}, {b12, c12}, ..., {b1m, c1m}},
 {{b21, c21}, {b22, c22}, ..., {b2m, c2m}},
 ...,
 {{bn1, cn1}, {bn2, cn2}, ..., {bnm, cnm}}}
```

and returns the list

```
{{a1, {b11, c11}, {b12, c12}, ..., {b1m, c1m}},
 {a2, {b21, c21}, {b22, c22}, ..., {b2m, c2m}},
 ...,
 {an, {bn1, cn1}, {bn2, cn2}, ..., {bnm, cnm}}}
```

for any positive integers *n* and *m*. For example,

```
list1 = Table[a[i], {i, 1, 3}]
```

```
{a[1], a[2], a[3]}
```

```
list2 = Table[{b[i, j], c[i, j]}, {i, 1, 3}, {j, 1, 2}]
```

```
{{{b[1, 1], c[1, 1]}, {b[1, 2], c[1, 2]}},
  {{b[2, 1], c[2, 1]}, {b[2, 2], c[2, 2]}},
  {{b[3, 1], c[3, 1]}, {b[3, 2], c[3, 2]}}}
```

```
func[list1, list2]
```

```
{{a[1], {b[1, 1], c[1, 1]}, {b[1, 2], c[1, 2]}},
  {a[2], {b[2, 1], c[2, 1]}, {b[2, 2], c[2, 2]}},
  {a[3], {b[3, 1], c[3, 1]}, {b[3, 2], c[3, 2]}}}
```

13. Show that

$$\nabla \times (\mathbf{A} \times \mathbf{B}) = (\mathbf{B} \cdot \nabla)\mathbf{A} - (\mathbf{A} \cdot \nabla)\mathbf{B} + \mathbf{A}(\nabla \cdot \mathbf{B}) - \mathbf{B}(\nabla \cdot \mathbf{A})$$

where **A** and **B** are any two vector point functions.

14. **(a)** Add rules for the built-in function `Sqrt` to simplify

```
Sqrt[x^2]
```

when `x` is declared real and either positive or negative.

   (b) `ComplexExpand[`*expr*`]` expands *expr*, assuming that all variables are real. `ComplexExpand[`*expr*`, {`*x1, x2, ...*`}]` expands *expr* assuming that variables matching any of the *xi* are complex. The option `TargetFunctions` can be given as a list of functions from the set {`Re, Im, Abs, Arg, Conjugate, Sign`}. `ComplexExpand` will try to give results in terms of functions specified. Let `a` be declared real and positive and `b` be declared real and negative. Simplify the expressions

```
Abs[b]
Abs[(3^(3/2))/(a^(2/3)) + (a^2/b^3) I]
Abs[((2/3) a - 3 b^2 a^(-7) I) b u^2/v^(2/3)]
Abs[(a^3 - (b^2/a^4) I)(u + a)^2/((2 + 3 a I)^5 v^3)]
```

where `u` and `v` are complex variables. Compare the results with those in Section 3.3.6 and explain any discrepancies. *Hint:* With a specification for the option `TargetFunctions`, apply `ComplexExpand` to `Abs[`*expr*`]`.

❀ **3.0 15.** Make the assignment

```
{v₀, θ₀} = {55 m/s, 30°};
```

Applying the function `Clear` to v_0 and θ_0 generates a couple of error messages:

```
Clear[v₀, θ₀]
Clear::ssym : v₀ is not a symbol or a string.
Clear::ssym : θ₀ is not a symbol or a string.
```

How can we clear the values for v_0 and θ_0 without using the `Unset` operator "`=.`"? How are these values stored in the global rule base? *Hint:* Use the function `FullForm` and the operator "`?`".

3.4 Procedures

A procedure, or compound expression, is a sequence of expressions separated by semicolons. *Mathematica* evaluates the expressions consecutively and returns the result from evaluating the last expression. For instance,

In[1] := `r = 3; s = Sqrt[4] + r; r^2 + s^2`

Out[1] = 34

If a procedure ends with a semicolon, *Mathematica* returns `Null`, which is a symbol representing "nothing" and is not printed as an output. For an example, define the function `func` to be a procedure ending with a semicolon:

In[2] := `func[y_] := func[y] = (Print["What a wonderful world!"];`
            `Plot[Sin[x], {x, 0, 2 Pi}];`
            `x = y;)`

Note that the parentheses are necessary for delimiting the body of the function and that the syntax for the definition allows the function to remember values that it finds, as explained in Section 3.3.3. Let us call the function:

In[3] := `func[6]`

What a wonderful world!

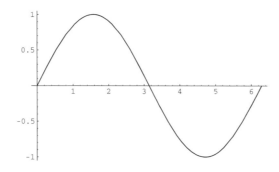

Mathematica displayed no output, that is, Out[*n*]. The value of func[6] is Null:

```
In[4] := ?func

    Global`func
    func[6] = Null

    func[y_] := func[y] =
        (Print["What a wonderful world!"]; Plot[Sin[x], {x, 0, 2*Pi}];
         x = y; )
```

If the last semicolon is removed from the definition of func, the value of func[6] would be 6. Printing an expression, generating a plot, and assigning value to a variable are just side effects. To retrieve the value of the last element in the compound expression, use the operator %*n*:

```
In[5] := %%^2 + 3

Out[5] = 39
```

Often, an unfortunate side effect of procedures is the introduction of global variables such as r, s, and x with assigned values:

```
In[6] := {r, s, x}

Out[6] = {3, 5, 6}
```

Section 3.4.1 will present a construct for avoiding global variables by declaring variables to be local to a particular procedure.

Mathematica normally evaluates the expressions in a procedure one after another. However, we can use conditional or iterative constructs to alter this sequential flow of control. Section 3.4.2 will examine conditionals for branching, and Section 3.4.3 will describe loops for iteration.

```
In[7] := ClearAll[func, r, s, x]
```

3.4.1   *Local Symbols*

Sections 2.2.4 and 2.2.13 warned against conflicts of global symbols, which have assigned values or function definitions, with those of the same names elsewhere in a *Mathematica* session and exalted the merit of clearing them as soon as they are no longer needed. To minimize the collisions of symbols, we can often use transformation rules and pure functions instead of temporary variables and auxiliary functions. *Mathematica* supports another construct for avoiding symbol collisions. Module[{$x1$, $x2$, $y1 = y1_0$, $x3$, $y2 = y2_0$, ...}, *body*] specifies that occurrences of the symbols xi, yi, ... in *body* should be treated as local and that yi should be assigned the initial value yi_0. Module declares symbols as local to the module by giving them new and unique names within its body.

As an example, let us create a global variable p with the value 10:

```
In[8] := p = 10

Out[8] = 10
```

This variable has no effects on evaluations within `Module` if `p` is specified as a local variable within the module:

```
In[9] := Module[{q, p}, Print[q]; q = p^2; Print[q, "\n"];
             p = 50; Print[p, "   ", q]]

    q$1

    p$1^2

    50    2500
```

Also, evaluations within `Module` have not affected the global variable `p`:

```
In[10] := p

Out[10] = 10
```

Furthermore, the value of the local variable `q` is not visible outside the module:

```
In[11] := q

Out[11] = q
```

The local variables `p` and `q` within `Module` are given, respectively, the names `p$n` and `q$n` to distinguish them from the corresponding global variables and variables in other modules. Thus, local variables are completely independent of the global variables and variables in other modules with identical names. The positive integer n in *symbol$n* increases by 1 each time `Module` is called in a *Mathematica* session. Thus, names of local variables are always unique to a particular module.

```
In[12] := Module[{q, p}, Print[q]; q = p^2; Print[q, "\n"];
             p = 50; Print[p, "   ", q]]

    q$2

    p$2^2

    50    2500
```

Note that `p$1` and `q$1` have become `p$2` and `q$2`. (With *Mathematica* version 2.2, `p$3` and `q$3` have become `p$4` and `q$4`.)

A few words of caution are in order. Even if variables are specified as local to a module, their global values will be captured if they appear in the expressions of initial values:

```
In[13] := u = 5;

In[14] := Module[{u = 120, v = u^2 + 1}, Print[v]]

    26

In[15] := ClearAll[p, u]
```

3.4.2 Conditionals

A conditional function returns the value of one of several alternative expressions only if a condition is met. The conditional functions of *Mathematica* are `If`, `Which`, and `Switch`:

If [*condition*, *t*]	give *t* if *condition* evaluates to **True**, and **Null** if it evaluates to **False**
If [*condition*, *t*, *f*]	give *t* if *condition* evaluates to **True**, and *f* if it evaluates to **False**
If [*condition*, *t*, *f*, *u*]	give *u* if *condition* evaluates to neither **True** nor **False**
Which [*condition1*, *value1*, *condition2*, *value2*, ...]	evaluate each of the *conditioni* in turn, returning the value of the *valuei* corresponding to the first one that yields **True**
Switch [*expr*, *form1*, *value1*, *form2*, *value2*, ...]	evaluate *expr*, then compare it with each of the *formi* in turn, evaluating and returning the *valuei* corresponding to the first match found

The relational and logical expressions described in Section 2.2.14 and the built-in predicates introduced in Sections 2.4.3 and 3.1.3.1 can serve as conditions for these conditional expressions.

Section 3.2.3.3 showed that we can put conditions on patterns, transformation rules, and definitions. The pattern object *pattern* /; *condition* stands for any expression matching *pattern* provided that the evaluation of *condition* yields **True**. The transformation rule *lhs* :> *rhs* /; *condition* and the definition *lhs* := *rhs* /; *condition* apply only if the evaluation of *condition* yields **True**.

To illustrate the use of conditional constructs, let us consider three examples.

EXAMPLE 3.4.1 Write several functions for determining the maximum of a list of numbers.

We begin by defining the function **ourMax** for finding the maximum of a list of numbers to be a procedure containing the conditional function **If**:

```
In[16] := ourMax[x_List] := Module[
                  {maxNum = x[[1]]},
                  Do[If[x[[i]] > maxNum, maxNum = x[[i]]],
                    {i, 2, Length[x]}];
                  maxNum
                                        ]
```

The function **Do** was introduced in Section 2.1.18 and will be explained further in Section 3.4.3.2.

To extract the maximum, we can also define the recursive function **myMax** with a multiclause definition using patterns:

```
In[17] := myMax[{x_, y_, z___}] /; x > y] := myMax[{x, z}]
          myMax[{x_, y_, z___}] /; x <= y] := myMax[{y, z}]
          myMax[{x_}] := x
```

Conditional constructs may not be necessary. Here is a simple function for determining the maximum of a list of numbers:

```
In[20] := newMax[x_List] := Last[Sort[x]]
```

Of course, we can just use the built-in function **Max**.

Let us apply these functions to a list of random integers from 0 to 100 and of random length from 2 to 15:

In[21] := `Table[Random[Integer, 100], {Random[Integer, {2, 15}]}]`

Out[21] = `{81, 76, 97, 56}`

In[22] := `{ourMax[%], myMax[%], newMax[%], Max[%]}`

Out[22] = `{97, 97, 97, 97}`

All these functions return the same result.

In[23] := `ClearAll[ourMax, myMax, newMax]`   ∎

EXAMPLE 3.4.2 Write a function for the Kronecker delta δ_{nm} defined by

$$\delta_{nm} = \begin{cases} 1 & n = m \\ 0 & n \neq m \end{cases}$$

where n and m are integers.

We can write a single-clause definition with the function `If`:

In[24] := `kDel[n_Integer, m_Integer] := If[n == m, 1, 0]`

Thus,

In[25] := `{kDel[1, 1], kDel[1, 0]}`

Out[25] = `{1, 0}`

We can also give a multiclause definition:

In[26] := `newkDel[n_Integer, m_Integer] := 1 /; n == m`
          `newkDel[n_Integer, m_Integer] := 0 /; n != m`

Therefore,

In[28] := `{newkDel[1, 1], newkDel[1, 0]}`

Out[28] = `{1, 0}`

Often multiclause definitions are easier to understand and modify than single-clause definitions.

In[29] := `ClearAll[kDel,newkDel]`   ∎

EXAMPLE 3.4.3 Write a function for the step potential

$$v(x) = \begin{cases} 0 & x < -3a \\ -2b & -3a \leq x < -a \\ -b & -a \leq x < a \\ -2b & a \leq x < 3a \\ 0 & 3a \leq x \end{cases}$$

Let $a = 1$ and $b = 1$.

The conditional function `Which` allows us to give a single-clause definition:

```
In[30] := v[x_] := Which[x < -3 || x >= 3, 0,
              (x >= -3 && x < -1)||(x >= 1 && x < 3), -2,
              x >= -1 && x < 1, -1]
```

In the definition, we have used the logical operators `&&` and `||`. Let us plot this function:

```
In[31] := Plot[v[x], {x, -5, 5}, PlotStyle -> Thickness[0.01],
              AxesLabel -> {" x", "v"}];
```

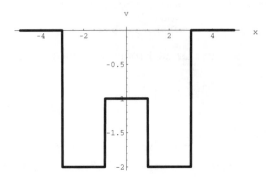

We can also give a multiclause definition:

```
In[32] := newv[x_ /; x < -3 || x >= 3] := 0
         newv[x_ /; (x >= -3 && x < -1)||(x >= 1 && x < 3)] := -2
         newv[x_ /; x >= -1 && x < 1] := -1
```

Plotting this function produces the same graph as before.

Whereas multiclause definitions are not differentiable, we can differentiate single-clause definitions involving conditional functions. For example,

```
In[35] := newv'[x]
```

```
Out[35] = newv'[x]
```

```
In[36] := v'[x]
```

```
Out[36] = Which[x < -3 || x >= 3, 0,
              x >= -3 && x < -1 || x >= 1 && x < 3, 0,
              x >= -1 && x < 1, 0]
```

We must exercise caution in using this feature of conditional functions because *Mathematica* may return flawed derivatives. For instance,

```
In[37] := v'[1]
```

```
Out[37] = 0
```

The result is wrong as the potential v is in fact not differentiable at 1.

```
In[38] := ClearAll[v, newv]
```

∎

If an evaluated condition yields neither `True` nor `False` in a call to the function `Which`, *Mathematica* returns all or part of the conditional expression:

```
In[39] := ClearAll[x, y]
```

```
In[40] := x = 2
```

```
Out[40] = 2
```

```
In[41] := Which[x < 1, 1, y < 1, 2, x > 1, 3, y > 1, 4, True, "Wonderful!"]
```

❀ **3.0**

```
Which[y < 1, 2, x > 1, 3, y > 1, 4, True, Wonderful!]
```

❀ **2.2**

```
Which[x < 1, 1, y < 1, 2, x > 1, 3, y > 1, 4, True, Wonderful!]
```

The evaluation of `y < 1` yielded neither `True` nor `False`.

```
In[42] := ClearAll[x]
```

When a function with more than one definition (i.e., with a multiclause definition) is called, *Mathematica*, in principle, starts with the most specific definition, proceeds from the more specific definitions to the more general ones, and uses the first definition that matches the call. As pointed out in Section 3.3.2, "Selecting a Definition," *Mathematica*'s notion of generality, however, is not always scrutable. Consider the example cited in that section:

```
In[43] := ClearAll[f]
          f[x_Integer] := x^2
          f[x_Integer?Positive] := 1/x
          f[x_Integer /; x > 50] := -x
```

What would *Mathematica* return for `f[60]`? Since the last definition is the most specific, we expect `-60` to be the value of `f[60]`.

```
In[47] := f[60]
```

```
Out[47] = 3600
```

Contrary to expectation, *Mathematica* used the first definition. The function `Switch` allows us to give a single-clause definition in which the patterns are always tested in the specified order:

```
In[48] := ClearAll[f]
          f[y_] := Switch[y, x_Integer /; x > 50, -y,
                             x_Integer?Positive, 1/y,
                             x_Integer, y^2]
```

```
In[50] := f[60]
```

```
Out[50] = -60
```

There is no ambiguity this time. Since 60 matches the first pattern, *Mathematica* returns -60 for f[60], as expected.

```
In[51] := ClearAll[f]
```

3.4.3 Loops

3.4.3.1 CHANGING VALUES OF VARIABLES

In using looping constructs, we often wish to change the values of variables. *Mathematica* provides special notations for such modifications:

x++	increase the value of x by 1, returning the old value of x
++x	increase the value of x by 1, returning the new value of x
x--	decrease the value of x by 1, returning the old value of x
--x	decrease the value of x by 1, returning the new value of x
x += dx	add dx to x and return the new value of x
x -= dx	subtract dx from x and return the new value of x
x *= c	multiply x by c and return the new value of x
x /= c	divide x by c and return the new value of x
{x, y} = {y, x}	interchange the values of x and y
PrependTo[s, *elem*]	prepend *elem* to the value of s, and reset s to the result
AppendTo[s, *elem*]	append *elem* to the value of s, and reset s to the result

For example, consider

```
In[52] := x = 3b
```

$Out[52] = 3 b$

```
In[53] := x *= 4 b^2
```

$Out[53] = 12 b^3$

```
In[54] := x
```

$Out[54] = 12 b^3$

Note that prior to the modification of a variable, it must already have an assigned value.

After increasing the value of x by 1, ++x returns the new value of x whereas x++ yields the old one. For instance,

```
In[55] := x = 3
```

$Out[55] = 3$

```
In[56] := ++x
```

$Out[56] = 4$

In[57] := **x**

Out[57] = 4

In[58] := **x++**

Out[58] = 4

In[59] := **x**

Out[59] = 5

The functions **PrependTo** and **AppendTo** modify lists. For an example, consider

In[60] := **s = {1, 2, 3, 4}**

Out[60] = {1, 2, 3, 4}

In[61] := **AppendTo[s, r]**

Out[61] = {1, 2, 3, 4, r}

In[62] := **s**

Out[62] = {1, 2, 3, 4, r}

In[63] := **ClearAll[x, s]**

3.4.3.2 DO, WHILE, and FOR

The functions **Do**, **While**, and **For** evaluate expressions or procedures repeatedly. The value returned by these functions is **Null**, which, as mentioned earlier, is simply a symbol indicating the absence of an expression or a result and is not printed as an output. These functions are used only for their side effects, such as assigning values to variables, generating graphics, and printing expressions.

The function **Do** evaluates an expression or a procedure repetitively a specified number of times. The iterative constructs are

Do[*expr*, {*imax*}]	evaluate *expr* *imax* times
Do[*expr*, {*i*, *imax*}]	evaluate *expr* with the variable *i* successively taking on the values 1 through *imax* (in steps of 1)
Do[*expr*, {*i*, *imin*, *imax*}]	start with *i* = *imin*
Do[*expr*, {*i*, *imin*, *imax*, *di*}]	use steps *di*
Do[*expr*, {*i*, *imin*, *imax*}, {*j*, *jmin*, *jmax*}, ...]	evaluate *expr* looping over different values of *j*, etc. for each *i*

Do has no output, as it always returns **Null**. For instance, consider

In[64] := **ClearAll[f]**
        f[x_] := x^2
        Do[f[i], {i, 3}]

To display `f[i]` with `i = 1, 2,` and `3,` we include the function `Print`:

```
In[67] := Do[Print[f[i]], {i, 3}]
```

```
1
4
9
```

The number of times `Do[expr, {i, imin, imax, di}]` evaluates *expr* is

$$\frac{imax - imin}{di} + 1$$

rounded off to the nearest smaller integer. Consider determining

$$\sum_{n=1}^{\infty} \frac{1}{n^k}$$

for $k = 2, 4, 6,$ and 8.

```
In[68] := Do[Print["k = ", k, "    ",
              Sum[1/(n^k), {n, 1, Infinity}]//N
             ],
           {k, 2, 8, 2}
          ]
```

```
k = 2      1.64493
k = 4      1.08232
k = 6      1.01734
k = 8      1.00408
```

The number of iterations is $(8-2)/2 + 1 = 4$. We can obtain the same result with, for example, the iterator `{k, 2, 9.3, 2}`.

Instead of constructing a `Do` loop, we can often use the function `Nest` discussed in Section 3.3.4, "Functional Iterations." Example 3.3.2 introduced the logistic map defined by the equation

$$x_{n+1} = f(x_n)$$

where $n = 0, 1, \ldots$ and

$$f(x) = \mu x(1 - x)$$

with $0 \leq x \leq 1$ and $0 \leq \mu \leq 4$. With a `Do` loop, we can determine x_{120} for the map parameter $\mu = 3.64$ and the initial point $x_0 = 0.500$:

```
In[69] := x = 0.500; Do[x = 3.64 x (1 - x), {120}]; x
```

```
Out[69] = 0.371626
```

We can also compute x_{120} with the function `Nest`:

```
In[70] := Nest[(3.64 # (1 - #))&, 0.500, 120]
```

```
Out[70] = 0.371626
```

Nest is perhaps more polished than **Do**.

In[71] := **ClearAll[f, x]**

The functions **While** and **For** evaluate expressions or procedures repeatedly as long as certain conditions are satisfied. The looping constructs are

While[*condition*, *body*]	evaluate *condition*, then *body*, repetitively, until *condition* first fails to give **True**
For[*start*, *condition*, *incr*, *body*]	execute *start*, then repeatedly evaluate *body* and *incr* until *condition* fails to give **True**

To illustrate the use of **While**, determine x such that $x = 1 + 1/x$:

In[72] := **x = 1.0;**
While[1 + 1/x =!= x, x = 1 + 1/x];
x

Out[74] = 1.61803

where **UnsameQ**[*lhs*, *rhs*], with the special input form *lhs* =!= *rhs*, yields **True** if the expression *lhs* is not identical to *rhs*, and yields **False** otherwise. The number $(1 + \sqrt{5})/2 \approx 1.61803$ is called the golden ratio. Using the function **FixedPoint** introduced in Section 3.3.4 to determine the golden ratio is perhaps more elegant than using the looping construct:

In[75] := **FixedPoint[(1 + 1/#)&, 1.0]**

Out[75] = 1.61803

Of course, this result agrees with the approximate numerical value of the built-in mathematical constant **GoldenRatio**:

In[76] := **N[GoldenRatio]**

Out[76] = 1.61803

In[77] := **ClearAll[x]**

As an example on using **For**, generate a list of the prime numbers less than 40:

In[78] := **For[{n, v} = {1, {}}, Prime[n] < 40, n++,**
AppendTo[v, Prime[n]]];
v

Out[79] = {2, 3, 5, 7, 11, 13, 17, 19, 23, 29, 31, 37}

Prime[*n*] gives the *n*th prime number, the special form *n++* increments the value of *n* by 1, and **AppendTo**[*v*, *elem*] is equivalent to $v = $ **Append**[*v*, *elem*].

In[80] := **ClearAll[n, v]**

After executing *start*, the order of evaluation in a **For** loop is *condition*, *body*, and *incr*. For instance,

```
In[81] := For[n = 1, n < 3, n++, Print[n]]
1
2
```

Here n is first set to 1. The test $n < 3$ gives `True`, 1 is printed, and n is increased to 2; $n < 3$ again yields `True`, 2 is printed, and n is increased to 3; $n < 3$ now gives `False` and the loop stops. A side effect is that n retains the value 3:

```
In[82] := n
```

```
Out[82] = 3
```

For both `While` and `For`, *body* may not be evaluated at all if *condition* yields `False` at the outset.

```
In[83] := For[n = 1, n < 1, n++, Print[n]]
```

Here n is first set to 1, $n < 1$ gives `False`, the loop terminates immediately, and n still has the value 1:

```
In[84] := n
```

```
Out[84] = 1
```

```
In[85] := ClearAll[n]
```

3.4.4 *Named Optional Arguments*

Mathematica functions allow two kinds of arguments: positional arguments and named optional arguments or, simply, options. Whereas the meanings of the positional arguments are determined by their positions in the argument sequence, those of the named optional arguments are specified by their names.

Built-in *Mathematica* plotting functions, for example, accept many options that, as we have seen, are expressed as transformation rules and can be entered in any order after the positional arguments in the argument sequences of the functions. To illustrate how to set up options in user-defined functions, consider a problem from introductory electrostatics: For points on a plane, write a function that returns the electric potential and electric field produced by a system of point charges confined to the plane.

At position $\mathbf{r}$, the electric potential due to a system of N point charges $q_1, q_2, \ldots, q_N$ at positions $\mathbf{r}_1, \mathbf{r}_2, \ldots, \mathbf{r}_N$, respectively, can be written as

$$\varphi(\mathbf{r}) = k \sum_{i=1}^{N} \frac{q_i}{|\mathbf{r} - \mathbf{r}_i|} \tag{3.4.1}$$

and the electric field is given by

$$\mathbf{E}(\mathbf{r}) = -\nabla \varphi(\mathbf{r}) \tag{3.4.2}$$

where k is the Coulomb constant. In the SI system of units, $k = 8.98755 \times 10^9$ volt meter/coulomb and the units of charge, electric potential, and electric field are coulomb, volt, and volt/meter, respectively; in the Gaussian system, $k = 1$ and the units of charge, electric potential, and electric field are statcoulomb (or esu), statvolt, and statvolt/centimeter, respectively. In a two-dimensional Cartesian coordinate system, Equations 3.4.1 and 3.4.2 imply

$$\varphi(x, y) = k \sum_{i=1}^{N} \frac{q_i}{\sqrt{(x - x_i)^2 + (y - y_i)^2}} \tag{3.4.3}$$

$$E_x = -\frac{\partial \varphi}{\partial x} \tag{3.4.4}$$

$$E_y = -\frac{\partial \varphi}{\partial y} \tag{3.4.5}$$

where E_x and E_y are the x and y components of the electric field, respectively.

For determining the electric potential and electric field at a point (x, y), the function electrostaticField takes two positional arguments. The first argument is for a list of lists, each of which has three elements: the charge, the x coordinate, and the y coordinate of a charge in the charge distribution producing the field. The pattern *patt* .. is a pattern repeated one or more times. The pattern q:{{_, _, _}..} stands for a list of trios, and this list is to be named q in the body of the function. The second argument is for a list of x and y coordinates of the point at which the electric potential and electric field are evaluated. The pattern *patt*:v or Optional[*patt*, v], introduced in Section 3.2.6, "Defaults," represents an expression that, if omitted, has default value v. The pattern Optional[p:{_, _}, {0, 0}] matches any two-element list to be named p on the *rhs* of the function definition; if we do not explicitly give a second positional argument to the function electrostaticField when it is called, p assumes the default value {0, 0}. That is, unless otherwise specified, the origin is taken as the point at which the electric potential and electric field are determined. With the inclusion of opts___Rule, options may be given as trailing arguments for specifying the system of units and the unit for the angle that the electric field makes counterclockwise from the $+x$ axis. The value for the option systemOfUnits can be either SI or Gaussian, and that for the option angle can be either radian or degree. Comments on the function are included in the body of the function.

With the function Options, we first name the options and specify their default values. Options[*f*] gives the list of default options assigned to *f*.

```
In[86] := Options[electrostaticField] =
                {systemOfUnits -> SI, angle -> radian};
```

Let us present the definition for the function and then analyze how options work.

```
electrostaticField[
      q:{{_, _, _}..}, Optional[p:{_, _}, {0, 0}], opts___Rule
            ] :=
Module[(* declare local variables *)
         {phi, x, y, ex, ey, e,
          electricPotential, electricField, theta,

          (* determine option values and assign them
             to local variables as initial values *)
```

```
      units =
      systemOfUnits/.{opts}/.Options[electrostaticField],
      dir = angle/.{opts}/.Options[electrostaticField]
      },

 (* from Equation 3.4.3; electric potential at a point
    (x, y) in units of (k*unit of charge/unit of length) *)
 phi = Sum[
      q[[i, 1]]/Sqrt[(x - q[[i, 2]])^2 + (y - q[[i, 3]])^2],
      {i, 1, Length[q]}
          ];

 (* from Equations 3.4.4 and 3.4.5; x and y components
    of the electric field at the point p in units
    of (k*unit of charge/unit of length^2) *)
 ex = N[-D[phi, x]/.{x -> p[[1]], y -> p[[2]]}];
 ey = N[-D[phi, y]/.{x -> p[[1]], y -> p[[2]]}];

 (* magnitude of the electric field at the point p
    in units of (k*unit of charge/unit of length^2) *)
 e = Sqrt[ex^2 + ey^2];

 (* the electric potential at the point p in units
    of (k*unit of charge/unit of length) *)
 phi = N[phi/.{x -> p[[1]], y -> p[[2]]}];

 (* determine the specified system of units, electric
    potential, and magnitude of the electric field *)
 Which[units === SI,
       electricPotential =
         ((phi 8.98755 10^9 volts)//ScientificForm);
       electricField =
         ((e 8.98755 10^9 volts/meter)//ScientificForm),
       units === Gaussian,
       electricPotential =
          ((phi statvolts)//ScientificForm);
       electricField =
          ((e statvolts/centimeter)//ScientificForm),
       True,
       Return["The value of the option systemOfUnits
 must be either SI or Gaussian."]
       ];

 (* print the values of the electric potential and
    magnitude of the electric field *)
 Print["electric potential = ", electricPotential,
 "\n\n"];
```

```
    Print["magnitude of electric field = ",
    electricField, "\n\n\n"];

    (* determine the direction that the electric field
       makes counterclockwise from the +x axis *)
    If[(ex == 0)&&(ey == 0), Return[],
       theta = ArcTan[Abs[ey/ex]]
      ];
    Which[(ex > 0)&&(ey >= 0)||(ex == 0)&&(ey > 0), Null,
          (ex < 0)&&(ey >= 0), theta = (Pi - theta),
          (ex <= 0)&&(ey < 0), theta = (Pi + theta),
          (ex > 0)&&(ey < 0), theta = (2Pi - theta),
          True, Return["Determination of angle failed!"]
         ];

    (* assign appropriate unit to the angle *)
    Which[dir === radian, theta = (N[theta] rad),
          dir === degree, theta = (N[theta/Degree] deg),
          True, Return["The value of the option angle
  must be either radian or degree."]
         ];

    (* print the value for the angle *)
    Print["angle of electric field from +x axis = ", theta];
       ]
```

The function `Return` appears at several places in the function definition. `Return[`*expr*`]` returns the value *expr* from a function and exits all procedures and loops in the function; `Return[]` returns the value `Null`, which is not printed, and exits the function.

3.0 We can embellish the previous definition with special characters and two-dimensional forms. To proceed safely and efficiently, paste a copy of the definition below, click the bracket of the new cell, choose **StandardForm** in the Convert To sub-menu of the Cell menu, and edit the result. An adorned definition for the function `electrostaticField` is

```
In[87] := electrostaticField[
            q : {{_, _, _}..}, Optional[p : {_, _}, {0, 0}], opts___Rule
                              ] :=

          Module[
           {(* declare local variables *)
            φ, x, y, E, ℰ, electricPotential, electricField, θ,

             (* determine option values and assign them
                to local variables as initial values *)
```

```
        units = systemOfUnits/.{opts}/.Options[electrostaticField],
        dir = angle/.{opts}/.Options[electrostaticField]
    },


    (* from Equation 3.4.3; electric potential at a point
        (x, y) in units of (k*unit of charge/unit of length) *)
```

$$\varphi = \sum_{i=1}^{Length[q]} \frac{q[\![i, 1]\!]}{\sqrt{(x - (q[\![i, 2]\!]))^2 + (y - (q[\![i, 3]\!]))^2}};$$

```
    (* from Equations 3.4.4 and 3.4.5; x and y components
        of the electric field at the point p in units
        of (k*unit of charge/unit of length^2) *)
    Eₓ = N[-∂ₓφ /. {x → (p[[1]]), y → (p[[2]])}];
    E_y = N[-∂_yφ /. {x → (p[[1]]), y → (p[[2]])}];


    (* magnitude of the electric field at the point p
        in units of (k*unit of charge/unit of length^2) *)
```

$$\mathcal{E} = \sqrt{E_x^2 + E_y^2};$$

```
    (* the electric potential at the point p in units
        of (k*unit of charge/unit of length) *)
    φ = N[φ/.{x → (p[[1]]), y → (p[[2]])}];


    (* determine the specified system of units, electric
        potential, and magnitude of the electric field *)
    Which[units === SI,
            electricPotential =
                ScientificForm[φ 8.98755 10⁹ volts];
```

$$electricField = ScientificForm\left[\frac{\mathcal{E}\,8.98755\,10^9\,volts}{meter}\right],$$

```
            units === Gaussian,
            electricPotential = ScientificForm[φ statvolts];
```

$$electricField = ScientificForm\left[\frac{\mathcal{E}\,statvolts}{centimeter}\right],$$

```
            True,
            Return["The value of the option
    systemOfUnits must be either SI or Gaussian."]];


    (* print the values of the electric potential and
        magnitude of the electric field *)
    Print["electric potential = ", electricPotential, "\n\n"];
    Print["magnitude of electric field = ",
    electricField, "\n\n\n"];
```

```
    (* determine the direction that the electric field
       makes counterclockwise from the +x axis *)
    If[Ex == 0 ∧ Ey == 0, Return[], θ = ArcTan[Abs[Ey/Ex]]];
    Which[Ex > 0 ∧ Ey ≥ 0 ∨ Ex == 0 ∧ Ey > 0, Null,

           Ex < 0 ∧ Ey ≥ 0, θ = π - θ,

           Ex ≤ 0 ∧ Ey < 0, θ = π + θ,

           Ex > 0 ∧ Ey < 0, θ = 2 π - θ,
           True, Return["Determination of angle failed!"]];

    (* assign appropriate unit to the angle *)

    Which[dir === radian, θ = N[θ] rad,
          dir === degree, θ = N[θ/Degree] deg,
          True, Return["The value of the
    option angle must be either radian or degree."]];

    (* print the value for the angle *)
    Print["angle of electric field from +x axis = ", θ];]
```

where φ, E_x, E_y, ε, and θ have replaced phi, ex, ey, e, and theta, respectively, and we have reentered all the comments that were lost in the conversion. Beware that the Greek letter ε used here appears similar to, but is in fact different from, the ordinary keyboard letter E, which represents the exponential constant in *Mathematica*. Furthermore, E, rather than E_x and E_y, is included in the list of local variables, which must be symbols. Whereas E is a symbol, E_x and E_y are not, as their full forms are Subscript[E, x] and Subscript[E, y], respectively. (Could we have replaced e with the Greek letter ε instead of the script letter ε?)

Let us examine how *Mathematica* extracts the option values. With Options, we name the options and specify their default values:

```
Options[electrostaticField] =
           {systemOfUnits -> SI, angle -> radian};
```

The option systemOfUnits allows the choice between the SI and Gaussian systems of units, and the option angle permits the selection between radian and degree for the unit of angle. The default settings are SI and radian for systemOfUnits and angle, respectively. When the function electrostaticField is called, *Mathematica* evaluates

```
systemOfUnits/.{opts}/.Options[electrostaticField]
```

and

```
angle/.{opts}/.Options[electrostaticField]
```

to determine the option values. Consider, for example, the evaluation of

```
systemOfUnits/.{opts}/.Options[electrostaticField]
```

The replacement operator `/.` associates to the left. That is, the rules in `{opts}` are tried before those in `Options[electrostaticField]`. If the rule `systemOfUnits -> ` *optionValue* appears explicitly in the actual argument sequence of the function `electrostaticField` when it is called, this rule is passed to `{opts}`, and `systemOfUnits/.{opts}` becomes *optionValue* after the replacement. The rules in `Options[electrostaticField]` are not used, because *optionValue* does not match the *lhs* of any of the rules. Thus, `systemOfUnits/.{opts}/.Options[electrostaticField]` is evaluated to *optionValue*. If a rule for `systemOfUnits` has not been specified explicitly in the actual argument sequence of the function `electrostaticField`, that is, in `{opts}`, `systemOfUnits/.{opts}` simply becomes `systemOfUnits`. The rule `systemOfUnits -> SI` in `Options[electrostaticField]` is then applied, and the entire expression `systemOfUnits/.{opts}/.Options[electrostaticField]` is evaluated to `SI`. In both cases, the evaluation results in an appropriate option value that is then assigned as an initial value to the local variable `units`. Similarly, proper option value is determined and assigned as an initial value to the local variable `dir`. Instead of recycling the symbols `systemOfUnits` and `angle`, we assign the option values to the new symbols `units` and `dir` because assigning values to symbols with the option names can be perilous. (See Exercises 7 and 8 of Section 3.4.6.) It is not necessary to assign the option values as initial values to the local variables `units` and `dir`. We can make the assignments to these local variables later in the execution of the function after the local variables declaration. If we choose to assign the option values as initial values in the manner indicated in the local variable list of `Module`, we should remember that initial values are always evaluated before the module is executed. That is, the global values of local variables are captured if they appear in the expressions of initial values, as explained in Section 3.4.1.

To illustrate the use of the function `electrostaticField`, consider the charge distribution: A charge of $+500$ esu is located at $(-10, 30)$ and another charge of -200 esu is located at $(30, 20)$, where the coordinates are in centimeters. Let us determine the electric potential and the electric field at the origin:

```
In[88] := electrostaticField[{{500, -10, 30}, {-200, 30, 20}},
                             systemOfUnits -> Gaussian,
                             angle -> degree]
```

electric potential = 1.02644×10^1 statvolts

magnitude of electric field = $\dfrac{4.82897 \times 10^{-1} \text{ statvolts}}{\text{centimeter}}$

angle of electric field from +x axis = 306.335 deg

For another example, consider a column of 101 equally spaced point charges, each of 2 μC, distributed along the y axis from $y = -50$ to $+50$, where lengths are in meters. To determine the electric potential and electric field at the point $(100, 0)$, call `electrostaticField`:

```
In[89] := electrostaticField[
              Table[{2 10^-6, 0, i}, {i, -50, 50}], {100, 0}
              ]
```

$$\text{electric potential} = 1.74603 \times 10^4 \text{ volts}$$

$$\text{magnitude of electric field} = \frac{1.62058 \times 10^2 \text{ volts}}{\text{meter}}$$

$$\text{angle of electric field from +x axis} = 0$$

Now consider a uniform line charge of length b placed along the y axis, as in Figure 3.4.1, where the center of the line charge is at the origin.

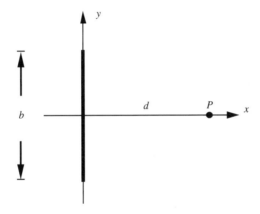

Figure 3.4.1 A uniform line charge of length b positioned along the y axis.

The electric potential at a point P on the x axis at a distance d from the origin is given by

$$\varphi = \frac{2kQ}{b} \ln\left(\frac{b/2 + \sqrt{(b/2)^2 + d^2}}{d}\right) \tag{3.4.6}$$

where Q is the total charge (see [Ser87]). The magnitude of the electric field at point P can be expressed as

$$E = \frac{2kQ}{d}\frac{1}{\sqrt{b^2 + 4d^2}} \tag{3.4.7}$$

and the electric field is in the positive x direction (see [RHK92]). For $Q = 101 \times 2\ \mu C$, $b = 100$ m, $d = 100$ m, Equations 3.4.6 and 3.4.7 give

$$\varphi = 1.74727\ 10^4 \text{ volts}$$

$$E = 1.62382\ 10^2\ \frac{\text{volts}}{\text{meter}}$$

Comparing these results with those obtained earlier for the column of point charges suggests that a column of closely spaced (i.e., distance between two charges $\ll d$) equal point charges approximates rather well a continuous uniform line charge. The difference between the values for the electric potential is less than 0.1%, and that for the electric field is less than 0.2%.

The function `SetOptions` allows us to change the default values for options. `SetOptions[f, opt1->value1, opt2->value2, ...]` resets the specified default options for f. Let `Gaussian` be the default value for `systemOfUnits`:

```
In[90] := SetOptions[electrostaticField, systemOfUnits -> Gaussian];
```

Consider again the earlier example with a distribution of two point charges: A charge of $+500$ esu at $(-10, 30)$ and another charge of -200 esu at $(30, 20)$, where the coordinates are in centimeters. In calling `electrostaticField` to determine the electric potential and electric field at the origin, we do not need to specify the `systemOfUnits` option explicitly anymore:

```
In[91] := electrostaticField[{{500, -10, 30}, {-200, 30, 20}},
                             angle -> degree]
```

electric potential = 1.02644×10^1 statvolts

magnitude of electric field = $\dfrac{4.82897 \times 10^{-1} \text{ statvolts}}{\text{centimeter}}$

angle of electric field from +x axis = 306.335 deg

In setting up the function `electrostaticField`, we made the assignment

```
Options[electrostaticField] =
               {systemOfUnits -> SI, angle -> radian};
```

before defining the function. If this expression is located within the body of the function, we cannot assign the option values as initial values to local variables. Furthermore, we cannot use `SetOptions` to change the default options, because every call to `electrostaticField` returns immediately the default options to their original settings and overrides the effects of `SetOptions`.

```
In[92] := ClearAll[electrostaticField]
```

3.4.5 *An Example: Motion of a Particle in One Dimension*

The equation of motion of a particle with mass m moving along the x axis can be written as

$$x''(t) = \frac{1}{m} F(x(t), x'(t), t) \tag{3.4.8}$$

where $F(x(t), x'(t), t)$ is the resultant force acting on the particle and $x(t)$ is its position at time t. This equation simply states that the acceleration of the particle equals the net force per unit mass.

To illustrate the use of several constructs presented earlier in this section, that is, Section 3.4, we write a function for plotting the position, velocity, and acceleration versus time of the particle and for displaying these plots together. The function takes four arguments: the net force per unit mass, the initial position, the initial velocity, and the time interval from $t = 0$. Options can be entered as trailing arguments for specifying the plots to be omitted and the labels to be placed at the ends of the axes or on the edges of the frame of each plot. Two-dimensional graphics options that affect all the plots can also be included for passing to the functions `Plot` and `Show` with the exceptions: `PlotLabel`, `AxesLabel`, `Ticks`, `FrameLabel`, and `FrameTicks`. With the specifications `velocityPlot -> False` and `combinationPlot -> False`, for example, the function does not produce the velocity versus time plot nor display together the plots that are generated. With the inclusion of `accelerationAFLabel -> {`*xlabel*, *ylabel*`}`, for instance, the function places the specified labels at the ends of the x and y axes of the acceleration versus time plot if the option value of `Axes` is `True` and on the bottom and left-hand edges of the frame around the plot if the option value of `Frame` is `True`; the default settings of `Axes` and `Frame` are `False` and `True`, respectively. The description of the function is embedded in its body.

```
In[93] := (* name the options and specify their default values *)
          Options[motion1DPlot] =
           {positionPlot -> True,
            velocityPlot -> True,
            accelerationPlot -> True,
            combinationPlot -> True,
            positionAFLabel -> {"t (s)", "x (m)"},
            velocityAFLabel -> {"t (s)", "v (m/s)"},
            accelerationAFLabel -> {"t (s)", "a (m/s^2)"},
            combinationAFLabel -> {"t (s)", None}};


In[95] := Needs["Utilities`FilterOptions`"]


In[96] := motion1DPlot[a_, x0_, v0_, tmax_, opts___Rule] :=          ←
          Module[{
          (* declare local variables *)
          sol, curves = {}, plotx, plotv, plota,

          (* determine option values and assign them
             as initial values to local variables *)
          position =
            positionPlot/.{opts}/.Options[motion1DPlot],
          velocity =
            velocityPlot/.{opts}/.Options[motion1DPlot],
          acceleration =
            accelerationPlot/.{opts}/.Options[motion1DPlot],
          combination =
            combinationPlot/.{opts}/.Options[motion1DPlot],
          positionLabel =
            positionAFLabel/.{opts}/.Options[motion1DPlot],
```

```
        velocityLabel =
          velocityAFLabel/.{opts}/.Options[motion1DPlot],
        accelerationLabel =
          accelerationAFLabel/.{opts}/.Options[motion1DPlot],
        combinationLabel =
          combinationAFLabel/.{opts}/.Options[motion1DPlot],

        (* select valid options for Plot and Show and assign
           them as initial values to local variables *)
        optPlot = FilterOptions[Plot, opts],
        optShow = FilterOptions[Graphics, opts]
              },

        (* set text of a warning message *)
        motion1DPlot::argopt = "Each of the values for the
    options positionPlot, velocityPlot, accelerationPlot, and
    combinationPlot must be either True or False.";

        (* verify option specifications *)
        If[Count[{position, velocity, acceleration, combination},
            True|False] =!= 4,
            Message[motion1DPlot::argopt]; Return[$Failed]
          ];

        (* solve the equation of motion numerically *)
        sol =
          NDSolve[{x''[t] == a, x[0] == x0, x'[0] == v0},
                  x, {t, 0, tmax}
                ];

        (* plot position vs. time *)
        If[position,
            plotx = Plot[Evaluate[x[t] /. sol], {t, 0, tmax},
                        PlotLabel -> "position vs. time",
                        AxesLabel -> positionLabel,
                        Ticks -> Automatic,
                        FrameLabel -> positionLabel,
                        FrameTicks -> Automatic,
                        Evaluate[optPlot],
                        PlotRange -> All,
                        Axes -> False,
                        Frame -> True
                      ];
          AppendTo[curves, plotx]
          ];
```

```
            (* plot velocity vs. time *)
            If[velocity,
               plotv = Plot[Evaluate[x'[t] /. sol], {t, 0, tmax},
                           PlotLabel -> "velocity vs. time",
                           AxesLabel -> velocityLabel,
                           Ticks -> Automatic,
                           FrameLabel -> velocityLabel,
                           FrameTicks -> Automatic,
                           Evaluate[optPlot],
                           PlotStyle -> Dashing[{0.03, 0.03}],
                           PlotRange -> All,
                           Axes -> False,
                           Frame -> True
                           ];
               AppendTo[curves, plotv]
              ];

            (* plot acceleration vs. time *)
            If[acceleration,
               plota = Plot[Evaluate[a /. sol], {t, 0, tmax},
                           PlotLabel -> "acceleration vs. time",
                           AxesLabel -> accelerationLabel,
                           Ticks -> Automatic,
                           FrameLabel -> accelerationLabel,
                           FrameTicks -> Automatic,
                           Evaluate[optPlot],
                           PlotStyle -> RGBColor[1, 0, 0],
                           PlotRange -> All,
                           Axes -> False,
                           Frame -> True
                           ];
               AppendTo[curves, plota]
              ];

            (* combine the plots *)
            If[(combination)&&(Length[curves] > 1),
               Show[curves,
                    PlotLabel -> "combination",
                    AxesLabel -> combinationLabel,
                    Ticks -> {Automatic, None},
                    FrameLabel -> combinationLabel,
                    FrameTicks -> {Automatic, None},
                    optShow];
              ]

                  ]
```

In verifying option specifications, `motion1DPlot` uses the function `Message` to print a warning message if any of the options `position`, `velocity`, `acceleration`, and `combination` have values other than `True` or `False`. `Message[`*symbol* `::`*tag*`]` prints the message previously assigned to *symbol* `::` *tag*.

For passing options to `Plot` and `Show`, `motion1DPlot` employs the function `FilterOptions` in the package `Utilities`FilterOptions`` because `Plot` and `Show` do not accept user-defined options such as `positionPlot` and `velocityAFLabel`. `FilterOptions[`*f*, *opt1*, *opt2*, ...`]` returns a sequence of options from *opt1*, *opt2*, ... that are valid options for *f*. We assign the sequence of specified options that are valid for `Plot` as an initial value to the local variable `optPlot`, and the sequence of specified options that are valid for `Show` as an initial value to `optShow`.

To illustrate the use of `motion1DPlot`, let us consider several examples. For the first example, a pebble is thrown vertically upward with an initial speed of 25 m/s. If air resistance is negligible, the force acting on the pebble is

$$F = -mg$$

where g is the magnitude of the acceleration due to gravity and the positive x direction is upward. The function `motion1DPlot` generates the graphs:

In[97] := `motion1DPlot[-9.80, 0, 25, 5.1]`

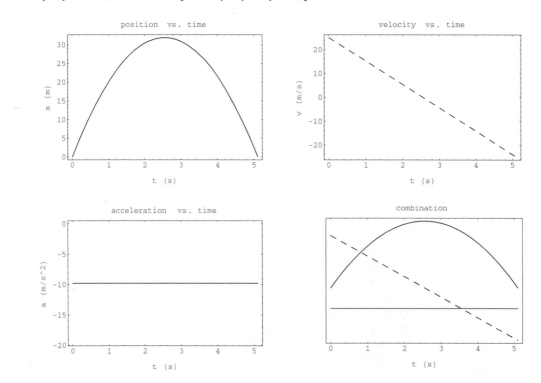

To approximate the effect of air resistance, we include a retarding drag force proportional to the square of the velocity and express the resultant force as

$$F = -mg - kx'(t)|x'(t)|$$

where the parameter k is about 10^{-4} kg/m for a pebble of radius 0.01 m (see [GT88] for more information). Let the pebble's mass $m = 10^{-2}$ kg. The function `motion1DPlot` produces a different set of graphs:

```
In[98] := motion1DPlot[-9.80 - 10^(-2) x'[t] Abs[x'[t]],
                       0, 25, 4.49]
```

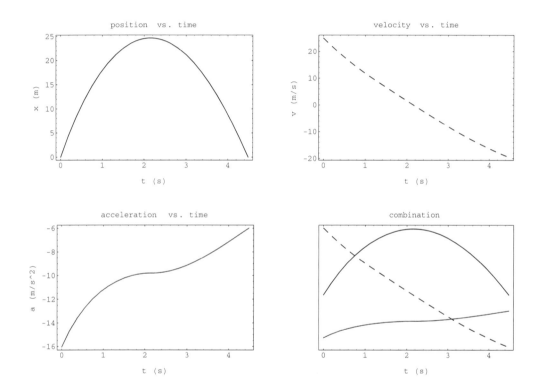

The maximum height is about 32 m and the acceleration is constant if air resistance is absent, whereas the actual height reached is less than 25 m and acceleration in fact increases during the flight. Furthermore, when the pebble returns to its initial position, the speed is the same as the initial speed of 25 m/s if air resistance is neglected, whereas the realistic speed is less than 20 m/s.

For another example, a particle initially at rest has mass $m = 1$ kg and is subjected to a force

$$F(t) = kte^{-\alpha t}$$

where $k = 1$ N/s and $\alpha = 0.5$ s^{-1}. (This example is extracted from a problem in [MT95].) Let us call `motion1DPlot`:

```
In[99] := motion1DPlot[t Exp[-0.5 t], 0, 0, 14,
                       positionPlot -> False,
```

```
combinationPlot -> False,
Axes -> True,
Frame -> False]
```

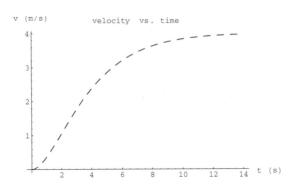

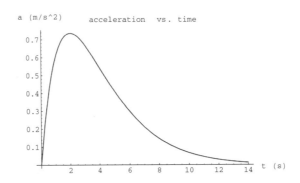

We conclude this section with an example from molecular physics. Solution of the time-independent Schrödinger equation for a diatomic molecule is difficult. For an approximation, we can separate the equation into three equations for different kinds of motion: motion of the electrons, rotation of the nuclei, and vibration of the nuclei. The equation for the electronic motion is still complex because it must include the interactions of the electrons with the nuclei as well as their mutual interactions [Gas74]. However, the solution of the equation for rotational motion is straightforward because the nuclei rotate as a rigid body about the center of mass [ER85]. The two-body problem of the vibrational motion, which is our focus here, can be transformed into two one-body problems: the motion of the center of mass and the one-dimensional motion of either nucleus as viewed from the other. The motion of the center of mass is just that of a free particle. The analysis of the relative motion can be further simplified with the observation that semiclassical approximation is often justified because of the large mass of the nuclei. In Equation 3.4.8, which is the classical equation of motion, the mass m is now the reduced mass and the resultant force is the force on the moving nucleus due to the observing one [Sym71]. The vibrational motion is reduced to the motion of a particle in a one-dimensional potential that is often taken to be the Lennard-Jones or 6–12 potential

$$V(x) = 4b\left[\left(\frac{d}{x}\right)^{12} - \left(\frac{d}{x}\right)^6\right]$$

where b and d are the empirical parameters. If $V(x)$ and x are measured, respectively, in units of $4b$ and d, this potential becomes

$$V(x) = \frac{1}{x^{12}} - \frac{1}{x^6}$$

Let us plot this potential:

```
In[100] := Plot[(1/x^12 - 1/x^6), {x, 0.4, 3},
              PlotRange -> {-0.25, 0.3},
              AxesLabel -> {" x (d)", " V (4b)"}];
```

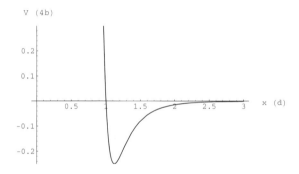

It is steep and repulsive for $x < 1$, has a minimum of $-1/4$ at $x = 2^{1/6}$, and approaches zero as x goes to infinity. The force on the moving nucleus is

```
In[101] := -D[1/x^12 - 1/x^6, x]
```

$$Out[101] = \frac{12}{x^{13}} - \frac{6}{x^7}$$

For $x_0 = 1.32$ and $v_0 = 0$, the energy is between $-1/4$ and 0 and the motion is bounded and periodic but asymmetric about $x = 2^{1/6}$. Let the unit of mass be m. The function `motion1DPlot` gives the graphs describing the motion of either nucleus as viewed from the other:

```
In[102] := motion1DPlot[(12/x[t]^13 - 6/x[t]^7), 1.32, 0, 6.8,
              positionAFLabel -> {"t ((d/2)(m/b)^(1/2))", "x (d)"},
              velocityAFLabel ->
                  {"t ((d/2)(m/b)^(1/2))", "v ((4b/m)^(1/2))"},
              accelerationAFLabel ->
                  {"t ((d/2)(m/b)^(1/2))", "a (4 b/(m d))"},
              combinationAFLabel -> {"t ((d/2)(m/b)^(1/2))", None}]
```

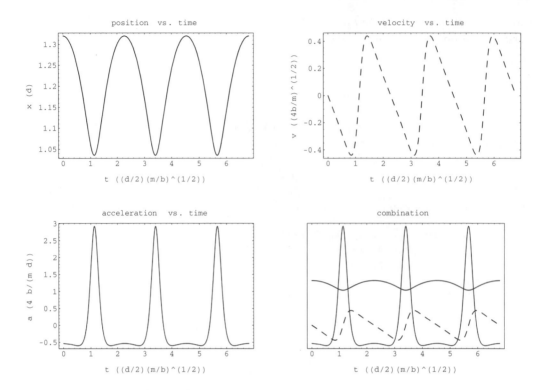

Though these graphs, based on Newton's laws, offer insight into the vibration of the nuclei, we must be leery of the Newtonian notion that all energies between $-1/4$ and 0 are possible for the bound states. The vibrational energy is in fact quantized. In a semiclassical approximation, we impose on the classical results the Sommerfeld-Wilson quantization rule in order to isolate the discrete energies. (Refer to [KM90] for a discussion of semiclassical quantization of molecular vibrations.)

For the H_2 molecule, the Lennard-Jones potential unfortunately yields vibrational energies that are inconsistent with experimental observations. A better choice leading to results that are in reasonable agreement with experimental data is the Morse potential

$$V(x) = V_0\left\{\left(1 - e^{-\alpha(x/x_{min}-1)}\right)^2 - 1\right\}$$

where the dissociation energy $V_0 = 4.73$ eV, the equilibrium internuclear separation $x_{min} = 0.74$ Å, and the parameter $\alpha = 1.44$ for the H_2 molecule. Let us plot the potential in units of V_0 as a function of x in units of x_{min}:

```
In[103] := Plot[(1 - Exp[-1.44 (x - 1)])^2 - 1, {x, 0, 6},
                AxesLabel -> {" x (xmin)", " V (Vo)"}];
```

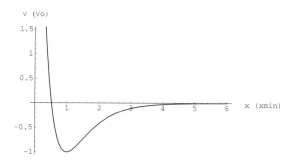

While the Lennard-Jones and Morse potentials have the same essential features, the latter allows an additional parameter α for adjusting the curvature of the minimum. We will not invoke the function `motion1DPlot` to generate the graphs because they resemble those produced with the Lennard-Jones potential.

In[104] := **ClearAll[motion1DPlot]**

3.4.6 Exercises

1. Write a single-clause definition and also a multiclause definition for the function
$$f(x) = |x|$$
What is the derivative of f at 0?

*2. Write a single-clause definition for a function that takes a list of numbers as its argument and returns a list of their inverses, except the function should only report the number of zeros present when there are zeros in the list. For example,

```
func[{1, 2, 3, 4, a}]
func[{1, 2, 3, 4, a}]
```

```
func[{2, 3, 4, 10}]
```
$$\{\frac{1}{2}, \frac{1}{3}, \frac{1}{4}, \frac{1}{10}\}$$

```
func[{0, 3, 0, 0}]
3 zeros in list
```

Hint: Use **Map**, **If**, **Count**, **VectorQ**, **NumberQ**, **Print**, and a pure function.

3. Write a single-clause definition and also a multiclause definition for the potential
$$v(x) = \begin{cases} 0 & x < -2a \\ -2b & -2a \le x < -a \\ 0 & -a \le x < a \\ -b & a \le x < 3a \\ b & 3a \le x \end{cases}$$

Let $a = 1$ and $b = 1$.

4. A particle of mass m moves in the potential
$$V(x) = \frac{-V_0 a^2 (a^2 + x^2)}{8a^4 + x^4}$$

(a) Plot the potential and the force on the particle as a function of the position of the particle.

(b) Choose three sets of initial conditions: one resulting in the energy of the particle greater than zero, one the energy of the particle between $-V_0/8$ and zero, and one the energy of the particle between $-V_0/4$ and $-V_0/8$. For each set of initial conditions, plot the particle's position, velocity, and acceleration as a function of time. *Hint:* Adopt an appropriate system of units.

5. A particle of mass m moves in the potential
$$V(x) = V_0 \left(\left(\frac{x}{a}\right)^2 - \beta \left(\frac{x}{a}\right)^3 \right)$$

where V_0 and a are positive constants. Adopt a system of units in which $m = 1$, $V_0 = 1$, and $a = 1$. Let $\beta = 1/12$.

(a) Plot the potential and also the force on the particle.

(b) Choose two sets of initial conditions: $x_0 = -3$ and v_0 resulting in the energy of the particle being greater than 64/3, and $x_0 = 5$ and v_0 resulting in the energy of the particle lying between 64/3 and zero. For each set of initial conditions, plot the particle's position, velocity, and acceleration as a function of time.

*6. Modify the function `motion1DPlot` in Section 3.4.5 so that it takes as its first argument the entire Equation 3.4.8 rather than just the *rhs* of the equation. Also, the function should print an error message if any expression other than an equation is entered as the first argument, the initial position and velocity are not real numbers, or the time interval is not a positive real number. Real numbers include all approximate real numbers, integers, and rational numbers. Implement the error message feature of the function two ways: one with the function `If` and another with the pattern object *pattern/;condition*.

7. Consider the function `binomialExpansion`:

```
Options[binomialExpansion] = {caption -> False, exp -> 2};

binomialExpansion[x_Symbol, y_Symbol, opts___] :=
    Module[{caption, exp},
       caption = caption/.{opts}/.Options[binomialExpansion];
       exp = exp/.{opts}/.Options[binomialExpansion];
       If[caption === True,
         Print["Expansion of ", (x + y)^exp]
         ];
       Expand[(x + y)^exp]
          ]
```

Why doesn't this function work as expected? For example,

```
binomialExpansion[a, b]
```
$(a + b)$ exp$70

```
binomialExpansion[a, b, exp -> 10, caption -> True]
```
$(a + b)$ exp$71

8. Consider the function `myBinomialExpansion`:

```
ClearAll[myBinomialExpansion, caption]
```

```
Options[myBinomialExpansion] = {caption -> False};
```

```
myBinomialExpansion[
  x_Symbol, y_Symbol, exp_, opts___Rule] :=
  Module[{},
    caption = caption/.{opts}/.Options[myBinomialExpansion];
    If[caption === True,
      Print["Expansion of ", (x + y)^exp]
    ];
    Expand[(x + y)^exp]
      ]
```

Why do two identical calls to the function give different results? The first call prints the caption, whereas the second does not:

```
myBinomialExpansion[a, b, 2, caption -> True]
```

Expansion of $(a + b)^2$
$a^2 + 2\,a\,b + b^2$

```
myBinomialExpansion[a, b, 2, caption -> True]
```
$a^2 + 2\,a\,b + b^2$

3.5 Graphics

Section 2.3 discussed the interactive use of *Mathematica*'s graphical capabilities. This section considers graphics programming that extends these capabilities. Section 3.5.1 introduces the concept of graphics objects, which are *Mathematica* expressions containing graphics elements. There are two kinds of graphics elements: graphics primitives and graphics directives. Graphics primitives are the building blocks of graphics objects, and graphics directives determine how they are rendered. The function **Show** displays the graphics objects. Graphics options can be specified for **Show** or inserted into graphics objects. Whereas graphics directives affect the rendering of individual graphics primitives, graphics options influence the overall appearance of the graphics. Sections 3.5.2 and 3.5.3 describe how to produce two- and three-dimensional graphics, respectively.

The availability of color is one reason that *Mathematica* graphics are so enticing. Unfortunately, this book is not printed in color; in the printing of this book, all color specifications are converted to gray levels.

3.5.1 Graphics Objects

Built-in plotting functions generate plots of lists of data or mathematical functions. To make a plot, the function creates a graphics object and then calls for its rendering. The function **InputForm** allows us to see the graphics object. **InputForm**[*expr*] prints as a version of *expr* suitable for input to *Mathematica*.

```
In[1] := InputForm[ListPlot[Table[Random[], {5}],
                   PlotStyle -> PointSize[0.02],
                   DisplayFunction -> Identity]]

Out[1]//InputForm=
 Graphics[{PointSize[0.02], {Point[{1, 0.7642066534997113}],
     Point[{2, 0.9185176242220508}], Point[{3, 0.1582315364042221}],
     Point[{4, 0.7302390083601948}], Point[{5, 0.5385308951521238}]}}\
 , {PlotRange -> Automatic, AspectRatio -> GoldenRatio^(-1),
   DisplayFunction :> Identity, ColorOutput -> Automatic,
   Axes -> Automatic, AxesOrigin -> Automatic, PlotLabel -> None,
   AxesLabel -> None, Ticks -> Automatic, GridLines -> None,
   Prolog -> {}, Epilog -> {}, AxesStyle -> Automatic,
   Background -> Automatic, DefaultColor -> Automatic,
   DefaultFont :> $DefaultFont, RotateLabel -> True, Frame -> False,
   FrameStyle -> Automatic, FrameTicks -> Automatic,
   FrameLabel -> None, PlotRegion -> Automatic,
   ImageSize -> Automatic, TextStyle :> $TextStyle,
   FormatType :> $FormatType}]
```

(The *Mathematica* version 3.0 output shown here differs slightly from that of *Mathematica* version 2.2.) We have suppressed the rendering of the plot by setting the option **DisplayFunction** to **Identity**. The graphics object is an expression with the head **Graphics** and elements consisting of graphics elements and graphics options. The graphics elements are the graphics primitives **Point**[{*x*, *y*}] and the graphics directive **PointSize**[*n*]. The graphics options have the form *OptionName* -> *OptionValue*.

While the plotting functions are excellent for plotting lists of data and mathematical functions, graphics programming must be invoked to produce more complex graphics. In these cases, we need to construct the graphics objects and display them with the function **Show**.

There are several kinds of graphics objects:

         Graphics[*elements*, *options*]   two-dimensional graphics

         DensityGraphics[*array*]   density plot

`ContourGraphics[`*array*`]`	contour plot
`SurfaceGraphics[`*array*`]`	three-dimensional surface
`Graphics3D[`*elements, options*`]`	three-dimensional graphics
`GraphicsArray[{{`*g11, g12, ...*`}, ...}]`	array of graphics objects

This book focuses only on `Graphics` and `Graphics3D`.

Unless there is only one graphics primitive, the first argument of `Graphics` or `Graphics3D` must be a list or nested list of graphics elements that are either graphics primitives or graphics directives. The first argument takes the form

{prim1, dir1, dir2, prim2, sublist1, prim3, dir3, prim4, prim5, sublist2, ...}

where the *primj* are the graphics primitives, the *dirj* are the graphics directives, and the *sublistj* have a similar form as that of the list. Examples of graphics primitives are `Point`, `Line`, `Polygon`, and `Text`; examples of graphics directives are `GrayLevel`, `RGBColor`, `PointSize`, and `Thickness`.

The function `Show`, introduced in Section 2.3, displays graphics objects:

`Show[`*graphics, options*`]`	display two- and three-dimensional graphics, using the options specified
`Show[`*g1, g2, ...*`]`	display several graphics objects combined

Section 2.3.1.2 considered graphics options. Option specifications can be included as trailing arguments of the `Show` function. Until they are overridden, these graphics options remain with the graphics object that `Show` displays and returns.

In a list, each graphics directive modifies the subsequent graphics primitives in the list or sublists of the list:

```
In[2] := Show[Graphics[
              {Point[{1, 1}],
               PointSize[0.15],
               {GrayLevel[0.5], Point[{2, 2}]},
               Point[{3, 3}]}
                 ],
           AxesOrigin -> {0, 0},
           Axes -> True,
           PlotRange -> {{0, 4}, {0, 4}}
           ];
```

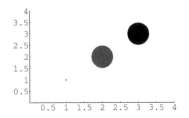

In the preceding graphics, `GrayLevel` affects only the gray-level intensity of the point at {2, 2} and `PointSize` specifies the radius of the points at {2, 2} and {3, 3}. The graphics directives have no influence on the almost invisible point at {1, 1}.

If there is another graphics directive with the same name but different specification further down the list, the latter prevails:

```
In[3] := Show[Graphics[{
                PointSize[0.05], Point[{1, 1}],
                {PointSize[0.10], GrayLevel[0.5], Point[{2, 2}]},
                Point[{3, 3}], PointSize[0.175], Point[{4, 4}]
                   },
                AxesOrigin -> {0, 0},
                Axes -> True,
                PlotRange -> {{0, 5}, {0, 5}}
                   ]
          ];
```

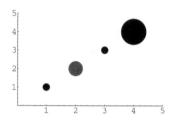

Here, `PointSize[0.05]` specifies the radius of the points at {1, 1} and {3, 3}. `PointSize[0.10]` determines the radius of the point at {2, 2}, and `PointSize[0.175]` stipulates the size of the point at {4, 4}. Note that `PointSize[0.10]` and `GrayLevel[0.5]` do not affect points outside their list.

Whereas graphics directives modify the rendering of graphics primitives in the list, graphics options have global effects on the plot. Specifications of graphics options cause *Mathematica* to construct a collection of graphics elements. The function `FullGraphics` allows us to see these elements. `FullGraphics[g]` takes a graphics object and generates a new one in which graphics options, if any, are given as explicit lists of graphics elements.

```
In[4] := Short[
            InputForm[
                FullGraphics[Graphics[Point[{1, 1}], Axes -> True]]
                   ], 7
               ]
```

Out[4]//Short=
Graphics[{Point[{1, 1}], {{GrayLevel[0.], AbsoluteThickness[0.25],
       Line[{{0.5, 0.}, {0.5, 0.02123669610234236}}]},
       Text[0.5, {0.5, -0.04247339220468473}, {0., 1.}],

```
{GrayLevel[0.], AbsoluteThickness[0.25],
 Line[{{1., 0.}, {1., 0.02123669610234236}}]}, <<46>>,
{GrayLevel[0.], AbsoluteThickness[0.25],
 Line[{{0., -0.05}, {0., 2.049999999999999}}]}}}]
```

With the function **short**, we have reduced the length of the output. (For *Mathematica* version 2.2, specify 9 instead of 7 lines long in the function **short** here. Also, the *Mathematica* version 3.0 output shown here differs somewhat from that of *Mathematica* version 2.2.)

3.5.2 *Two-Dimensional Graphics*

3.5.2.1 TWO-DIMENSIONAL GRAPHICS PRIMITIVES

Mathematica recognizes many two-dimensional graphics primitives:

`Point[{x, y}]`	point at coordinates $\{x, y\}$
`Line[{{x1, y1}, {x2, y2}, ...}]`	line joining points $\{x1, y1\}, \{x2, y2\}, \ldots$
`Rectangle[{xmin, ymin}, {xmax, ymax}]`	filled rectangle, oriented parallel to the axes
`Polygon[{{x1, y1}, {x2, y2}, ...}]`	filled polygon with the specified list of corners
`Circle[{x, y}, r]`	circle of radius r centered at the point $\{x, y\}$
`Circle[{x, y}, {rx, ry}]`	ellipse with semiaxes rx and ry
`Circle[{x, y}, r, {theta1, theta2}]`	circular arc where *theta1* and *theta2* are, respectively, the starting and finishing angles measured counterclockwise in radians from the positive x direction
`Circle[{x, y}, {rx, ry}, {theta1, theta2}]`	elliptical arc
`Disk[{x, y}, r]`	filled disk of radius r centered at the point $\{x, y\}$
`Disk[{x, y}, {rx, ry}]`	elliptical disk with semiaxes rx and ry
`Disk[{x, y}, r, {theta1, theta2}]`	segment of a disk
`Text[expr, {x, y}]`	text corresponding to the printed form of *expr*, centered at $\{x, y\}$
`Text[expr, {x, y}, {-1, 0}]`	text with its left-hand end at $\{x, y\}$
`Text[expr, {x, y}, {1, 0}]`	text with its right-hand end at $\{x, y\}$
`Text[expr, {x, y}, {0, -1}]`	text centered above $\{x, y\}$
`Text[expr, {x, y}, {0, 1}]`	text centered below $\{x, y\}$

Coordinates of graphics primitives may be given in terms of the original coordinates $\{x, y\}$ that we have been using in this book or the display coordinates relative to the display area for the graphics. The display, or scaled, coordinates are given in the form `Scaled[{sx, sy}]`. These coordinates, $\{sx, sy\}$, vary from 0 to 1 along the x and y directions, and the origin is located at the lower-left corner of the display area. As an example, let us display two rectangles, one specified with the original coordinates and another specified with the scaled or display coordinates:

```
In[5] := Show[Graphics[
                Rectangle[{0.5, 0.5}, {1, 1}]
                    ],
            Graphics[
                Rectangle[Scaled[{0.5, 0.5}], Scaled[{1, 1}]]
                    ],
            PlotRange -> {{0, 2}, {0, 2}},
            Frame -> True,
            FrameTicks -> {{0, 0.5, 1, 2}, {0, 0.5, 1, 2}}
            ];
```

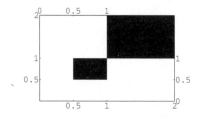

In the graphics object, the lower-left and upper-right corners of the smaller rectangle are specified in the original coordinates, whereas those of the larger rectangle are given in the scaled coordinates.

To illustrate the use of some of the graphics primitives in setting up graphics objects, we select an example from geometrical optics.

EXAMPLE 3.5.1 Construct a ray diagram for a thin converging lens.

To begin, we generate a list of graphics primitives representing the lines. The function **Map** with the special input form `/@` applies the function **Line** to each list of points joined by a line:

```
In[6] := g1 = Line /@ {{{-13, 0}, {17, 0}},
                        {{0, -4}, {0, 4}},
                        {{-10, 0}, {-10, 2}},
                        {{15, 0}, {15, -3}},
                        {{-10, 2}, {0, 2}, {15, -3}},
                        {{-10, 2}, {15, -3}}
                        }
```

```
Out[6] = {Line[{{-13, 0}, {17, 0}}], Line[{{0, -4}, {0, 4}}],
          Line[{{-10, 0}, {-10, 2}}], Line[{{15, 0}, {15, -3}}],
          Line[{{-10, 2}, {0, 2}, {15, -3}}], Line[{{-10, 2}, {15, -3}}]}
```

Map and **Polygon** give a list of graphics primitives for the triangles depicting the arrowheads:

```
In[7] := g2 = Polygon /@ {
                    {{-10, 2}, {-9.6, 1.6}, {-10.4, 1.6}},
                    {{15, -3}, {15.4, -2.6}, {14.6, -2.6}},
                    {{-5, 2}, {-5.4, 2.4}, {-5.4, 1.6}},
                    {{-5, 1}, {-5.314, 1.471}, {-5.471, 0.6862}}
                          }

Out[7] = {Polygon[{{-10, 2}, {-9.6, 1.6}, {-10.4, 1.6}}],
          Polygon[{{15, -3}, {15.4, -2.6}, {14.6, -2.6}}],
          Polygon[{{-5, 2}, {-5.4, 2.4}, {-5.4, 1.6}}],
          Polygon[{{-5, 1}, {-5.314, 1.471}, {-5.471, 0.6862}}]}
```

The functions **Thread** and **Circle** provide a list of two graphics primitives representing the two circular arcs of the lens:

```
In[8] := g3 = Thread[Circle[{{13, 0}, {-13, 0}}, 13.60,
                            {{2.843, 3.440}, {5.985, 6.582}}
                           ]
                    ]

Out[8] = {Circle[{13, 0}, 13.6, {2.843, 3.44}],
          Circle[{-13, 0}, 13.6, {5.985, 6.582}]}
```

Using **Thread** is elegant but optional. We may enter the list of graphics primitives directly. **Thread** and **Disk** yield a list of two graphics primitives for the two focal points:

```
In[9] := g4 = Thread[Disk[{{-6, 0}, {6, 0}}, 0.25]]

Out[9] = {Disk[{-6, 0}, 0.25], Disk[{6, 0}, 0.25]}
```

Finally, we produce a list of graphics primitives representing the labels:

✿ **2.2**

```
g5 = {Text["F", {-6, -0.5}, {0, 1}],
      Text["F", {6, 0.5}, {0, -1}],
      Text[FontForm["Ray Diagram for a\nThin Converging Lens",
                    {"Helvetica-Bold", 9}],
           {3.5, 3.0}, {-1, 0}]
     };
```

The function **FontForm** determines the font and size of the title for the diagram. **FontForm**[*expr*, {"*font*", *size*}] specifies that *expr* should be printed in the specified font and size. In a text string, \n indicates a new line.

❋ **3.0**

```
In[10] := g5 = {
            Text["F", {-6, -0.12}, {0, 1}],
            Text["F", {6, 0.12}, {0, -1}],
            Text[StyleForm["Ray Diagram for a\nThin Converging Lens",
                    FontFamily -> "Helvetica",
                    FontWeight -> "Bold",
                    FontSize -> 9],
                {3.5, 2.8}, {-1, 0}]
          };
```

The function `StyleForm` determines the style of the title for the diagram. `StyleForm[`*expr,* `options]` prints *expr* using the specified style options. Some common style options are `FontSize, FontWeight, FontSlant, FontFamily, FontColor,` and `Background.` `StyleForm[`*expr,* `"`*style*`"]` prints *expr* using the specified cell style in the current notebook. Some typical cell styles are `Title, Section, Subsection, Text, Input,` and `Output.` In a text string, \n indicates a new line.

The function `Show` displays the diagram:

```
In[11] := Show[Graphics[Join[g1, g2, g3, g4, g5]],
              AspectRatio -> Automatic];
```

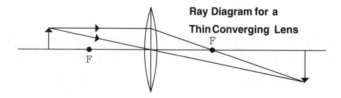

With the option `AspectRatio -> Automatic`, the ratio of height to width of the display area is determined from the original coordinate system; that is, a unit in the *x* direction and a unit in the *y* direction in the original coordinate system have the same size in the final display. The diagram retains its natural shape, and the arcs of the lens are truly circular.

```
In[12] := ClearAll[g1, g2, g3, g4, g5]
```                                                                          ∎

3.5.2.2 TWO-DIMENSIONAL GRAPHICS DIRECTIVES

Mathematica supports many two-dimensional graphics directives:

`GrayLevel [`*level*`]`	gray level between 0 (black) and 1 (white)
`RGBColor[`*r,* *g,* *b*`]`	color with specified red, green, and blue components, each between 0 and 1

`Hue[h]`	color with hue h between 0 and 1
`Hue[h, s, b]`	color with specified hue, saturation, and brightness, each between 0 and 1
`PointSize[r]`	give all points a radius r as a fraction of the total width of the graph
`AbsolutePointSize[d]`	give all points a radius d measured in absolute units
`Thickness[r]`	give all lines a thickness r as a fraction of the total width of the graph
`AbsoluteThickness[d]`	give all lines a thickness d measured in absolute units
`Dashing[{r1, r2, ...}]`	show all lines as a sequence of dashed segments of lengths $r1, r2, \ldots$ (repeated cyclically) with ri given as a fraction of the total width of the graph
`AbsoluteDashing[{d1, d2, ...}]`	use absolute units to measure dashed segments

Here are `RGBColor` specifications for some common colors:

Beige	`RGBColor[0.640004, 0.580004, 0.5]`
Black	`RGBColor[0., 0., 0.]`
Blue	`RGBColor[0., 0., 1.]`
Brown	`RGBColor[0.5, 0.164693, 0.164693]`
Chocolate	`RGBColor[0.823496, 0.411802, 0.117603]`
Cyan	`RGBColor[0., 1., 1.]`
Gold	`RGBColor[1., 0.843104, 0.]`
Green	`RGBColor[0., 1., 0.]`
Ivory	`RGBColor[1., 1., 0.941206]`
Lavender	`RGBColor[0.902005, 0.902005, 0.980407]`
Magenta	`RGBColor[1., 0., 1.]`
Olive	`RGBColor[0.230003, 0.370006, 0.170003]`
Orange	`RGBColor[1., 0.5, 0.]`
Peach	`RGBColor[0.44, 0.26, 0.26]`
Pink	`RGBColor[1., 0.752907, 0.796106]`
Purple	`RGBColor[0.627506, 0.125507, 0.941206]`
Red	`RGBColor[1., 0., 0.]`
Tomato	`RGBColor[1., 0.388195, 0.278405]`
Turquoise	`RGBColor[0.250999, 0.878399, 0.815699]`

Violet	RGBColor[0.559999, 0.370006, 0.599994]
White	RGBColor[1., 1., 1.]
Yellow	RGBColor[1., 1., 0.]

To illustrate the use of some of the graphics directives, we set up and display a graphics object representing a favorite lecture demonstration.

EXAMPLE 3.5.2 Construct a diagram showing a lecture demonstration in which a projectile is aimed at a target that falls from rest at the moment when the projectile is fired.

To begin, we plot the trajectory of the projectile. The option `DisplayFunction ->` `Identity` prevents the display of the plot.

```
In[13] := gp1 = Plot[(10/15) x - (7/225) x^2, {x, 0, 15},
                PlotStyle -> Thickness[0.01],
                DisplayFunction -> Identity];
```

We then enter a list of graphics elements for the lines; two of them are dashed lines:

```
In[14] := gp2 = {Line[{{0, 0}, {5, 0}}],
                {Dashing[{0.04, 0.02}],
                 Line[{{0, 0}, {15, 10}}],
                 Line[{{15, 10}, {15, 3}}]}
                };
```

The package `Graphics`Arrow`` furnishes the graphics primitive `Arrow[`*start*, *finish*`]` representing an arrow from coordinates *start* to *finish*. Let us load the package and enter a list of graphics elements representing the arrows in the diagram:

```
In[15] := Needs["Graphics`Arrow`"]
          gp3 = {{Thickness[0.01],
                  Arrow[{0, 0}, {5, 5 10/15}],
                  Arrow[{15, 10}, {15, 7}],
                  Arrow[{11.8956, 3.5332}, {12.3375, 3.48956}]},
                 Arrow[{12.5, 1}, {14.5, 2.5}]
                };
```

We enter a list of elements for the projectile and the target; the projectile is colored red:

```
In[17] := gp4 = {{RGBColor[1, 0, 0], Disk[{15, 3}, 0.5]},
                 Disk[{0, 0}, 0.5],
                 Disk[{15, 10}, 0.5]
                };
```

Finally, we give a list of `Text` primitives representing the labels:

```
(* Mathematica Version 2.2 Specifics *)

gp5 = {Text[FontForm["v", {"Helvetica-Bold", 10}],
            {3, 4}],
       Text[FontForm["0", {"Helvetica", 8}],
            {3.5, 3.75}],
```

```
        Text[FontForm["q", {"Symbol", 10}],
             {2.5, 0.75}],
        Text[FontForm["0", {"Helvetica", 8}],
             {3, 0.475}],
        Text[FontForm["Point of\ncollision",
                      {"Helvetica", 9}],
             {10.75, 1}]};
```

```
In[18] := (* ❋ 3.0 *)

        gp5 = {Text[StyleForm[v₀, "Subsubsection"],
                    {3.3, 3.8}],
               Text[θ₀, {2.5, 0.6}],
               Text[StyleForm["Point of\ncollision",
                              FontFamily -> "Helvetica",
                              FontSize -> 9],
                    {10.75, 1}]};
```

The function **Show** displays the combined graphics objects:

```
In[19] := Show[gp1, Graphics[Join[gp2, gp3, gp4, gp5]],
              PlotRange -> {{-1, 17}, {-1, 12}},
              Axes -> False,
              DisplayFunction -> $DisplayFunction,
              AspectRatio -> Automatic];
```

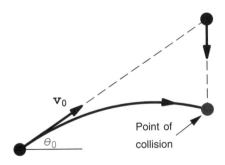

Note that **Show** can combine graphics objects created by built-in plot functions with those constructed explicitly with graphics elements. Furthermore, **Show** renders graphics objects in the order they appear as its arguments, and later objects cover earlier ones in the final display. **Show** also renders the graphics primitives in a graphics object in the order they appear in the list, and later primitives are similarly drawn over earlier ones.

```
In[20] := ClearAll[gp1, gp2, gp3, gp4, gp5]
```

3.5.2.3 TWO-DIMENSIONAL GRAPHICS OPTIONS

Options can be specified for **Show** or inserted directly into graphics objects. If an option appears with different values in the graphics objects and in the **Show** function, the latter specification prevails in the rendering of a piece of graphics.

For two-dimensional graphics, _Mathematica_ permits the specification of a large number of options:

Option Name	Default Value	Description
AspectRatio	1/GoldenRatio	the ratio of height to width for the display area; **Automatic** sets it from the original coordinate system, namely, one unit in the x direction of the original coordinate system has the same size in the display area as one unit in the y direction
Axes	False	whether to include axes; **True** draws all axes and **False** draws no axes
AxesLabel	None	labels to be put on axes; _ylabel_ specifies a label for the y axis; {_xlabel, ylabel_} puts labels on both axes
AxesOrigin	Automatic	the point at which axes cross; {_x, y_} specifies the point where they cross; **Automatic** uses an internal algorithm to determine where the axes should cross
AxesStyle	Automatic	the graphics directive or list of graphics directives with which all axes are to be rendered; {{_xstyle_}, {_ystyle_}} specifies individual styles for axes
Background	Automatic	the background color for a plot; a setting must be a **GrayLevel**, **Hue**, or **RGBColor** directive; **Automatic** produces a white background
ColorOutput	Automatic	the type of color output to produce; **Automatic** uses color specifications as given; **GrayLevel** converts all color specifications to gray levels
DefaultColor	Automatic	the default color to use for lines, points, and so forth; the setting must be a **GrayLevel**, **Hue**, or **RGBColor** directive; **Automatic** gives a color

Option Name	Default Value	Description
		complementary to the background specified
DisplayFunction	$DisplayFunction	the function to apply to graphics objects in order to display them; Identity causes the objects to be returned, but no display to be generated
Epilog	{}	a list of graphics elements to be rendered after the main part of the graphics is rendered
Frame	False	whether to draw a frame around the plot
FrameLabel	None	labels to be placed on the edges of the frame around a plot; {xmlabel, ymlabel} specifies labels for the bottom and left-hand edges of the frame; {xmlabel, ymlabel, xplabel, yplabel} specifies labels for each of the edges of the frame
FrameStyle	Automatic	the graphics directive or list of graphics directives with which the edges of a frame are to be rendered; Automatic uses a default style; {{xmstyle}, {ymstyle}, {xpstyle}, {ypstyle}} specifies different styles for each edge of the frame
FrameTicks	Automatic	what tick marks to draw if there is a frame; None gives no tick marks; {xmticks, ymticks, xpticks, ypticks} specifies tick mark options separately for each edge of the frame; Automatic places tick marks automatically
GridLines	None	what grid lines to include; None gives no grid lines; Automatic places a grid line for every major tick mark; {xgrid, ygrid} specifies the position of grid lines in each direction
PlotLabel	None	an overall label for a plot
PlotRange	Automatic	the range of coordinates to include in the plot; All includes all points;

Option Name	Default Value	Description
		Automatic drops outlying points; $\{min, max\}$ specifies limits for y; $\{\{xmin, xmax\}, \{ymin, ymax\}\}$ gives limits for each coordinate
PlotRegion	Automatic	the region that the plot should fill in the final display area; $\{\{sxmin, sxmax\}, \{symin, symax\}\}$ specifies the region in scaled coordinates; Automatic fills the final display area with the plot
Prolog	{}	a list of graphics elements to be rendered before the main part of the graphics is rendered
RotateLabel	True	labels on vertical frame axes are rotated to be vertical
Ticks	Automatic	what tick marks to draw if there are axes; None gives no tick marks; Automatic places tick marks automatically; $\{xticks, yticks\}$ specifies tick mark options separately for each axis; tick mark option $\{coord1, coord2, \ldots\}$ draws tick marks at the specified positions; $\{\{coord1, label1\}, \ldots\}$, with the specified labels; $\{\{coord1, label1, len1\}, \ldots\}$, with the specified scaled lengths; $\{\{coord1, label1, \{plen1, mlen1\}\}, \ldots\}$, with the specified lengths in the positive and negative directions; $\{\{coord1, label1, len1, style1\}, \ldots\}$, with the specified styles

⌘ 2.2

DefaultFont	$DefaultFont	the default font to use for text in a plot; $\{$ "*font*" , *size*$\}$ specifies the name and size of the font to use; $DefaultFont gives the default font to use; its default value is {"Courier", 10} for both Macintosh and Windows

---

❀ **3.0**

FormatType	$FormatType	the default format type to use for text in a graphic; common settings are OutputForm, StandardForm, and TraditionalForm; the default value of $FormatType is StandardForm; FormatType is also an option for the function Text
TextStyle	$TextStyle	the default style to use for text in a graphic; possible forms of settings are "*style*", which specifies a cell style in the current notebook, {*opt1 -> val1,...*}, which gives a list of option specifications, or {"*style*", *opt1 -> value1,...*}, which stipulates a style modified by option settings; common cell styles are Title, Section, Subsection, Text, Input, and Output; common options are FontSize, FontSlant, FontWeight, FontFamily, FontColor, and Background; setting for FontSlant can be "Plain", "Italic", and "Oblique"; setting for FontWeight can be "Plain" and "Bold"; common choices for FontFamily are "Courier", "Times", and "Helvetica"; setting for FontColor and Background must be a GrayLevel, Hue, or RGBColor directive; TextStyle is also an option for the function Text
ImageSize	Automatic	the absolute size at which to render a graphic; {*x, y*} specifies a width of *x* printer's points and a height of *y* printer's points

Section 2.3.1.2 discussed many of these options. In what follows, we provide examples for several new possibilities. Here is a plot of $I = [\sin\alpha/\alpha]^2$:

```
(* Mathematica Version 2.2 Specifics *)

Plot[(Sin[alpha]/alpha)^2, {alpha, - 3Pi, 3Pi},
     PlotRange -> {0, 1.0},
     Frame -> True,
     PlotLabel -> "DemoPlot",
     FrameLabel -> {FontForm["a", {"Symbol", 10}], "I"}
     ];
```

```
In[21] := (* ❀ 3.0 *)

Plot[(Sin[α]/α)^2, {α, -3Pi, 3Pi},
     PlotRange -> {0, 1.0},
     Frame -> True,
     PlotLabel -> "DemoPlot",
     FrameLabel -> {"α", "I"}
     ];
```

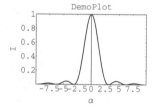

The **FrameTicks** option allows new tick mark specification for each edge of the frame:

```
(* Mathematica Version 2.2 Specifics *)

Show[%%, FrameTicks ->
           {{{-3Pi, FontForm["-3p", ff]},
             {-Pi, FontForm["-p", ff]},
             {Pi, FontForm["p", ff]},
             {3Pi, FontForm["3p", ff]}} /.
                 ff -> {"Symbol", 10},
             {0, 1}}
     ];
```

```
In[23] := (* ❀ 3.0 *)

Show[%, FrameTicks ->
           {{-3 Pi, -Pi, Pi, 3 Pi}, {0, 1}, None, None}];
```

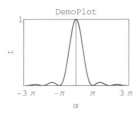

Specification of the **Background** option changes the background color of the plot:

```
In[25] := Show[%, Background -> GrayLevel[0.4]];
```

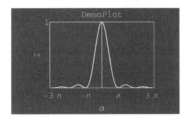

Show[*g1*, *g2*, ...] displays several graphics objects combined. **Epilog** and **Prolog** provide other ways to combine graphics objects for display.

```
In[26] := Plot[Sin[x], {x, 0, 2Pi},
              Epilog -> {{GrayLevel[0.8],
                          Polygon[{{2, 0}, {3, 1}, {4, 0}}]}}
          ];
```

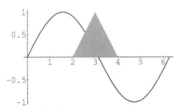

With **Epilog**, the main part of the graphics is rendered first and, therefore, the triangle covers part of the curve.

```
In[27] := Plot[Sin[x], {x, 0, 2Pi},
              Prolog -> {{GrayLevel[0.8],
                          Polygon[{{2, 0}, {3, 1}, {4, 0}}]}}
          ];
```

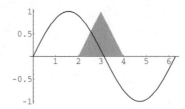

With `Prolog`, the triangle is behind the curve, which is rendered last.

❈ **2.2**   There are three ways to change the fonts for text in graphics: `$DefaultFont`, `DefaultFont` option, and `FontForm`, in increasing order of precedence. Resetting the global variable `$DefaultFont` changes the default font for all subsequent graphical output in a _Mathematica_ session; specifying the `DefaultFont` option affects the default font for a particular plot; using the `FontForm` function imposes the font for a specific piece of text.

❈ **3.0**   There are several ways to change the formats for text in graphics: global variables `$TextStyles` and `$FormatType`, options `TextStyle` and `FormatType`, and functions `StyleForm` and `Text`, in increasing order of precedence. Resetting the global variables `$TextStyles` and `$FormatType` changes the default style and format type for all subsequent graphical output in a _Mathematica_ session; specifying the options `TextStyle` and `FormatType` affects the default style and format type for a particular graphic; using the `StyleForm` and `Text` functions imposes the style and format type for a specific piece of text.

Here is a plot of the charge on the capacitor versus time for the _RC_ circuit:

```
In[28] := Plot[{1, 1 - Exp[-t]}, {t, 0, 4},
            PlotRange -> {0, 1.2},
            PlotStyle -> {Dashing[{0.015, 0.015}],
                          Thickness[0.009]},
            AxesLabel -> {"t", "q"},
            PlotLabel -> "Charge vs. time",
            Ticks -> {{{1, "RC"}}, {{0.632, "Q'"}, {1, "Q"}}},
            Epilog -> {Dashing[{0.015, 0.015}],
                       Line[{{1, 0}, {1, 0.632}}],
                       Line[{{0, 0.632}, {1, 0.632}}]}
          ];
```

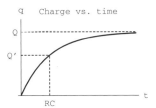

where after one time constant, RC, the charge reaches the value Q', which is 0.632 of the maximum value Q. Let us change the text styles in the plot:

```
(* Mathematica Version 2.2 Specifics *)

Show[%%, DefaultFont -> {"Times-Italic", 10},
        PlotLabel -> FontForm["Charge vs. time",
                                {"Helvetica-Bold", 10}]
    ];
```

```
In[29] := (* ❀ 3.0 *)

Show[%, TextStyle -> {FontFamily -> "Times",
                      FontSlant -> "Italic"},
        PlotLabel ->
            StyleForm["Charge vs. time",
                      FontFamily -> "Helvetica",
                      Fontslant -> "Plain",
                      FontWeight -> "Bold"]
    ];
```

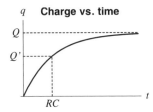

❀ **2.2**   Note that font specifications in **FontForm** have higher precedence than those in **DefaultFont**.

❀ **3.0**   Note that style specifications in **StyleForm** have higher precedence than those in **TextStyle**.

3.5.2.4   WAVE MOTION

This section applies the two-dimensional graphics programming capabilities of *Mathematica* to an introductory physics problem.

■ **The Problem**

A simple harmonic or sine wave travels in a long string stretched along the x direction. Animate the motion of a particle in the string, and demonstrate that the frequency and period of the simple harmonic motion of the particle equal those of the wave motion.

■ **Physics of the Problem**

The equation of the wave is

$$y = A \sin 2\pi(\frac{x}{\lambda} - \frac{t}{T}) \qquad (3.5.1)$$

where

y: the displacements of the particles in the y direction away from their equilibrium positions
x: the distance along the direction of propagation of the wave
t: time
A: amplitude, the magnitude of the maximum y displacement
λ: wavelength, the distance between two adjacent points having the same phase
T: period, the time for one complete vibration to pass a given point

The frequency f of the wave is defined as the number of vibrations per second passing a given point. It is related to the period by

$$f = \frac{1}{T} \qquad (3.5.2)$$

Equation 3.5.1 implies that each particle of the string oscillates in the y direction with simple harmonic motion having the same period and, according to Equation 3.5.2, the same frequency as those of the wave motion.

If y, x, and t are in units of A, $\lambda/2\pi$, and $T/2\pi$, respectively, then Equation 3.5.1 can be written as

$$y = \sin(x - t) \qquad (3.5.3)$$

Without loss of generality, we focus on the motion of the particle at $x = 0$. From Equation 3.5.3, the displacement of the particle as a function of time is

$$y = \sin(-t) \qquad (3.5.4)$$

■ **Solution with *Mathematica***

To animate the motion of the particle as the wave passes by, we create a sequence of graphs of Equations 3.5.3 and 3.5.4 together for a number of instants:

```
In[31] := Do[Show[Graphics[{{RGBColor[0, 0, 1], Line[{{0, -1}, {0, 1}}]},
                    PointSize[0.05], RGBColor[1, 0, 0],
                    Point[{0, Sin[- i (2 Pi/8)]}]}]
                  ],
          Plot[Sin[x - i (2 Pi/8)], {x, -4 Pi, 4 Pi},
               DisplayFunction -> Identity
               ],
          Frame -> True, FrameTicks -> None,
          Background -> GrayLevel[0.7],
          Axes -> False,
          DisplayFunction -> $DisplayFunction
          ];,
    {i, 0, 7}
    ]
```

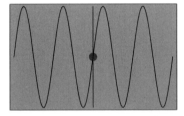

(Only the head cell of the closed group of eight graphics cells is shown.) A vertical line is included in each graph to show that the direction of motion of the particle is transverse to the direction of wave propagation.

To animate the graphics, click the group bracket and press Command-Y for the Macintosh or Control+Y for Windows. The speed of the animation can be controlled with the two buttons on the right of the pallette of buttons at the bottom of the notebook window.

Note that the particle completes a cycle of motion as one wave vibration passes by it. In other words, the frequency and period of the motion of the particle equal those of the wave motion.

3.5.3 *Three-Dimensional Graphics*

3.5.3.1 THREE-DIMENSIONAL GRAPHICS PRIMITIVES

Several three-dimensional graphics primitives are available in *Mathematica*:

`Point[{x, y, z}]`	point with coordinates $\{x, y, z\}$
`Line[{{x1, y1, z1}, {x2, y2, z2}, ...}]`	line through the points $\{x1, y1, z1\}$, $\{x2, y2, z2\}$, ...
`Polygon[{{x1, y1, z1}, {x2, y2, z2}, ...}]`	filled polygon with the specified list of corners

`Cuboid[{`*xmin, ymin, zmin*`}]`	unit cube with edges oriented parallel to the axes; two opposite corners have coordinates {*xmin, ymin, zmin*} and {*xmin* + 1, *ymin* + 1, *zmin* + 1}
`Cuboid[{`*xmin, ymin, zmin*`}, {`*xmax, ymax, zmax*`}]`	cuboid (rectangular parallelepiped) with edges oriented parallel to the axes; two opposite corners have the specified coordinates
`Text[`*expr*`, {`*x, y, z*`}]`	text corresponding to the printed form of *expr*, centered at the point {*x, y, z*}

Mathematica encloses each three-dimensional figure within a cuboidal box. The *x*, *y*, and *z* axes are oriented parallel to the edges of the box and form a right-handed coordinate system. If the option `Boxed -> False` is included in `Show` or `Graphics3D`, the box is not shown explicitly.

Coordinates for graphics primitives can be specified in either original or scaled coordinates. Original coordinates are given in the form `{`*x, y, z*`}`; scaled coordinates, running from 0 to 1 in each dimension, are given in the form `Scaled[{`*sx, sy, sz*`}]`.

```
In[32] := Show[Graphics3D[{
              Cuboid[{0.5, 0.5, 0.5}, {0.95, 0.95, 0.95}],
              Cuboid[Scaled[{0.5, 0.5, 0.5}],
                     Scaled[{0.95, 0.95, 0.95}]],
              Dashing[{0.02, 0.02}],
              Line /@ {{{0.5, 0.5, 0.5}, {0.5, 0.5, 0}},
                       {{0.5, 0.5, 0}, {0.5, 0, 0}},
                       {{0.5, 0.5, 0}, {0, 0.5, 0}}}
                      }
                    ],
              PlotRange -> {{0, 2}, {0, 2}, {0, 2}},
              Axes -> True,
              AxesLabel -> {"x", "y", "z"}
          ];
```

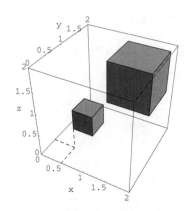

The opposite corners of the smaller cuboid are specified in original coordinates, whereas those of the larger one are given in scaled coordinates. To facilitate visualization, three dashed lines are drawn for locating a corner of the smaller cuboid.

With a sufficient number of appropriate polygons, we can build any three-dimensional object. The package `Graphics`Shapes`` contains functions that generate lists of polygons for some common three-dimensional shapes. `Sphere[]` provides a list of 262 polygons for a sphere:

```
In[33] := Needs["Graphics`Shapes`"]
          Short[Sphere[], 6]
```

```
Out[34]//Short=
    {Polygon[{{0.207912, 0, -0.978148}, {0.406737, 0, -0.913545},
        {0.38683, 0.125689, -0.913545}, {0.197736, 0.0642482, -0.978148}}
        ], Polygon[{{0.406737, 0, -0.913545}, {0.587785, 0, -0.809017},
        {0.559017, 0.181636, -0.809017}, {0.38683, 0.125689, -0.913545}}]
        , Polygon[{{0.587785, 0, -0.809017}, <<3>>}], <<258>>,
    Polygon[<<1>>]}
```

`Show` renders the sphere:

```
In[35] := Show[Graphics3D[Sphere[]]];
```

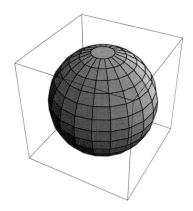

The corners of a polygon must lie in a plane.

```
In[36] := Show[Graphics3D[
                Polygon[{{-1, -1, 0}, {-1, 1, 0}, {1, 1, 0},
                    {1, -1, 0}, {0.5, 0, 1}, {-0.5, 0, 1}}]
                  ],
            Axes -> True
          ];
```

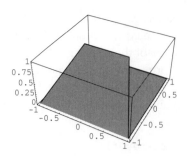

This is not a polygon because the last two corners are off the plane. The corners of a polygon must also form a convex figure.

```
In[37] := Show[Graphics3D[
              Polygon[{{-1, -1, 0}, {-1, 1, 0}, {1, 1, 0},
                      {1, -1, 0}, {0.5, 0, 0}, {-0.5, 0, 0}}]
                  ],
          Axes -> True
          ];
```

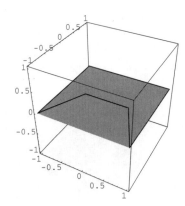

Although the corners are coplanar, this is still not a polygon because it is folded.

In the rendering of two-dimensional graphics, graphics objects that appear later as arguments of Show cover the earlier ones. Similarly, graphics primitives that come later in the list for Graphics hide the earlier primitives. In the displaying of three-dimensional graphics, the order of appearance of the graphics objects in Show or graphics primitives in Graphics3D is irrelevant. In this case, parts of the figure in front obscure those behind them.

```
In[38] := Show[Graphics3D[
              Polygon /@ {{{-1, -1, 0}, {-1, 1, 0},
                          {1, 1, 0}, {1, -1, 0}},
                          {{-1, -1, -1.5}, {1, 1, -1.5},
                          {0, 0, 1}}}
                  ]
          ];
```

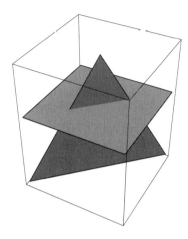

The triangle and square hide portions of each other.

To demonstrate the use of three-dimensional graphics primitives, we choose a problem from introductory electricity and magnetism.

EXAMPLE 3.5.3 Construct a diagram depicting a snapshot of a linearly polarized, sinusoidal, and plane electromagnetic wave propagating in the positive y direction.

We begin with the plot of the electric and magnetic fields as a function of y at a particular instance of time:

```
In[39] := gr1 = ParametricPlot3D[
                  {{0, t, Sin[t]}, {Sin[t], t, 0}},
                  {t, 0, 4.2 Pi}, DisplayFunction -> Identity
                       ];
```

`ParametricPlot3D[{{fx, fy, fz}, {gx, gy, gz}, ...}, {t, tmin, tmax}]` produces together several three-dimensional space curves parameterized by a variable t, which runs from *tmin* to *tmax*. The following list contains the `Line` primitives representing the axes:

```
In[40] := gr2 = Line /@ {{{0, 0, 0}, {0, 4.5Pi, 0}},
                          {{0, 0, 0}, {0, 0, 1}},
                          {{0, 0, 0}, {1, 0, 0}}};
```

The `Line` and `Polygon` primitives in the following list represent the arrows for the electric and magnetic field vectors:

```
In[41] := gr3 = Join[
                 Line /@ Flatten[
                        Table[{{{0, i, 0}, {0, i, Sin[i]}},
                               {{0, i, 0}, {Sin[i], i, 0}}},
                          {i, 0, 4Pi, Pi/4}
                          ], 1
                      ],
```

```
Polygon /@ (Flatten[
        Table[{
    {{0, i, Sin[i]},
     {0, i + del, Sin[i] - Sign[Sin[i]] del},
     {0, i - del, Sin[i] - Sign[Sin[i]] del}
    },
    {{Sin[i], i, 0},
     {Sin[i] - Sign[Sin[i]] del, i + del, 0},
     {Sin[i] - Sign[Sin[i]] del, i - del, 0}
    }
        },
    {i, 0, 4Pi, Pi/4}
        ], 1
        ] /. del -> 0.2)
    ];
```

The Text primitives are for the labels:

```
(* Mathematica Version 2.2 Specifics *)

gr4 = MapThread[Text, {
        MapThread[FontForm, {
            {"x", "y", "z", "E", "B"},
            {{"Times-Italic", 10},
             {"Times-Italic", 10},
             {"Times-Italic", 10},
             {"Times-Bold", 10},
             {"Times-Bold", 10}
            }
                }
            ],
        {{1 + 0.25, 0, 0},
         {0, 4.5Pi + 0.5, 0},
         {0, 0, 1 + 0.25},
         {0, 3 Pi/4 + 0.25, Sin[3 Pi/4] + 0.5},
         {Sin[3 Pi/4] + 0.5, 3 Pi/4 + 0.25, 0}
        }
        }
    ];
```

```
In[42] := (* ❀ 3.0 *)

gr4 = MapThread[Text, {
        MapThread[StyleForm, {
            {"x", "y", "z", "E", "B"},
```

```
                    {opts1, opts1, opts1, opts2, opts2}
                                    }
                        ],
                {{1 + 0.2, 0.5, 0},
                 {0, 4.5Pi + 0.5, 0},
                 {0, 0, 1 + 0.25},
                 {0, 3 Pi/4 + 0.25, Sin[3 Pi/4] + 0.5},
                 {Sin[3 Pi/4] + 0.5, 3 Pi/4 + 0.4, 0}
                }
                    }
        ] /.{opts1 ->
                Sequence[FontFamily -> "Times",
                         FontSlant -> "Italic"],
                opts2 ->
                Sequence[FontFamily -> "Times",
                         FontWeight -> "Bold"]};
```

We have taken the opportunity to illustrate the use of the function `MapThread`. `MapThread[f, {{a1, a2, ...}, {b1, b2, ...}, ...}]` gives $\{f[a1, b1, ...], f[a2, b2, ...], ...\}$. Its use is elegant but optional. Perhaps, it may be clearer just to specify each piece of text explicitly in the form `Text[FontForm[`*expr*`, {"font", size}], {x, y, z}]` for *Mathematica* version 2.2, or in the form `Text[StyleForm[`*expr*`, options], {x, y, z}]` or `Text[`*expr*`, {x, y, z}, options]` for *Mathematica* version 3.0. Finally, `Show` displays the graphics objects combined:

```
In[44] := Show[gr1, Graphics3D[Join[gr2, gr3, gr4]],
            BoxRatios -> {0.3, 1, 0.3},
            Boxed -> False,
            Axes -> False,
            PlotRange -> {{-1, 1}, {0, 4.5 Pi}, {-1, 1}},
            Lighting -> False,
            ViewPoint -> {2.647, 1.137, 1.775},
            DisplayFunction -> $DisplayFunction
            ];
```

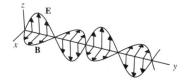

Note that we can combine graphics objects produced by built-in plotting functions with those constructed explicitly from graphics elements.

```
In[45] := ClearAll[gr1, gr2, gr3, gr4, gr5]
```

3.5.3.2 THREE-DIMENSIONAL GRAPHICS DIRECTIVES

The graphics directives introduced in Section 3.5.2.2 for two dimensions remain efficacious in three dimensions: `PointSize`, `AbsolutePointSize`, `Thickness`, `AbsoluteThickness`, `Dashing`, `AbsoluteDashing`, `GrayLevel`, `RGBColor`, and `Hue`. *Mathematica* provides several additional directives for three dimensions: `EdgeForm`, `FaceForm`, and `SurfaceColor`.

Two- and three-dimensional graphics have the same graphics directives for the `Point` and `Line` primitives:

```
In[46] := Show[Graphics3D[{PointSize[0.25],
                  Point[{0, 0, 0}],
                  Hue[1],
                  AbsoluteThickness[8],
                  Line[{{0, 0, -1}, {0, 0, 1}}]}
              ],
          PlotRange -> {{-1, 1}, {-1, 1}, {-1, 1}}
          ];
```

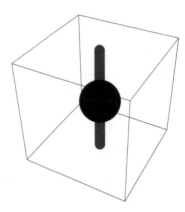

The radius of the point is 25 percent of the total width of the display area; the thickness of the line is about eight printer's points, or $8 \times 1/72$ of an inch.

Just as in two dimensions, color directives affect the rendering of `Text` primitives in three-dimensional graphics:

```
(*   Mathematica Version 2.2 Specifics *)

Show[Graphics3D[{RGBColor[1, 0, 0],
              Text[FontForm["RED", {"Times-Bold", 40}],
                  {0.5, 0.5, 0}]}
          ],
      PlotRange -> {{0, 1}, {0, 1}, {0, 1}}
      ];
```

```
In[47] := (* ☂ 3.0 *)

Show[Graphics3D[{RGBColor[1, 0, 0],
                Text[StyleForm["RED", FontFamily -> "Times",
                                      FontWeight -> "Bold",
                                      FontSize -> 40],
                     {0.5, 0.5, 0}]}
               ],
      PlotRange → {{0, 1}, {0, 1}, {0, 1}}
      ];
```

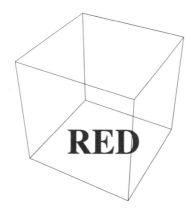

EdgeForm determines how the edges of polygons and cuboids are drawn. With **EdgeForm[]**, no lines are drawn; with **EdgeForm[g]**, lines are drawn according to the specified graphics directive or list of graphics directives *g*.

```
In[48] := Show[Graphics3D[{
                EdgeForm[{RGBColor[1, 0, 0], Thickness[0.03]}],
                Polygon[{{-1, -1, -1.5}, {1, 1, -1.5}, {0, 0, 1}}],
                EdgeForm[],
                Cuboid[{0, -1, -1.5}, {0.75, 0, -0.5}]
                         }
                         ],
              Boxed -> False
              ];
```

No edges are drawn for the cuboid; the edges of the triangle have a thickness of three percent of the total display area width, and their color is red.

How the polygons and cuboids are colored depends on the specification of the option `Lighting` for `Show` or `Graphics3D`. With `Lighting -> False`, the color directives determine their colors.

```
In[49] := Show[Graphics3D[{
               RGBColor[1, 0, 0],
               Polygon[{{-1, -1, -1.5}, {1, 1, -1.5}, {0, 0, 1}}],
               GrayLevel[1],
               Cuboid[{0, -1, -1.5}, {0.75, 0, -0.5}]
                    }
                   ],
             Boxed -> False,
             Lighting -> False
           ];
```

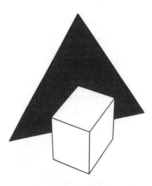

The color of the triangle is red and that of the cuboid is white.

`FaceForm` allows us to give different color specifications to the front and back faces of polygons. `FaceForm[gf, gb]` specifies the directive *gf* for the front faces and the directive *gb* for the back faces. What is a front face? When we look at the front face of a polygon, the consecutive points in `Polygon[{point1, point2, point3, ...}]`, which represents the polygon, proceed in a counterclockwise direction.

```
In[50] := Show[Graphics3D[{
            FaceForm[RGBColor[1, 0, 0], GrayLevel[1]],
            Polygon[{{-1, -1, 0}, {1, -1, 0},
                     {1, 1, 0}, {-1, 1, 0}}],
            Polygon[{{-1, -1, -1.5}, {0, 0, 1},
                     {1, 1, -1.5}}]
                }
              ],
         Axes -> True,
         AxesLabel -> {"x", "y", "z"},
         Lighting -> False
         ];
```

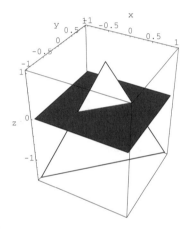

We are looking at the front of the square and the back of the triangle. **FaceForm** allows us to create some interesting visual effects:

```
In[51] := Needs["Graphics`Shapes`"]
          Show[Graphics3D[{
              FaceForm[RGBColor[1, 1, 0], GrayLevel[0.35]],
              Drop[Sphere[], -54]
                   }
                 ],
             Lighting -> False,
             Boxed -> False
             ];
```

The front and back faces of the polygons have different colors; the inside of the sphere glows in bright yellow, and the outside is shrouded with dark gray.

With the default option setting `Lighting -> True` for `Graphics3D`, _Mathematica_ uses a simulated lighting model to color the faces of polygons and cuboids in three-dimensional graphics. The color directives are ignored. The lighting model has three components: the graphics directive `SurfaceColor` and the graphics options `AmbientLight` and `LightSources`. `AmbientLight` and `LightSources` control how surfaces are illuminated, and `SurfaceColor` determines how light is reflected from the surfaces. The graphics options will be discussed in the next section.

With a varying between 0 and 1, `SurfaceColor[GrayLevel[a]]` specifies that the polygons should act as diffuse, white reflectors of light and that the intensity of reflected light is the product of a, $\cos(\alpha)$, and the intensity of the incident light, where α is the angle the incident light makes with the normal to the surface. With `SurfaceColor[GrayLevel[0]]`, there is no reflection, and the surfaces appear black. If no `SurfaceColor` directives are specified, _Mathematica_ assumes the default setting `SurfaceColor[GrayLevel[1]]`. With `SurfaceColor[RGBColor[r, g, b]]`, the intensities of the red, green, and blue components of the reflected light should be respectively r, g, and b times the intensities of the corresponding components of the incident light, multiplied by $\cos(\alpha)$. `FaceForm` permits different `SurfaceColor` specifications for the front and back faces of polygons.

In what follows, the illumination is the same for the three triangles. The `SurfaceColor` directive is omitted for the first triangle, whereas the second triangle is given `SurfaceColor[GrayLevel[0]]` and the third is assigned `SurfaceColor[RGBColor[1, 0, 0]]`. Consequently, the first triangle has the color of the default illumination, the second shows black, and the third displays red.

```
In[53] := Show[GraphicsArray[{
            Graphics3D[{
              Polygon[{{-1, -1, -1.5}, {1, 1, -1.5}, {0, 0, 1}}]
                  }
                ],
            Graphics3D[{
              SurfaceColor[GrayLevel[0]],
              Polygon[{{-1, -1, -1.5}, {1, 1, -1.5}, {0, 0, 1}}]
                  }
                ],
            Graphics3D[{
              SurfaceColor[RGBColor[1, 0, 0]],
              Polygon[{{-1, -1, -1.5}, {1, 1, -1.5}, {0, 0, 1}}]
                  }
                ]
                  }
                ]
          ];
```

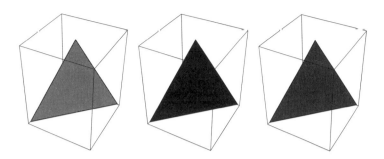

To illustrate the use of three-dimensional graphics directives, we select a problem from special relativity.

EXAMPLE 3.5.4 An element of volume in the form of a small cube is at rest in the O' reference frame that is moving at a velocity **v** with respect to the O reference frame along the positive direction of the common y axis. The two frames overlap when their origins coincide. Construct a figure depicting the situation.

Here are the graphics elements for the reference frame O:

```
In[54] := gra1 = {
              EdgeForm[{Thickness[0.008]}], GrayLevel[0.85],
              Polygon /@ {
                  {{0, 0, 0}, {2, 0, 0}, {2, 2, 0}, {0, 2, 0}},
                  {{0, 0, 0}, {2, 0, 0}, {2, 0, 2}, {0, 0, 2}},
                  {{0, 0, 0}, {0, 2, 0}, {0, 2, 2}, {0, 0, 2}}
                          }
              };
```

Here are those for the reference frame O':

```
In[55] := gra2 = {
              EdgeForm[{Thickness[0.008]}], GrayLevel[1],
              Polygon /@ {
                  {{0, 1, 0}, {1.5, 1, 0}, {1.5, 2.5, 0},
                   {0, 2.5, 0}},
                  {{0, 1, 0}, {1.5, 1, 0}, {1.5, 1, 1.5},
                   {0, 1, 1.5}},
                  {{0, 1, 0}, {0, 2.5, 0}, {0, 2.5, 1.5},
                   {0, 1, 1.5}}
                          }
              };
```

The graphics elements for the cuboid are

```
In[56] := gra3 = {GrayLevel[0.75],
                  Cuboid[{0.5, 1.75, 0.75}, {0.75, 2, 1}]};
```

The following `Line` primitives are for the lines indicating the position of the cuboid:

```
In[57] := gra4 = Line /@ {
              {{1.25/2, 3.75/2, 1.75/2}, {1.25/2, 3.75/2, 0}},
              {{1.25/2, 3.75/2, 0}, {1.25/2, 0, 0}},
              {{1.25/2, 3.75/2, 0}, {0, 3.75/2, 0}}
                         };
```

The graphics elements for the velocity vector are

```
In[58] := gra5 = {
             {Thickness[0.008], Line[{{0, 1, 1}, {0, 1.7, 1}}]},
             Polygon[{{0, 1.75, 1}, {0, 1.75 - 0.1, 1 + 0.075},
                 {0, 1.75 - 0.1, 1 - 0.075}}]
                  };
```

The following `Text` primitives are for the labels:

```
(* Mathematica Version 2.2 Specifics *)

gra6 = ({
    Text[FontForm["v", {"Helvetica-Bold", 10}],
         {0, 1.6, 1.2}],
    Text[FontForm["x", fn], {2 - 0.2, 0.2, 0}],
    Text[FontForm["z", fn], {0, 0.2, 2 - 0.2}],
    Text[FontForm["x'", fn], {1.5 - 0.2, 1 + 0.2, 0}],
    Text[FontForm["y'", fn], {0, 2.5 - 0.2, 0.2}],
    Text[FontForm["z'", fn], {0, 1 + 0.2, 1.5 - 0.2}],
    Text[FontForm["O", fn], {0, 0.125, 0.125}],
    Text[FontForm["O'", fn], {0, 1 + 0.15, 0.15}]
         } /. fn -> {"Helvetica-Italic", 8});
```

```
In[59] := (* ※ 3.0 *)

gra6 = ({
    Text[StyleForm["v", fn1], {0, 1.6, 1.2}],
    Text[StyleForm["x", fn2], {2 - 0.2, 0.2, 0}],
    Text[StyleForm["z", fn2], {0, 0.2, 2 - 0.2}],
    Text[StyleForm["x'", fn2], {1.5 - 0.2, 1 + 0.2, 0}],
    Text[StyleForm["y'", fn2], {0, 2.5 - 0.2, 0.2}],
    Text[StyleForm["z'", fn2], {0, 1 + 0.2, 1.5 - 0.2}],
    Text[StyleForm["O", fn2], {0, 0.125, 0.125}],
    Text[StyleForm["O'", fn2], {0, 1 + 0.15, 0.15}]
         } /.
    {fn1 -> Sequence[FontFamily -> "Helvetica",
                     FontWeight -> "Bold",
                     FontSize -> 10],
```

```
        fn2 -> Sequence[FontFamily -> "Helvetica",
                        FontSlant -> "Italic",
                        FontSize -> 8]});
```

Finally, **Show** displays the graphics object:

```
In[61] := Show[Graphics3D[{gra1, gra2, gra3, gra4, gra5, gra6}],
              ViewPoint -> {2.647, 1.137, 1.775},
              Lighting -> False,
              Boxed -> False];
```

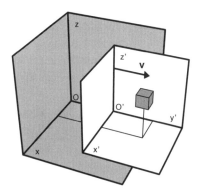

```
In[62] := ClearAll[gra1, gra2, gra3, gra4, gra5, gra6]
```

3.5.3.3   THREE-DIMENSIONAL GRAPHICS OPTIONS

Whereas graphics directives control the rendering of particular graphics primitives in a graphics object, graphics options affect globally the display of the plot. *Mathematica* offers many graphics options for three-dimensional graphics.

Several graphics options, discussed in Section 3.5.2.3, work the same in three dimensions as in two dimensions: **Background**, **ColorOutput**, **DisplayFunction**, **PlotLabel**, **PlotRegion**, and, in *Mathematica* version 2.2, **DefaultFont**, or, in *Mathematica* version 3.0, **TextStyle** as well as **FormatType**.

A number of the options, introduced in Section 3.5.2.3 for two dimensions, can be generalized to three dimensions:

Option Name	Default Value	Description
AspectRatio	Automatic	the ratio of height to width for the display area; **Automatic** derives the ratio from **BoxRatios**, that is, the ratios of side lengths for the bounding box; if a numerical value is explicitly specified, the values of **BoxRatios** are forced to adjust so that the plot may fill the display area

Axes	**False**	whether to include axes; **True** draws x, y, and z axes on the edges of the cuboidal bounding box; **False** draws no axes
AxesLabel	**None**	labels to be put on axes; **None** gives no axis labels; *zlabel* specifies a label for the z axis; {*xlabel, ylabel, zlabel*} puts labels on all three axes
AxesStyle	**Automatic**	the graphics directive or list of graphics directives with which all axes are to be rendered; {{*xstyle*}, {*ystyle*}, {*zstyle*}} specifies separate styles for each axis
PlotRange	**Automatic**	the ranges of x, y, and z coordinates to include in the cuboidal bounding box; **All** includes all parts; **Automatic** tries to use an internal algorithm to include the interesting parts of a plot, but to drop outlying parts; {*min, max*} specifies limits for z; {{*xmin, xmax*}, {*ymin, ymax*}, {*zmin, zmax*}} gives explicit limits for each coordinate
Ticks	**Automatic**	what tick marks to draw if there are axes; **None** draws no tick marks; **Automatic** places tick marks automatically; {*xtick, ytick, ztick*} specifies tick mark options separately for each axis; tick mark option {*coord1, coord2, …*} draws tick marks at the specified positions; {{*coord1, label1*}, …}, with the specified labels; {{*coord1, label1, len1*}, …}, with the specified scaled lengths; {{*coord1, label1, {plen1, mlen1}*}, …}, with the specified lengths in the positive and negative directions; {{*coord1, label1, len1, style1*}, …}, with the specified styles

To illustrate the use of some of these graphics options, let us consider an example.

```
In[63] := Show[Plot3D[Sin[Sqrt[x^2 + y^2]]/Sqrt[x^2 + y^2],
               {x, -8Pi, 8Pi}, {y, -8Pi, 8Pi},
               PlotPoints -> 50,
               DisplayFunction -> Identity
             ],
          Graphics3D[Text["Demo Plot", {14, 20, 2.0}]],
          PlotRange -> {-0.5, 1.2},
          Axes -> True,
          AxesLabel -> {"x", "y", "z"},
          DisplayFunction -> $DisplayFunction
        ];
```

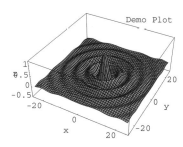

PlotRange specifies the range of z coordinates to include in the cuboidal box; **Axes** stipulates the drawing of the x, y, and z axes on the edges of the bounding box.

```
(* Mathematica Version 2.2 Specifics *)

Show[%%,
     DefaultFont -> {"Times", 10},
     Ticks ->
      {spec, spec, {{0, FontForm["0", fs]},
                    {1, FontForm["1", fs]}}
     } //.
        {spec ->
           ({#, FontForm[g[#], fs]}& /@
            Table[(3i)Pi, {i, -2, 2}]),
         fs -> {"Symbol", 9},
         g[n_Integer Pi|n:0] -> n p}
     ];
```

```
In[64] := (* ✳ 3.0 *)

   Show[%,
        TextStyle -> {FontFamily -> "Times", FontSize -> 10},
        Ticks -> ({spec, spec,
                   {{0, StyleForm["0", fs]},
                    {1, StyleForm["1", fs]}}
                  } //.
                  {spec ->
                      ({#, StyleForm[#, fs]}& /@
                       Table[(3 i)Pi, {i, -2, 2}]),
                   fs -> (FontSize -> 9)})
       ];
```

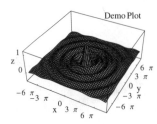

Tick marks are drawn at the specified positions with the given labels. In *Mathematica* version 2.2, `DefaultFont` changes the default font for the plot. In *Mathematica* version 3.0, `TextStyle` changes the default text style for the plot. The `ReplaceRepeated` operator `//.` repeatedly performs replacements until the result no longer changes. Using the pure function is elegant but unnecessary. We can enter the tick mark specifications explicitly:

```
(*  Mathematica Version 2.2 Specifics *)

Show[%%, DefaultFont -> {"Times", 12},
        Ticks -> {
        {
         {-6Pi, FontForm["-6p", fn]},
         {-3Pi, FontForm["-3p", fn]},
         {0, FontForm["0", fn]},
         {3Pi, FontForm["3p", fn]},
         {6Pi, FontForm["6p", fn]}
        },
        {
         {-6Pi, FontForm["-6p", fn]},
         {-3Pi, FontForm["-3p", fn]},
         {0, FontForm["0", fn]},
         {3Pi, FontForm["3p", fn]},
         {6Pi, FontForm["6p", fn]}
        },
        {
         {0, FontForm["0", fn]},
         {1, FontForm["1", fn]}
        }
                } /. fn -> {"Symbol", 9}

    ];
```

```
(* ❋ 3.0  *)

Show[%, TextStyle -> {FontFamily -> "Times", FontSize -> 12},
    Ticks -> {
            {
             {-6Pi, StyleForm["-6π", fn]},
```

```
                {-3Pi, StyleForm["-3π", fn]},
                {0, StyleForm["0", fn]},
                {3Pi, StyleForm["3π", fn]},
                {6Pi, StyleForm["6π", fn]}
            },
            {
                {-6Pi, StyleForm["-6π", fn]},
                {-3Pi, StyleForm["-3π", fn]},
                {0, StyleForm["0", fn]},
                {3Pi, StyleForm["3π", fn]},
                {6Pi, StyleForm["6π", fn]}
            },
            {
                {0, StyleForm["0", fn]},
                {1, StyleForm["1", fn]}

            }
        } /. fn -> (FontSize -> 9)
    ];
```

The choice is a matter of temperament.

Here are some of the graphics options that *Mathematica* provides specifically for three dimensions:

Option Name	Default Value	Description
AmbientLight	GrayLevel[0.]	the level of simulated ambient illumination of polygons and cuboids; the setting must be a **GrayLevel**, **Hue**, or **RGBColor** directive; the default setting **GrayLevel[0.]** gives no ambient lighting; with the option setting **Lighting -> False**, *Mathematica* does not use simulated illumination and ignores the **AmbientLight** specification
AxesEdge	Automatic	where to draw axes; **Automatic** uses an internal algorithm to decide on which exposed box edges axes should be drawn; {{*ydir, zdir*}, {*xdir, zdir*}, {*xdir, ydir*}} specifies on which three edges of the bounding box axes are drawn; the *idir* must be either $+1$ or -1, and specify whether axes are drawn on the edge of the box with a larger or smaller value of coordinate *i*, respectively; any pair {*idir, jdir*} can be replaced by **Automatic** or **None**; **None** draws no axis

Option Name	Default Value	Description
Boxed	True	whether to draw a cuboidal bounding box around the graphics; **True** draws the edges of the box; **False** omits the box
BoxRatios	Automatic	the ratios of side lengths for the bounding box; **Automatic** determines the ratios using the range of actual coordinate values in the plot; if a numerical value is explicitly specified for the option **AspectRatio**, the values of **BoxRatios** are forced to adjust so that the plot may fill the display area
BoxStyle	Automatic	the graphics directive or list of graphics directives with which the bounding box is to be rendered; **Automatic** uses a default style
DefaultColor	Automatic	the default color to use for the edges of the cuboidal bounding box around the graphics; the setting must be a **GrayLevel**, **Hue**, or **RGBColor** directive; **Automatic** gives a color complementary to the background specified
FaceGrids	None	whether to draw grid lines on the faces of the bounding box; **All** draws grid lines on all faces; **None** draws no grid lines; {*face1*, *face2*, ...} draws grid lines on the specified faces; faces are specified as {*xdir*, *ydir*, *zdir*}, where two of the *idir* must be 0, and the third one must be either +1 or −1; {{*face1*, {*xgrid*, *ygrid*}}, ...} specifies an arrangement of grid lines; {*xgrid*, *ygrid*} specifies the position of grid lines in each direction
Lighting	True	whether simulated illumination of polygons and cuboids is used; **True** uses simulated illumination, **False** does not
LightSources	{{{1., 0., 1.}, RGBColor[1, 0, 0]}, {{1., 1., 1.}, RGBColor[0, 1, 0]}, {{0., 1., 1.}, RGBColor[0, 0, 1]}}}	positions and colors of point light sources for simulated illumination of polygons and cuboids; specifications are given in the form {{*pos1*, *color1*}, {*pos2*, *color2*}, ...}; a position is specified as {*x, y, z*} in the display coordinate system, and a color can be specified by **GrayLevel**, **Hue**, or

Option Name	Default Value	Description
		RGBColor; with the option setting Lighting -> False, *Mathematica* does not use simulated illumination and ignores the LightSources specification
RenderAll	True	whether to draw all polygons; True draws all polygons, starting from the back; False draws only those polygons or parts of polygons that are visible in the final image
Shading	True	whether surfaces should be shaded; True gives shaded surfaces; False yields white ones
SphericalRegion	False	whether the final image should be scaled so that a sphere drawn around the three-dimensional bounding box would fit in the display area specified; True scales the final image so that a sphere drawn around the bounding box always fits in the display area specified; the center of the sphere is at the center of the box; False scales the image to be as large as possible, given the specified display area
ViewCenter	Automatic	the point that appears at the center of the display area in the final plot; Automatic centers the whole bounding box in the final image area; $\{x, y, z\}$ specifies the point in scaled coordinates, which run from 0 to 1 across each dimension of the bounding box; SphericalRegion -> True always centers the circumscribing sphere, regardless of the setting for ViewCenter
ViewPoint	{1.3, -2.4, 2.}	the point in space from which the objects plotted are to be viewed; $\{x, y, z\}$ gives the position of the viewpoint in a special coordinate system in which the longest side of the bounding box has length 1 and the center of the box has coordinates $\{0, 0, 0\}$; viewpoint must lie outside the bounding box; see Section 2.3.2.2

Option Name	Default Value	Description
`ViewVertical`	`{0., 0., 1.}`	what direction should be vertical in the final image; $\{x, y, z\}$ specifies the direction in scaled coordinates, which run from 0 to 1 across each dimension of the bounding box; $\{x, y, z\}$ and $\{rx, ry, rz\}$ are equivalent specifications, where r is a real constant

With the default option setting `Lighting -> True`, *Mathematica* uses simulated lighting to color polygons and cuboids. The simulated illumination has three components: `SurfaceColor`, `AmbientLight`, and `LightSources`. Section 3.5.3.2 discussed the `SurfaceColor` directive; this section considers the other two components. The `AmbientLight` option assigns the color for diffuse isotropic lighting. The `LightSources` option specifies the positions and colors of a collection of point light sources. For both `AmbientLight` and `LightSources`, colors can be given by `GrayLevel`, `RGBColor`, or `Hue`. The position of a light source is given as $\{x, y, z\}$ in the display coordinate system where x increases horizontally from left to right and y increases vertically from low to high across the display area. The z axis is perpendicular to the display area and is directed toward the viewer. What matter are the directions of the light sources; their distances from the object are irrelevant, that is, $\{x, y, z\}$ and $\{rx, ry, rz\}$ are equivalent if r is a normalization factor. Let us demonstrate the effects of different settings for `AmbientLight` and `LightSources`. We begin with the default settings for `AmbientLight` and `LightSources`.

```
In[66] := Needs["Graphics`Shapes`"]
          Show[Graphics3D[Torus[]],
               Boxed -> False,
               RenderAll -> False];
```

`Boxed -> False` omits the cuboidal bounding box, and `RenderAll -> False` limits the rendering to only what is visible in the final image. Let us remove the light sources and allow only isotropic illumination:

```
In[68] := Show[Graphics3D[Torus[],
                   AmbientLight -> GrayLevel[0.5],
                   LightSources -> {},
```

```
                          Boxed -> False,
                          RenderAll -> False
                     ]
     ];
```

The torus is uniformly shaded with a gray level of 0.5. Finally, we place a single white light source on the right and use no ambient light:

```
In[69] := Show[Graphics3D[Torus[],
                LightSources -> {{{1, 0.10, 0.25}, GrayLevel[1]}},
                Boxed -> False,
                RenderAll -> False
                    ]
          ];
```

Parts of the torus are now in the shadow.

With the setting **Axes -> True**, the option **AxesEdge** specifies on which edges of the bounding box the axes should be drawn. For each axis, there are four choices. **AxesEdge -> {{ydirx, zdirx}, {xdiry, zdiry}, {xdirz, ydirz}}** gives separate specifications for each of the x, y, and z axes, where *idirj* must be either $+1$ or -1, depending on whether the j axis is to be drawn on the box edge with a larger or smaller value of coordinate i, respectively.

```
In[70] := Show[
             Graphics3D[{},
                PlotRange -> {{-10, 10}, {-10, 10}, {-10, 10}},
                Axes -> True,
                AxesLabel -> {"x", "y", "z"},
                AxesEdge -> {{+1, +1}, {+1, -1}, {-1, -1}}
                    ]
             ];
```

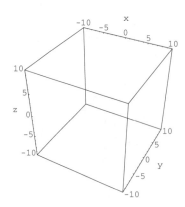

The x axis is on the box edge with $y = 10$ and $z = 10$; the y axis, with $x = 10$ and $z = -10$; the z axis, with $x = -10$ and $y = -10$.

With the default option setting `SphericalRegion -> False`, *Mathematica* scales the objects in a three-dimensional picture so that it is as large as possible, given the final display area. Consequently, the sizes of the objects may depend on the viewpoint. This dependence of sizes of objects on their orientations can cause problems in an animation sequence involving different viewpoints. Consider the rotation of a cuboid. The package `Graphics`Animation`` contains the function `SpinShow`, which generates sequences of three-dimensional graphics for rotational animations. For `SpinShow`, `SphericalRegion -> True` is the default option setting, which is different from that for `Graphics3D`.

```
In[71] := Needs["Graphics`Animation`"]
          SpinShow[Graphics3D[Cuboid[{0, 0, 0}, {2, 0.5, 0.5}]],
              PlotRange -> {{0, 2}, {0, 1}, {0, 1}},
              SphericalRegion -> False
          ];
```

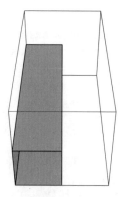

To animate, click the group bracket and press Command-Y for the Macintosh or Control+Y for Windows. (Only the head cell of the closed group of 24 graphics cells is shown here.) With the option setting `SphericalRegion -> False`, the cuboid wobbles as it rotates because its size depends on the orientation. Let us change the option setting to `SphericalRegion -> True`, which is the default setting for `SpinShow`:

```
In[73] := SpinShow[Graphics3D[Cuboid[{0, 0, 0}, {2, 0.5, 0.5}]],
              PlotRange -> {{0, 2}, {0, 1}, {0, 1}}
            ];
```

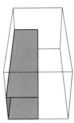

To animate, click the group bracket and press Command-Y for the Macintosh or Control+Y for Windows. (Only the head cell of the closed group of 24 graphics cells is shown here.) With the default option setting `SphericalRegion -> True` for `SpinShow`, the cuboid rotates in a manner intended because *Mathematica* scales the cuboid so that a sphere drawn around the bounding box always fits the display area. In this case, the cuboid remains consistent in size, regardless of its orientation.

3.5.3.4 CRYSTAL STRUCTURE

This section applies the three-dimensional graphics programming capabilities of *Mathematica* to an introductory solid state physics problem.

■ The Problem

Construct the sodium chloride structure, showing a conventional cubic cell.

■ Physics of The Problem

The sodium chloride structure consists of equal numbers of sodium and chlorine ions located at alternate sites of a simple cubic lattice. Since there are two kinds of ions, the structure no longer has the translational symmetry of the simple cubic Bravais lattice. It can be described as a face-centered cubic Bravais lattice with a basis of a sodium ion and a chlorine ion.

■ Solution with *Mathematica*

Comments describing the program are embedded in the input.

```
In[74] := (* a nested list of coordinates for some points
              of a simple cubic Bravais lattice *)
          coord = Table[{i, j, k}, {i, 0, 2}, {j, 0, 2}, {k, 0, 2}];
```

```
(* lines joining the nearest neighbors *)
lines = Line /@
    Join[Flatten[coord, 1],
        Map[RotateRight, Flatten[coord, 1], {2}],
        Map[RotateLeft, Flatten[coord, 1], {2}]
        ];

(* points representing the sodium ions *)
sodium = Point /@
        Select[Flatten[coord, 2], OddQ[Plus @@ #]&];

(* points representing the chlorine ions *)
chlorine = Point /@
        Select[Flatten[coord, 2], EvenQ[Plus @@ #]&];

(* showing the sodium chloride crystal structure *)
Show[Graphics3D[{
        lines,
        PointSize[0.06], chlorine,
        PointSize[0.04], RGBColor[1, 0, 0], sodium
                }
                ],
    Boxed -> False,
    ViewPoint -> {2.700, -1.825, 0.910},
    PlotRange -> {{-0.15, 2.15}, {-0.15, 2.15}, {-0.15, 2.15}}
    ];

Clear[coord, lines, sodium, chlorine]
```

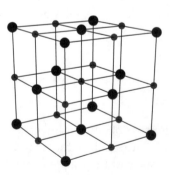

The smaller and larger points represent the sodium and chlorine ions, respectively.

3.5.4 Exercises

1. Construct a diagram, as shown, illustrating Kepler's second law stating that a line joining the Sun and any planet sweeps out equal areas in equal time intervals.

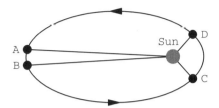

2. Construct a six-pointed star by placing two triangles together with one inverted over the other, place a point with a radius that is equal to 5% of the display width at each corner, and give a random color to each point.

3. Draw an Atwood's machine.

*4. Write a function that generates a sequence of graphs for animating the lecture demonstration discussed in Example 3.5.2. The sequence should end when the projectile hits the target. Also, the function should have two arguments: initial speed and elevation, that is, v_0 and θ_0.

*5. Animate the propagation of the electromagnetic wave shown in Example 3.5.3.

6. Construct the face-centered cubic Bravais lattice, as shown.

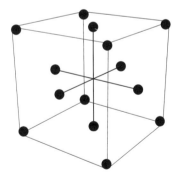

3.6   *Programming Styles*

Before focusing on programming styles, let us expose a common pitfall en route to effective use of *Mathematica*. Dazzled by the power of *Mathematica*, we may be awed into relinquishing our mind to it. We must keep thinking for several important reasons.

First, *Mathematica* cannot save us from bad physics. Before solving a problem with *Mathematica*, clearly state the problem and carefully summarize the physics for the problem.

As will be shown in Chapters 4, 5, and 6, each notebook should have three sections: The Problem, Physics of the Problem, and Solution with *Mathematica*.

Second, *Mathematica* can make mistakes. Vigilant users can catch the mistakes! For example, Wolfram [Wol91] points out that **NIntegrate**, with the default options, gives a wrong answer for the integral

$$\int_{-1000}^{1000} e^{-x^2}\, dx$$

Let us examine the error:

```
In[1] := NIntegrate[Exp[-x^2], {x, -1000, 1000}]

Out[1] = 1.34946 10^-26
```

Since $\exp(-x^2)$ is positive and is equal to 1 at $x = 0$, the integral cannot possibly be that small. The problem is that *Mathematica* made some invalid numerical assumptions about the integrand. To obtain the correct answer, simply use **Integrate**:

```
In[2] := Integrate[Exp[-x^2], {x, -1000, 1000}]//N

Out[2] = 1.77245
```

Some *Mathematica* errors are more subtle than this. Consider, for instance, a problem reported to MathGroup, an electronic forum. Let us give **f** a definition

```
In[3] := ClearAll[f]
         f[x_] := x^2
```

and later another definition

```
In[5] := f[y_] := y^4
```

Since the argument or pattern name, be it **x** or **y**, is irrelevant, the second definition should supersede the first and the value of, for example, **f[2]** should be **16**:

```
In[6] := f[2]

Out[6] = 4
```

Contrary to expectation, *Mathematica* used the first definition. The **?** operator reveals that *Mathematica* stores both definitions in the global rule base, giving higher priority to the first definition:

```
In[7] := ?f

      Global`f
      f[x_] := x^2

      f[y_] := y^4
```

(This problem cannot, perhaps, be called a bug because developers of *Mathematica* are well aware of this design flaw, which is quite difficult to mend. See [AHR94] for a discussion on the errors and subtleties of four widely used computer algebra systems: Derive, MACSYMA, Maple, and *Mathematica*.)

Finally, we must think in order to make efficient use of *Mathematica*. Improvement of seconds or even minutes on CPU time may be immaterial. Yet improvement of more than an hour is worthwhile. For example, consider plotting the series

$$v(x, y) = \frac{4}{\pi} \sum_{m=1}^{400} \frac{1}{(2m-1)} \sin(2m-1)\pi y\, e^{-(2m-1)\pi x}$$

which can be entered as

```
v[x_, y_] := ((4/Pi) Sum[(1/(2 m - 1)) Sin[(2 m - 1) Pi y]
                        Exp[-(2 m - 1) Pi x],
                    {m, 1, 400}])
```

Plotting `v[x, y]` with the command

```
Plot3D[v[x, y], {x, 0, 1.3}, {y, 0, 1.0},
       PlotRange -> {0, 1.2},
       ViewPoint -> {1.741, -2.565, 1.356},
       PlotPoints -> 30, BoxRatios -> {1, 1, 1},
       Boxed -> False, Axes -> False
    ];
```

takes hours on a Macintosh IIci or a Gateway 2000 4DX2-66V. By simply wrapping the function `Evaluate`, discussed in Section 2.3.1.1, around `v[x, y]` like

```
Plot3D[Evaluate[v[x, y]], {x, 0, 1.3}, {y, 0, 1.0},
       PlotRange -> {0, 1.2},
       ViewPoint -> {1.741, -2.565, 1.356},
       PlotPoints -> 30, BoxRatios -> {1, 1, 1},
       Boxed -> False, Axes -> False
    ];
```

we can reduce the plotting time to 163 seconds on a Macintosh IIci and 36 seconds on a Gateway 2000 4DX2-66V. (In this book, all timings have been performed with *Mathematica* version 2.2.)

Though *Mathematica*'s interactive capabilities are sufficient for simple problems, sooner or later we must write programs in order to solve more advanced problems. Computer programs are sets of instructions for computers to solve particular problems. This book has already covered the essential programming concepts of *Mathematica*. There are many ways in which we can use them to solve a problem and to interpret and visualize the result. Our particular approach depends on the problem and on our computational background as well as temperament.

There are three popular programming styles: procedural, functional, and rule-based programming. Section 3.6.1 discusses procedural programming, which is the approach of traditional languages such as FORTRAN, Pascal, and C. If we are already familiar with one or more of these languages, we are naturally prone to write all *Mathematica* programs in this style. Be aware that, for some problems, one of the other two programming styles may be more suitable. Section 3.6.2 examines functional programming and provides an example favoring this programming style. Similarly, Section 3.6.3 explicates rule-based programming.

A few words about clarity of programs are in order. The hallmark of computer programs is readability. Here is an opaque code for generating the crystal structure of sodium chloride:

```
Table[{i, j, k}, {i, 0, 2}, {j, 0, 2}, {k, 0, 2}];
Line /@ Join[Flatten[%, 1],
Map[RotateRight, Flatten[%, 1], {2}],
Map[RotateLeft, Flatten[%, 1], {2}]];
Point /@ Select[Flatten[%%, 2], OddQ[Plus @@ #]&];
Point /@ Select[Flatten[%%%, 2], EvenQ[Plus @@ #]&];
Show[Graphics3D[{%%%, PointSize[0.06], %,
PointSize[0.04], RGBColor[1, 0, 0], %%}],
Boxed -> False, ViewPoint -> {2.700, -1.825, 0.910}];
```

Such lack of clarity can be a major source of errors and reader frustration. Providing explanatory comments, following good formatting practices, and making assignments, if permitted, for temporary variables with descriptive names often add clarity to the program. Section 3.5.3.4, "Crystal Structure," shows an improved version of this code.

```
In[8] := ClearAll[f, v]
```

3.6.1 *Procedural Programming*

A procedural program consists of a procedure discussed in Section 3.4. For altering the sequential flow of control within a procedure, the conditional functions `If`, `Which`, and `Switch` permit branching and the iteration functions `Do`, `While`, and `For` allow looping. Procedural programs are often written as functions of the form

name[*arg1_, arg2_,* ...] := `Module`[{*name1, name2* = *val2,* ...}, *procedure*]

where the procedure is wrapped in a module. Within the procedure, there may be other modules or calls to other functions of this form. `Module` provides localization of symbols and thus sets up modules that are somewhat independent of each other.

The functions `electrostaticField` in Section 3.4.4 and `motion1DPlot` in Section 3.4.5 are procedural programs. This section presents an additional program. Consider again Example 2.1.7 on the motion of a planet under the gravitational attraction of the Sun. Whereas we used *Mathematica* interactively to plot the orbit of a planet before, we define here the function `planetMotion` to be a procedure that generates the graphics for animating the motion of a planet.

A planet moves in a plane defined by its initial position vector $\mathbf{r}_0$ from the Sun located at the origin and its initial velocity $\mathbf{v}_0$. Its energy is

$$E = \frac{1}{2}mv^2 - \frac{GMm}{r}$$

where m, v, and r are the planet's mass, speed, and distance from the Sun, respectively. Also,

G is the gravitational constant and M is the mass of the Sun. The angular momentum of the planet is

$$\mathbf{L} = m(\mathbf{r} \times \mathbf{v})$$

where $\mathbf{r}$ and $\mathbf{v}$ are, respectively, its position vector and velocity. If $L = 0$, the path of the planet is a straight line through the origin. If $L \neq 0$, its orbit is an ellipse, parabola, or hyperbola, depending on whether $E < 0, E = 0$, or $E > 0$, respectively. Since m is finite as well as positive and E together with L are constants of motion, we can express these conditions in terms of the parameters e and l defined by

$$e = \frac{E}{m} = \frac{v_{x0}^2 + v_{y0}^2}{2} - \frac{4\pi^2}{\sqrt{x_0^2 + y_0^2}}$$

and

$$l = \frac{L}{m} = \left| x_0 v_{y0} - y_0 v_{x0} \right|$$

where $x_0, y_0, v_{x0},$ and v_{y0} are the initial coordinates and components of the initial velocity and we have adopted a system of units in which length is in astronomical units and time is in years. In this system of units, $GM = 4\pi^2$. If $l = 0$, the path of the planet is a straight line. If $l \neq 0$, its orbit is an ellipse, parabola, and hyperbola, depending on whether $e < 0, e = 0$, or $e > 0$, respectively. For an elliptical orbit, the period of the motion is given by

$$\left| \frac{e}{2\pi^2} \right|^{-\frac{3}{2}}$$

(For a detailed discussion of the mechanics of planetary motion, see [Sym71].)

For producing the graphics that animate the motion of a planet, the function `planetMotion` takes two arguments: a list of initial coordinates and a list of components of the initial velocity. Two-dimensional graphics options and the option `frameNumber` can be included as trailing arguments. The option `frameNumber`, with a default value of 30, specifies the number of frames in the animation sequence. Explanatory comments on `planetMotion` are embedded in the body of the function.

```
In[9] := (* set up default value for the option frameNumber *)
         Options[planetMotion] = {frameNumber -> 30};

In[11] := (* load the package Utilities`FilterOptions`; see Section
          3.4.5 for a discussion of the function FilterOptions *)
         Needs["Utilities`FilterOptions`"]

In[13] := planetMotion[{x0_, y0_}, {vx0_, vy0_}, opts___Rule] :=
         Module[{(* determine the energy parameter and assign it
                   as initial value to a local variable *)
                 e = N[(vx0^2 + vy0^2)/2 -
                       (4 Pi^2)/Sqrt[x0^2 + y0^2]],
```

```
          (* determine the angular momentum parameter and
          assign it as initial value to a local variable *)
          l = N[Abs[x0 vy0 - y0 vx0]],

          (* determine option value of frameNumber and
          assign it as initial value to a local variable *)
          n = frameNumber/.{opts}/.Options[planetMotion],

          (* select valid options for Show and assign the
          sequence as initial value to a local variable *)
          optShow = FilterOptions[Graphics, opts],

          (* declare other local variables *)
          orbit, period, x, y, t
       },
Which[
(* verify that initial conditions are real numbers and
that frame number is a positive integer *)
((!(NumberQ[x0] && Im[x0]==0))||
 (!(NumberQ[y0] && Im[y0]==0))||
 (!(NumberQ[vx0] && Im[vx0]==0))||
 (!(NumberQ[vy0] && Im[vy0]==0))||
 (!(IntegerQ[n] && Positive[n]))),
Print["Initial conditions must be real numbers and
frame number must be a positive integer."],

(* check that the path is not a straight line *)
l == 0,
Print["The initial conditions result in a straight line
for the path whereas orbits of planets must be ellipses."],

(* check that the orbit is not a hyperbola *)
e > 0,
Print["The initial conditions result in a hyperbolic orbit
whereas orbits of planets must be ellipses."],

(* check that the orbit is not a parabola *)
e == 0,
Print["The initial conditions result in a parabolic orbit
whereas orbits of planets must be ellipses."],

(* for the elliptical orbit *)
e < 0,

(* determine the period of the motion *)
period = N[Abs[e/(2 (Pi^2))]^(-3/2)];

(* solve the equations of motion discussed in
Example 2.1.7 of Section 2.1.17 *)
```

```
orbit =
  NDSolve[
    {x''[t] + (4 Pi^2) x[t]/((x[t]^2 + y[t]^2)^(3/2)) == 0,
     y''[t] + (4 Pi^2) y[t]/((x[t]^2 + y[t]^2)^(3/2)) == 0,
     x[0] == x0,
     y[0] == y0,
     x'[0] == vx0,
     y'[0] == vy0
    },
    {x, y}, {t, 0, period}
        ];

(* generate the graphics for animation with a DO loop *)
Do[
  Show[

    (* the orbit *)
    ParametricPlot[Evaluate[{x[t], y[t]} /. orbit],
      {t, 0, period}, DisplayFunction -> Identity
                ],

    (* the Sun in yellow and the moving planet in red *)
    Graphics[
      {RGBColor[1, 1, 0], PointSize[0.1], Point[{0, 0}],
       RGBColor[1, 0, 0], PointSize[0.04],
       Point[{x[i], y[i]} /. orbit[[1]]]}
            ],

    (* option for preserving the true shape
    of the orbit *)
    AspectRatio -> Automatic,

    DisplayFunction -> $DisplayFunction,

    (* two-dimensional graphics options *)
    optShow

      ];,
  {i, 0, (1 - 1/n)period, period/n}
  ]
      ]
      ]
```

With this function, the sizes of the Sun and planet in the graphics are preset. For orbits of different sizes, we may wish that the function would allow us to adjust those of the Sun and planet. For adding this feature to **planetMotion**, see Exercise 1 in Section 3.6.4.

Let us call the function **planetMotion** with $x_0 = 1$, $y_0 = 0$, $v_{x0} = -2.5$, $v_{y0} = 5$, and several option specifications:

```
In[14] := planetMotion[{1, 0}, {-2.5, 5}, frameNumber -> 40,
                PlotRange -> {{-0.6, 1.21}, {-1.13, 0.6}},
                Axes -> None, Frame -> True,
                FrameTicks -> None,
                Background -> RGBColor[0.5, 0.85, 1]
              ]
```

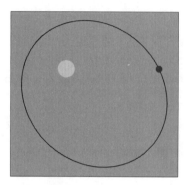

Only the head cell of the closed group of 40 graphics cells is shown. To animate, click the group bracket and press Command-Y for the Macintosh or Control+Y for Windows. That the speed of the planet varies around the orbit is consistent with Kepler's second law, which states that the position vector of a planet from the Sun sweeps out equal areas in equal times.

```
In[15] := ClearAll[planetMotion]
```

3.6.2 Functional Programming

A functional program, also known as a one-liner, consists of many nested functions, that is, a function acting on the results of applying other functions to the results of applying some other functions, and so forth, to some arguments. Functional programs are often written as functions of the form

> *name* [*arg1_*, *arg2_*, ...] := *rhs*

where *rhs* is a single, albeit sometimes very long, expression of nested functions. When the function *name* is invoked, the values of the functions at the deepest level of the nest serve as arguments of the functions at the next level, and so on, terminating at the top-level function, the value of which becomes the value of the function *name*. For one-liners, *rhs* is made up solely of *argi*, pure functions, and built-in functions and constants. Auxiliary function definitions and temporary variable assignments are forbidden. For iterations, functional programs often employ the functions **Nest**, **NestList**, **FixedPoint**, and **FixedPointList**. For manipulating expressions, they frequently utilize the functions **Map**, **MapAll**, **MapAt**, **Apply**, **Flatten**, and **FlattenAt**.

To illustrate functional programming, we generate a bifurcation diagram for the logistic map. As introduced in Example 3.3.2 of Section 3.3, the logistic map is defined by the one-dimensional difference equation

$$x_{n+1} = f(x_n)$$

where $n = 0, 1, \ldots$ and

$$f(x) = \mu x (1 - x)$$

with $0 \leq x \leq 1$ and $0 \leq \mu \leq 4$.

A bifurcation diagram shows the long-term iterates x_n for each value of the parameter μ that varies over a range of values. Let us write a function for producing a bifurcation diagram. The function **bifurcation** takes five arguments: the initial value, minimum value of μ, maximum value of μ, increment of μ, and number of iterations. The function also accepts two-dimensional graphics options as trailing arguments.

```
In[16] := bifurcation[x0_, mumin_, mumax_, del_,
                  nmax_, opts___Rule] := (
         ListPlot[
             Flatten[
               Table[
                 Thread[
                   List[
                     mu,
                     Drop[
                       NestList[(mu # (1 - #))&, x0, nmax],
                       200
                         ]
                       ]
                     ],
                 {mu, mumin, mumax, del}
                   ],
               1
                 ],
             opts
               ];
                                              )
```

For clarity, we formatted the code to show the successive levels of the nest. In *Mathematica* version 3.0, the symbols **μ**, **μmin**, **μmax**, and **Δ**, which are perhaps more familiar, can replace mu, mumin, mumax, and del in the function definition.

Let us dissect this function. For a given μ with the alias mu, NestList[(mu # (1 - #))&, x0, nmax] generates the list

$$\{x_0, x_1, x_2, \ldots, x_{nmax}\}$$

Drop discards the first 200 elements of this list to ensure omission of transient behavior. Thus, nmax must be an integer greater than or equal to 200. Thread then produces the list

$$\{\{\mu, x_{200}\}, \{\mu, x_{201}\}, \{\mu, x_{202}\}, \dots, \{\mu, x_{nmax}\}\}$$

because **Thread**$[f[a, \{b1, b2, \dots, bn\}]]$ returns the list

$$\{f[a, b1], f[a, b2], \dots, f[a, bn]\}$$

and, in our case, f is **List**, a is μ, and bi is x_{199+i}. **Table** gives a nested list with μ taking on a sequence of values ranging from $\mu_{\min}$ to $\mu_{\max}$ in steps of del:

$$\{\{\{\mu_{\min}, x_{200}\}, \{\mu_{\min}, x_{201}\}, \{\mu_{\min}, x_{202}\}, \dots, \{\mu_{\min}, x_{nmax}\}\},$$

$$\{\{\mu + del, x_{200}\}, \{\mu + del, x_{201}\}, \{\mu + del, x_{202}\}, \dots, \{\mu + del, x_{nmax}\}\},$$

$$\{\{\mu + 2del, x_{200}\}, \{\mu + 2del, x_{201}\}, \{\mu + 2del, x_{202}\}, \dots, \{\mu + 2del, x_{nmax}\}\},$$

$$\dots,$$

$$\{\{\mu_{\max}, x_{200}\}, \{\mu_{\max}, x_{201}\}, \{\mu_{\max}, x_{202}\}, \dots, \{\mu_{\max}, x_{nmax}\}\}\}$$

Flatten, with 1 as its second argument, flattens out the nested list to level 1 and yields a list of lists of coordinates for the points in the bifurcation diagram:

$$\{\{\mu_{\min}, x_{200}\}, \{\mu_{\min}, x_{201}\}, \{\mu_{\min}, x_{202}\}, \dots, \{\mu_{\min}, x_{nmax}\},$$

$$\{\mu + del, x_{200}\}, \{\mu + del, x_{201}\}, \{\mu + del, x_{202}\}, \dots, \{\mu + del, x_{nmax}\},$$

$$\{\mu + 2del, x_{200}\}, \{\mu + 2del, x_{201}\}, \{\mu + 2del, x_{202}\}, \dots, \{\mu + 2del, x_{nmax}\},$$

$$\dots,$$

$$\{\mu_{\max}, x_{200}\}, \{\mu_{\max}, x_{201}\}, \{\mu_{\max}, x_{202}\}, \dots, \{\mu_{\max}, x_{nmax}\}\}$$

Finally, **ListPlot** plots these points.

Let us call the function **bifurcation** with $x_0 = 0.2$, $\mu_{\min} = 2.9$, $\mu_{\max} = 4.0$, $del = 0.01$, $nmax = 500$, and several graphics option specifications:

```
(*   Mathematica Version 2.2 Specifics *)

bifurcation[0.2, 2.9, 4.0, 0.01, 500,
           AxesOrigin -> {2.8, 0},
           PlotRange -> {{2.8, 4}, {0, 1}},
           AxesLabel -> {FontForm[" 1", {"Symbol", 10}],
                        None},
           Epilog ->
             {Text[FontForm["x", {"Times", 10}],
                   {2.86, 0.96}],
              Text[FontForm["n", {"Times", 10}],
                   {2.90, 0.92}]}]
           ]
```

```
In[17] := (* ✿ 3.0 *)

    bifurcation[0.2, 2.9, 4.0, 0.01, 500,
                AxesOrigin -> {2.8, 0},
                PlotRange -> {{2.8, 4}, {0, 1}},
                AxesLabel -> {λ, xₙ}]
```

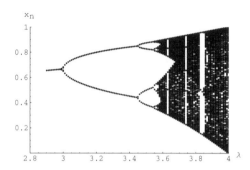

The bifurcation diagram shows period doublings, onset of chaotic behavior, and existence of windows of periodic behavior within the chaotic region. ([BG90] provides an excellent introduction to chaotic dynamics.)

Purists from the school of "strict" one-liners may be alarmed by the inclusion, on the *rhs* of the function definition for **bifurcation**, of the function **Table**, because it contains a named iteration variable **mu**, whereas the only permissible named variables on the *rhs* are the argument names of **bifurcation**. This posture also excludes pure functions of the form **Function[*var*, *body*]** from strict one-liners. It even rules out some other built-in functions such as **Sum**, **Integrate**, and **Solve**, unless their arguments are the argument names on the *lhs* of the function definition. Here is a strict one-liner for generating the bifurcation diagram for the logistic map:

```
bifurcationStrictOneLiner[x0_, mumin_, mumax_, del_,
                          nmax_, opts___Rule] := (
ListPlot[
  Flatten[
    Map[
      Thread,
      Transpose[
        {NestList[(# + del)&, mumin,
                  Round[(mumax - mumin)/del]
                 ],
```

```
          Map[
            Drop[#, 200]&,
            Map[
              NestList[#, x0, nmax]&,
              Map[
                Function,
                NestList[(# + del)&, mumin,
                         Round[(mumax - mumin)/del]
                     ] (# (1 - #))
                   ]
                 ]
               ]
           }
                 ]
          ],
      1
            ],
    opts
          ];
                                                           )
```

Choosing between `bifurcationStrictOneLiner` and `bifurcation` is a matter of aesthetics. The dissection of this function is left as an exercise in Section 3.6.4. (For further discussion of strict one-liners versus nonstrict or ordinary one-liners, see [Gra94].)

In[18] := `ClearAll[bifurcation, bifurcationStrictOneLiner]`

3.6.3   *Rule-Based Programming*

Rule-based programming is *Mathematica*'s forte. A rule-based program consists of a collection of user-defined rewrite rules such as assignment statements and function definitions. We create rewrite rules with the functions `Set`, `UpSet`, `SetDelayed`, or `UpSetDelayed` in the forms

`Set[`*lhs, rhs*`]`
`UpSet[`*lhs, rhs*`]`
`SetDelayed[`*lhs, rhs*`]`
`UpSetDelayed[`*lhs, rhs*`]`

The corresponding special input forms are

lhs `=` *rhs*
lhs `^=` *rhs*
lhs `:=` *rhs*
lhs `^:=` *rhs*

Often, *lhs* assumes the form

   name[*patt1*, *patt2*, ...]

where *patti* are patterns. We can also create rewrite rules with the syntax for dynamic programming so that *Mathematica* remembers the values it has found.

Rewrite rules reside in the global rule base, where more than one rule can be associated with a symbol. Whereas *Mathematica* stores the rules *lhs* = *rhs* and *lhs* ^= *rhs* with *rhs* evaluated, it stores the rules *lhs* := *rhs* and *lhs* ^:= *rhs* with *rhs* unevaluated. For example, let us create several rules:

```
In[19] := ClearAll[a, b, f, g, p, q]
```

```
In[20] := a = Random[Integer, 10]
Out[20] = 10
```

```
In[21] := f[n_] g[m_] ^= Random[Integer, {20, 30}]^(n m)
Out[21] = 20^m n
```

```
In[22] := b := Random[Integer, 10]
```

```
In[23] := p[n_] q[m_] ^:= Random[Integer, {20, 30}]^(n m)
```

The ? operator reveals how *Mathematica* stores these rules in the global rule base:

```
In[24] := ?a
        Global`a
        a = 10
```

```
In[25] := ?f
        Global`f
        f/: f[n_]*g[m_] = 20^(m*n)
```

```
In[26] := ?b
        Global`b
        b := Random[Integer, 10]
```

```
In[27] := ?p
        Global`p
        p/: p[n_]*q[m_] := Random[Integer, {20, 30}]^(n*m)
```

```
In[28] := ClearAll[a, b, f, g, p, q]
```

When *Mathematica* encounters an expression that matches *lhs* of a rewrite rule in the global rule base composed of built-in functions and user-defined rewrite rules, it applies the rule. If the rule has the form *lhs* = *rhs* or *lhs* ^= *rhs* and if *lhs* does not contain any patterns, it simply replaces the expression with *rhs*. If the rule has the form *lhs* := *rhs* or *lhs* ^:= *rhs* where *lhs,* as before, does not contain any patterns, it evaluates *rhs* and then replaces the expression with the result. If *lhs* contains patterns, it substitutes the actual values into the pattern names on *rhs* before applying the rule. If an expression matches *lhs* of several rules, *Mathematica* selects the rule according to the priority set forth in Section 3.3.2. That *Mathematica* uses user-defined rules before built-in ones bears emphasis. If the resulting expression or any of its parts again matches a rule, that rule is applied. The process continues until there are no more matching rules.

For implementing a set of mathematical relations, rule-based programming is usually the best choice. What are the formulas in mathematical tables? They are rewrite rules!

Let us illustrate rule-based programming with an example from quantum mechanics. With rules from the algebra of linear operators, we shall prove a number of commutator identities and simplify several commutators of functions of the one-dimensional position and momentum operators. ([Fea94] is the pioneering work on the application of *Mathematica* to quantum mechanics.)

The sum and product of two linear operators and the product of a linear operator with a number (more precisely, a scalar) are themselves linear operators. These operations of addition and multiplication obey all the rules of addition and multiplication of numbers, with one notable exception: multiplication of linear operators is, in general, not commutative. If A, B, and C are linear operators and a is a number, then

$$AU = UA = A \tag{3.6.1}$$

$$A(B + C) = AB + AC \tag{3.6.2}$$

$$(A + B)C = AC + BC \tag{3.6.3}$$

$$A(BC) = (AB)C \tag{3.6.4}$$

$$A(aB) = aAB \tag{3.6.5}$$

$$(aA)B = aAB \tag{3.6.6}$$

where U is the unit or identity operator. (Pedants may object to our omission of the rule for the zero operator. Letting the symbol 0 denote here both the number zero and the zero operator makes the rule unnecessary and offers convenience without creating difficulties in computations.)

To implement these rules in *Mathematica*, we introduce the built-in function with the name `NonCommutativeMultiply` because operator multiplication is often not commutative. `NonCommutativeMultiply[A, B]`, with the special input form A `**` B, is a general associative, but noncommutative, form of multiplication. The associative law, Equation 3.6.4, is already built into `**`:

```
In[29] := A ** (B ** C) == (A ** B) ** C
```

```
Out[29] = True
```

Equation 3.6.1 on the identity operator U can be entered as

```
In[30] := Unprotect[NonCommutativeMultiply];
```

In[31] := A_ ** U := A
            U ** A_ := A

where we have unprotected the built-in function `NonCommutativeMultiply` in order to specify rules for it. The distributive laws, Equations 3.6.2 and 3.6.3, can be entered as

In[33] := A_ ** (B_ + C_) := A ** B + A ** C
            (A_ + B_) ** C_ := A ** C + B ** C

To enter Equations 3.6.5 and 3.6.6, we naturally think of the pattern `a_?NumberQ` for representing numbers. Yet a less restrictive pattern is needed because the number a, in the application of these rules, can actually be an algebraic expression representing a number and *Mathematica* does not recognize as numbers the expressions consisting of constants, numerals, and operation signs. Let, for example, a and b be constants—symbols denoting particular numbers,

In[35] := NumberQ[a] ^= True;
            NumberQ[b] ^= True;

and evaluate

In[37] := Cases[{a, b, (2/3) b, I Sqrt[a], b^3, a b, a/b, a^b},
                  a_?NumberQ]

Out[37] = {a, b}

Mathematica recognizes only a and b as numbers. The less restrictive pattern (`x_. y_^n_. /; NumberQ[x]&&NumberQ[y]&&NumberQ[n]`) matches all the elements of the list in the previous example:

In[38] := Cases[{a, b, (2/3) b, I Sqrt[a], b^3, a b, a/b, a^b},
                  (x_. y_^n_. /; NumberQ[x]&&NumberQ[y]&&NumberQ[n])]

Out[38] = {a, b, $\frac{2\,b}{3}$, I Sqrt[a], b^3, a b, $\frac{a}{b}$, a^b}

This pattern is sufficiently general in representing numbers for most problems in quantum mechanics. With the function

In[39] := number3Q[x_, y_, n_] := NumberQ[x]&&NumberQ[y]&&NumberQ[n]

Equations 3.6.5 and 3.6.6 can be entered as

In[40] := A_ ** (B_ (x_. y_^n_. /; number3Q[x, y, n])) :=
                                          ((x y^n) A ** B)
            (A_ (x_. y_^n_. /; number3Q[x, y, n])) ** B_ :=
                                          ((x y^n) A ** B)

To avoid inadvertent modifications, we should restore the protection for the built-in function `NonCommutativeMultiply`:

In[42] := Protect[NonCommutativeMultiply];

Mathematica can now correctly simplify expressions consisting of linear operators and numbers. For example,

```
In[43] := {(3 A + B/5) ** C, ((2 a/Sqrt[5 a b]) U) ** A,
           ((a^2/b^3) U) ** (A + 2 B) ** C,
           (Sqrt[2 a] A) ** (B ** (3 C/(5 b^2)))
          }//PowerExpand//ExpandAll
```

$$Out[43] = \{3\ A\ **\ C\ +\ \frac{B\ **\ C}{5},\ \frac{2\ Sqrt[a]\ A}{Sqrt[5]\ Sqrt[b]},\ \frac{a^2\ A\ **\ C}{b^3}\ +\ \frac{2\ a^2\ B\ **\ C}{b^3},$$

$$\frac{3\ Sqrt[2]\ Sqrt[a]\ A\ **\ B\ **\ C}{5\ b^2}\}$$

With the rules (Equations 3.6.1 to 3.6.6) already in the global rule base, we can prove several commutator identities. The commutator of two linear operators is defined as

$$[A, B] = AB - BA$$

For *Mathematica*, we express this definition as

```
In[44] := commutator[A_, B_] := A ** B - B ** A
```

Let us prove the identities

$$[A, B] = -[B, A] \tag{3.6.7}$$

$$[aA, bB] = ab[A, B] \tag{3.6.8}$$

$$[A, B + C] = [A, B] + [A, C] \tag{3.6.9}$$

$$[A + B, C] = [A, C] + [B, C] \tag{3.6.10}$$

$$[A, BC] = [A, B]C + B[A, C] \tag{3.6.11}$$

$$[AB, C] = A[B, C] + [A, C]B \tag{3.6.12}$$

$$[[A, B], C] + [[C, A], B] + [[B, C], A] = 0 \tag{3.6.13}$$

For Equation 3.6.7,

```
In[45] := commutator[A, B] + commutator[B, A] == 0
```

```
Out[45] = True
```

For Equation 3.6.8,

```
In[46] := commutator[a A, b B] - a b commutator[A, B] == 0 //
          ExpandAll
```

```
Out[46] = True
```

For Equation 3.6.9,

```
In[47] := commutator[A, B + C] - commutator[A, B] -
          commutator[A, C] == 0
```

```
Out[47] = True
```

For Equation 3.6.10,

```
In[48] := commutator[A + B, C] - commutator[A, C] -
            commutator[B, C] == 0
```

Out[48] = True

For Equation 3.6.11,

```
In[49] := commutator[A, B ** C] - commutator[A, B] ** C -
            B ** commutator[A, C] == 0
```

Out[49] = True

For Equation 3.6.12,

```
In[50] := commutator[A ** B, C] - A ** commutator[B, C] -
            commutator[A, C] ** B == 0
```

Out[50] = True

For Equation 3.6.13, the Jacobi identity,

```
In[51] := commutator[commutator[A, B], C] +
            commutator[commutator[C, A], B] +
            commutator[commutator[B, C], A] == 0
```

Out[51] = True

The one-dimensional position and momentum operators x and p in quantum mechanics satisfy the commutator relation

$$[x, p] = i\hbar \tag{3.6.14}$$

Since the commutator of two operators is also an operator, the *rhs* of this commutation relation actually stands for

$$i\hbar\, U$$

where U is the identity operator. We can apply the commutation relation (Equation 3.6.14) to simplify a commutator with the function

```
In[52] := xpCommutator[expr_] :=
            ExpandAll[expr //. p ** x :> x ** p - I hb U]
```

where **hb**, the alias of Planck's constant divided by 2π, must be declared to be a number:

```
In[53] := NumberQ[hb] ^= True;
```

(In *Mathematica* version 3.0, the traditional symbol $\hbar$ can replace the alias **hb** as Planck's constant divided by 2π.)

To simplify, for example, the commutators

$$[x, p^4]$$
$$[p, x^2 p^2]$$
$$[x p^2, p x^2]$$

we have

In[54] := commutator[x, p ** p ** p ** p]//xpCommutator

Out[54] = 4 I hb p ** p ** p

In[55] := commutator[p, x ** x ** p ** p]//xpCommutator

Out[55] = -2 I hb x ** p ** p

In[56] := commutator[x ** p ** p, p ** x ** x]//xpCommutator

Out[56] = -6 hb^2 x ** p - 3 I hb x ** x ** p ** p

Let us conclude this section with the evaluation of the commutator

$$[H, p]$$

for the one-dimensional harmonic oscillator whose Hamiltonian H is

$$H = \frac{p^2}{2m} + \frac{1}{2}m\omega^2 x^2$$

where m and ω are the mass and frequency of the oscillator, respectively. After declaring m and ω (with the alias w) to be constants,

In[57] := **NumberQ[m]** ^= **True;**
               NumberQ[w] ^= **True;**

we have

In[59] := commutator[p**p/(2 m) + (m/2) w^2 x**x, p]//xpCommutator

Out[59] = I hb m w^2 x

(In *Mathematica* version 3.0, the use of the alias w for ω is unnecessary because ω is now a legitimate symbol. For further discussions of operators, commutators, and their algebras, see [DK67], [Has91], [Mes62], and [Mor90].)

In[60] := **ClearAll[a, b, m, w, hb, number3Q, xpCommutator]**

3.6.4 Exercises

1. Modify the function planetMotion in Section 3.6.1 to allow option specifications for the sizes of the Sun and planet in the graphics.

*2. Write a function that generates the graphics for animating the motion, as viewed from the laboratory frame, of two identical masses under their mutual gravitational attraction.

3. Modify the function bifurcation in Section 3.6.2 to generate a bifurcation diagram for the map

$$x_{n+1} = x_n \exp\left[\mu(1 - x_n)\right]$$

Produce a bifurcation diagram for $2.0 < \mu < 4.0$. *Hint:* Discard only the first 100 rather 200 elements, and let *nmax* $= 300$.

4. In Section 3.6.2, the function `bifurcationStrictOneLiner` is defined as a nested function call, that is, the application of a sequence of nested functions to some arguments. Following the dissection of the function `bifurcation` in the same section, take `bifurcationStrictOneLiner` apart and describe the result of each function application.

5. The orbital angular momentum operator of a three-dimensional particle is

$$l = r \times p$$

where **r** and **p** are the position and linear momentum operators, the components of which satisfy the commutation relations

$$[r_i, r_j] = 0$$

$$[p_i, p_j] = 0 \qquad (i, j = 1, 2, 3) \tag{3.6.15}$$

$$[r_i, p_j] = i\hbar\delta_{ij}$$

where each subscript takes on the values 1, 2, and 3, corresponding to the x, y, and z components respectively; that is, $r_1 = x$, $p_1 = p_x$, and so forth. The Kronecker delta δ_{ij} is defined by

$$\delta_{ij} = \begin{cases} 0 & i \neq j \\ 1 & i = j \end{cases}$$

With the rules (Equations 3.6.1 to 3.6.6) and the commutation relations (Equation 3.6.15), verify

$$[l_x, l_y] = i\hbar l_z$$

$$[l_y, l_z] = i\hbar l_x$$

$$[l_z, l_x] = i\hbar l_y$$

$$[l_z, l^2] = 0$$

where

$$l^2 = l_x^2 + l_y^2 + l_z^2$$

6. Consider a list of random number of random integers. Define a recursive function, that is, a function that calls itself, that returns a list of the elements at odd positions, that is, the first, third, fifth, ... elements. The recursive function must not contain the built-in function `Length`. For example,

```
f[{-59, 65, 4, 69, 83, -37, -96, -58}]

{-59, 4, 83, -96}
```

3.7   *Packages*

As discussed in Section 1.5, packages are files of definitions written in the *Mathematica* language, and a package must be loaded before its definitions are accessible. *Mathematica* comes with many standard packages, and we can write our own. What are the purposes of packages? Sections 2.2.4 and 2.2.13 warned that global symbols are potential pitfalls, and Section 3.4.1 showed how the function **Module** sets up local symbols. Using contexts, packages provide an encapsulation or localization mechanism for symbols in entire files so as to allow the use of the same symbol in a *Mathematica* session as well as in different packages. Packages also permit hiding implementations that are of no interest to users of the definitions or readers of the notebooks. Furthermore, they allow storing of definitions that are used often. Section 3.7.1 elaborates on the notion of contexts introduced in Section 1.5. Section 3.7.2 discusses the manipulation of contexts. Section 3.7.3 analyzes the construction of a sample package, and Section 3.7.4 provides a template for packages.

3.7.1   Contexts

Full names of *Mathematica* symbols have the form *context`short,* where *context`* is the context of a symbol and *short* is its short name. A context always ends with a backquote `, called a "context mark" in *Mathematica*. Symbols with the same short name but different contexts are different. For example, **physics`strangeness** and **psychology`strangeness** are different:

> *In[1] :=* **physics`strangeness === psychology`strangeness**

> *Out[1] =* False

SameQ[*lhs,* *rhs*], with the special input form *lhs* **===** *rhs,* yields **True** if the expression *lhs* is identical to *rhs,* and it yields **False** otherwise. Contexts are not symbols and must be enclosed by quotation marks when entered as arguments of commands. Symbols with common properties or functionalities or from one subject area are often put in the same context.

Contexts can be hierarchical. In other words, contexts can take the form *context1`context2`* ... *contextn`*. For example, the context of the symbol **mechanics`kinematics`velocity** is **mechanics`kinematics`**:

> *In[2] :=* **Context[mechanics`kinematics`velocity]**

> *Out[2] =* mechanics`kinematics`

Context[*s*] returns the context of symbol *s*.

At any point in a *Mathematica* session, there is a current context, which is the value of the variable **$Context**. When a new symbol is introduced, it is put in the current context. For example, enter **charm**:

> *In[3] :=* **charm**

> *Out[3] =* charm

The default current context is **Global`**:

In[4] := **$Context**

Out[4] = Global`

The context of **charm** is therefore **Global`**:

In[5] := **Context[charm]**

Out[5] = Global`

We can refer to symbols in the current context by their short names:

In[6] := **Global`charm === charm**

Out[6] = True

The symbol `` `subcontext` `` stands for *currentcontext*`` `subcontext` ``; that is, the full name of the context `` `subcontext` `` is *currentcontext*`` `subcontext` ``, where *currentcontext*`` ` `` is the current context. For example, if the current context is **Global`**, the context of the symbol `` `sound`wave `` is **Global`sound`**:

In[7] := **{$Context, Context[`sound`wave]}**

Out[7] = {Global`, Global`sound`}

Built-in *Mathematica* objects have the context **System`**. For example, **Plot3D** is in the context **System`**:

In[8] := **Context[Plot3D]**

Out[8] = System`

When we invoke a short name associated with several contexts, to which symbol does it refer? *Mathematica* first searches the current context for a symbol with that short name. If it cannot find a symbol with that short name, *Mathematica* searches the contexts in the context search path for a symbol with the short name. Let us create, for example, the symbols **physics`color** and **art`color**:

In[9] := **{physics`color, art`color}**

Out[9] = {physics`color, art`color}

As mentioned before, **$Context** returns the current context:

In[10] := **$Context**

Out[10] = Global`

The variable **$ContextPath** gives the context search path (i.e., the list of contexts, after the current context, that *Mathematica* searches for a symbol, in the order in which they appear in the list):

In[11] := **$ContextPath**

Out[11] = {Graphics`Animation`, Global`, System`}

(*Mathematica* version 2.2 does not include **Graphics`Animation`** in the default context search path.) Now add **physics`** and **art`** to the context search path:

In[12] := **$ContextPath = $ContextPath~Join~{"physics`", "art`"}**

Out[12] = {Graphics`Animation`, Global`, System`, physics`, art`}

Since there is no symbol with the short name **color** in the current context, the short name **color** now refers to the symbol **physics`color**, because **physics`** is the first context with a symbol having the short name **color** in the context search path:

In[13] := **Context[color]**

Out[13] = physics`

Contexts are system-independent. The contexts of symbols exported by packages are the context names of the packages. Section 1.5 discussed the relationship between the context name of a package and its file name, which is system-dependent. To use a symbol exported by a package, we must not invoke the symbol before loading the package. For example, let us introduce the symbol **DiracDelta**:

In[14] := **DiracDelta**

Out[14] = DiracDelta

This symbol is in the current context, which is **Global`**. Now load the package **Calculus`DiracDelta`**. The function **Needs["*context*`"]** loads a package with the context name *context`*.

In[15] := **Needs["Calculus`DiracDelta`"]**

```
DiracDelta::shdw:
    Symbol DiracDelta appears in multiple contexts
    {Calculus`DiracDelta`, Global`}; definitions in context
    Calculus`DiracDelta` may shadow or be shadowed by other
    definitions.
```

Mathematica warns that the definition for **DiracDelta** in the package may shadow or be shadowed by other definitions of this symbol. Let us evaluate the integral

In[16] := **Integrate[DiracDelta[x], {x, -Infinity, Infinity}]**

Out[16] = Integrate[DiracDelta[x], {x, -∞, ∞}]

Mathematica cannot find a definition for **DiracDelta** because it refers to the symbol in the context **Global`**:

In[17] := **Context[DiracDelta]**

Out[17] = Global`

For *Mathematica* to access the definition for **DiracDelta** in the package, we must first remove it from the current context:

In[18] := **Remove[DiracDelta]**

Mathematica now returns the correct value for the integral of a delta function:

In[19] := **Integrate[DiracDelta[x], {x, -Infinity, Infinity}]**

Out[19] = 1

3.7.2 *Context Manipulation*

Mathematica provides several commands for manipulating contexts:

`BeginPackage["`*context*`` `` `"]`	make *context*`` ` `` and `System`` ` `` the only active contexts
`BeginPackage["`*context*`` `` `",` `{"`*need1*`` `` `", "`*need2*`` `` `", ...}]`	call `Needs` on the *needi*
`EndPackage[ ]`	restore `$Context` and `$ContextPath` to their values before the preceding `BeginPackage`, and prepend the current context to the list `$ContextPath`
`Begin["`*context*`` `` `"]`	reset the current context
`End[ ]`	return the present context, and revert to the previous one

Let us illustrate the use of these commands. First, return the context search path to its default:

```
In[20] := $ContextPath = {"Graphics`Animation`", "Global`", "System`"}

Out[20] = {Graphics`Animation`, Global`, System`}
```

(Again, `Graphics`` ` ``Animation`` ` `` is not in the default context search path of *Mathematica* version 2.2.) Now enter

```
In[21] := BeginPackage["test`"]

Out[21] = test`
```

`BeginPackage["test`"]` resets the values of both `$Context` and `$ContextPath`:

```
In[22] := {$Context, $ContextPath}

Out[22] = {test`, {test`, System`}}
```

`EndPackage[ ]` restores the values of `$Context` and `$ContextPath` and prepends the current context to the list `$ContextPath`.

```
In[23] := EndPackage[ ]
```

The current context is once again `Global`` ` ``:

```
In[24] := $Context

Out[24] = Global`
```

The context search path is restored, but with context `test`` ` `` prepended:

```
In[25] := $ContextPath

Out[25] = {test`, Graphics`Animation`, Global`, System`}
```

Before proceeding, return the context search path to its default:

```
In[26] := $ContextPath = {"Graphics`Animation`", "Global`", "System`"}

Out[26] = {Graphics`Animation`, Global`, System`}
```

If **BeginPackage** has a second argument, the specified packages are loaded:

```
In[27] := BeginPackage["test`",
                        {"Calculus`Limit`", "Graphics`ContourPlot3D`"}]

Out[27] = test`
```

The current context becomes **test`**:

```
In[28] := $Context

Out[28] = test`
```

The context search path now includes the additional contexts:

```
In[29] := $ContextPath

Out[29] = {test`, Calculus`Limit`, Graphics`ContourPlot3D`, System`}
```

Without altering the context search path, **Begin["mycontext`"]** resets the current context to **mycontext`**:

```
In[30] := Begin["mycontext`"]

Out[30] = mycontext`
```

$Context shows the current context:

```
In[31] := $Context

Out[31] = mycontext`
```

The context search path remains the same:

```
In[32] := $ContextPath

Out[32] = {test`, Calculus`Limit`, Graphics`ContourPlot3D`, System`}
```

Without changing the context search path, **End[]** returns the current context and reverts to the previous one:

```
In[33] := End[]

Out[33] = mycontext`
```

The current context is once again **test`**:

```
In[34] := $Context

Out[34] = test`
```

The context search path is unaffected:

```
In[35] := $ContextPath

Out[35] = {test`, Calculus`Limit`, Graphics`ContourPlot3D`, System`}
```

Before proceeding to the next section, let us return the current context and context search path to their defaults:

```
In[36] := {$Context, $ContextPath} =
            {"Global`", {"Graphics`Animation`", "Global`", "System`"}}

Out[36] = {Global`, {Graphics`Animation`, Global`, System`}}
```

3.7.3 A Sample Package

With a sample package, this section shows how to use the commands introduced in Section 3.7.2 in setting up *Mathematica* packages.

3.7.3.1 THE PROBLEM

Write a package that provides the functions for the eigenenergies and normalized energy eigenfunctions of the hydrogen atom.

The normalized energy eigenfunctions of hydrogen can be written as

$$\psi_{nlm}(r, \theta, \phi) = R_{nl}(r) Y_{lm}(\theta, \phi)$$

where R_{nl} and Y_{lm} are the radial functions and the normalized spherical harmonics, respectively. In terms of the associated Laguerre polynomials L_p^q and the Bohr radius a, the radial functions are given by

$$R_{nl}(r) = a^{-3/2} \frac{2}{n^2} \sqrt{\frac{(n-l-1)!}{[(n+l)!]^3}} F_{nl}\left(\frac{2r}{na}\right)$$

with

$$F_{nl}(x) = x^l e^{-x/2} L_{n-l-1}^{2l+1}(x)$$

The associated Laguerre polynomials $L_p^q(x)$ are related to the generalized Laguerre polynomials `LaguerreL[p, q, x]` of *Mathematica* by

$$L_p^q(x) = (p+q)! \, \texttt{LaguerreL[p, q, x]}$$

The eigenenergies can be expressed as

$$E_n = -\frac{Ry}{n^2}$$

where Ry is the Rydberg constant.

The allowed quantum numbers n, l, and m are given by the rules

$$n = 1, 2, 3, \ldots$$

$$l = 0, 1, 2, \ldots, (n-1)$$

$$m = -l, -l+1, \ldots, 0, 1, 2, \ldots, +l$$

(For more information on the hydrogen atom, see [Lib92].)

3.7.3.2 THE PACKAGE

What follows is a package that exports functions for the eigenenergies and eigenfunctions. The unit of eigenenergy can be the electron volt (eV), the joule (J), or the erg.

```
BeginPackage["PT`Hydrogen`", "Miscellaneous`Units`"]

Hydrogen::usage = "Hydrogen.m is a package that
provides functions for the eigenenergies and
normalized energy eigenfunctions of the
hydrogen atom."

Psi::usage = "Psi[n, l, m, r, theta, phi] gives
the normalized energy eigenfunctions in the
spherical coordinates r, theta, and phi."

Energy::usage = "Energy[n] gives the eigenenergies
of hydrogen in electron volts (eV). Joule or Erg
may be entered as the value of the option Unit to
express the energies in the specified unit."

a::usage = "The symbol a stands for the Bohr radius."

Unit::usage = "Unit is an option of Energy[n]."

Options[Energy] = {Unit -> ElectronVolt}

Begin["`Private`"]

f[n_, l_, x_] := (x^l Exp[-x/2] (n + 1)! *
                  LaguerreL[n - l - 1, 2 l + 1, x])

radialFunction[n_, l_, r_] :=
                  (a^(-3/2) (2/(n^2)) *
                  Sqrt[(n - l - 1)!/(n + 1)!^3] *
                  f[n, l, 2r/(n a)])

Hydrogen::badarg = "You called `1` with `2`
argument(s)! It must have `3` argument(s)."

Psi::quannum = "Your set of quantum numbers
(`1`, `2`, `3`) is not allowed."

Energy::badarg = "The first argument must be a
positive integer. The second argument is optional
and, if specified, must be one of the rules:
Unit -> Joule or Unit -> Erg. You entered `1`."

Psi[n_Integer?Positive, l_Integer?NonNegative,
    m_Integer, r_, theta_, phi_] :=
                  (radialFunction[n, l, r] *
                  SphericalHarmonicY[l, m, theta, phi]) /;
                  (l <= (n - 1) && -1 <= m <= 1)

Psi[n_, l_, m_, r_, theta_, phi_] :=
                  Message[Psi::quannum, n, l, m]
```

```
Psi[arg___ /; Length[{arg}] != 6] :=
                Message[Hydrogen::badarg, Psi,
                      Length[{arg}], 6]

Energy[n_Integer?Positive, opts___Rule] :=
                (-Convert[Rydberg, Unit/.{opts}/.
                              Options[Energy]
                ]/n^2)

Energy[arg__ /; Length[{arg}] < 3] :=
                Message[Energy::badarg, {arg}]

Energy[arg___ /;
      (Length[{arg}] == 0 || Length[{arg}] > 2)] :=
                Message[Hydrogen::badarg, Energy,
                      Length[{arg}], "1 or 2"]


End[]

Protect[Psi, Energy]

EndPackage[]
```

3.7.3.3 ANALYSIS OF THE PACKAGE

This section analyzes the construction of the package in Section 3.7.3.2. In what follows, semicolons are put at the end of commands to suppress outputs that are normally absent in the actual loading of the package.

The package begins with the command

```
In[37] := BeginPackage["PT`Hydrogen`", "Miscellaneous`Units`"];
```

Mathematica imports, that is, reads in, the package `Miscellaneous`Units`` if it is not already loaded, changes the current context to `PT`Hydrogen``, and sets the context search path to `{"PT`Hydrogen`", "Miscellaneous`Units`", "System`"}`.

Next come the usage messages:

```
In[38] := Hydrogen::usage = "Hydrogen.m is a package that
          provides functions for the eigenenergies and
          normalized energy eigenfunctions of the
          hydrogen atom.";

          Psi::usage = "Psi[n, l, m, r, theta, phi] gives
          the normalized energy eigenfunctions in the
          spherical coordinates r, theta, and phi.";

          Energy::usage = "Energy[n] gives the eigenenergies
          of hydrogen in electron volts (eV). Joule or Erg
          may be entered as the value of the option Unit to
          express the energies in the specified unit.";
```

```
a::usage = "The symbol a stands for the Bohr radius.";

Unit::usage = "Unit is an option of Energy[n].";
```

After loading the package, we can access these usage messages with the ? operator just as we obtain those for the built-in objects. These messages serve another purpose of putting the symbols `Hydrogen`, `Psi`, `Energy`, `a`, and `Unit` in the context `PT`Hydrogen``. Since the context `PT`Hydrogen`` is included in the context search path after the package is loaded, these symbols, intended for export outside the package, are visible.

`Options[`*symbol*`]` gives the list of default options assigned to *symbol*. The package specifies a default option for the function `Energy` with the command

```
In[43] := Options[Energy] = {Unit -> ElectronVolt};
```

Then comes the command

```
In[44] := Begin["`Private`"];
```

The context search path remains `{"PT`Hydrogen`", "Miscellaneous`Units`", "System`"}`, whereas the current context becomes `PT`Hydrogen`Private``.

At this point in the package, there are definitions for two auxiliary functions `f` and `radialFunction`:

```
In[45] := f[n_, l_, x_] := (x^l Exp[-x/2] (n + 1)! *
                            LaguerreL[n - l - 1, 2 l + 1, x])

          radialFunction[n_, l_, r_] :=
                         (a^(-3/2) (2/(n^2)) *
                          Sqrt[(n - l - 1)!/(n + l)!^3] *
                          f[n, l, 2r/(n a)])
```

These symbols are put in the context `PT`Hydrogen`Private``. Since this context is excluded from the context search path after the package is loaded, these symbol are invisible outside the package as they are intended solely for internal uses in the package.

Several error messages for the functions intended for export are then defined:

```
In[47] := Hydrogen::badarg = "You called `1` with `2`
          argument(s)! It must have `3` argument(s).";

          Psi::quannum = "Your set of quantum numbers
          (`1`, `2`, `3`) is not allowed.";

          Energy::badarg = "The first argument must be a
          positive integer. The second argument is optional
          and, if specified, must be one of the rules:
          Unit -> Joule or Unit -> Erg. You entered `1`.";
```

There are three definitions for the function `Psi`:

```
In[50] := Psi[n_Integer?Positive, l_Integer?NonNegative,
              m_Integer, r_, theta_, phi_] :=
                         (radialFunction[n, l, r] *
                          SphericalHarmonicY[l, m, theta, phi]) /;
                         (l <= (n - 1) && -1 <= m <= 1)
```

```
Psi[n_, l_, m_, r_, theta_, phi_] :=
                Message[Psi::quannum, n, 1, m]

Psi[arg___ /; Length[{arg}] != 6] :=
                Message[Hydrogen::badarg, Psi,
                        Length[{arg}], 6]
```

The last two definitions trigger error messages when incorrect arguments are entered. **Message**[*symbol*::*tag*, *e1*, *e2*, ...] prints the message *symbol*::*tag*. The values of the *e1*, *e2*, ... replace `` `1` ``, `` `2` ``, ... in the message *symbol*::*tag* defined previously. The second definition returns an error message if unacceptable quantum numbers are specified:

In[53] := **Psi[-2, 1, 0, r, theta, phi]**

```
Psi::quannum:
    Your set of quantum numbers (-2, 1, 0) is not allowed.
```

The third definition gives an error message if there is a wrong number of arguments:

In[54] := **Psi[2, 1, 0, 2, r, theta, phi, z]**

```
Hydrogen::badarg:
    You called Psi with 8 argument(s)! It must have 6 argument(s).
```

There are also three definitions for the function **Energy**:

In[55] := **Energy[n_Integer?Positive, opts___Rule] :=**
```
                (-Convert[Rydberg, Unit/.{opts}/.
                                Options[Energy]
                    ]/n^2)

Energy[arg__ /; Length[{arg}] < 3] :=
                Message[Energy::badarg, {arg}]

Energy[arg___ /;
        (Length[{arg}] == 0 || Length[{arg}] > 2)] :=
                Message[Hydrogen::badarg, Energy,
                        Length[{arg}], "1 or 2"]
```

The second definition for **Energy** returns an error message if an unacceptable quantum number is entered or the option **Unit** is not specified in terms of a rule:

In[58] := **Energy[-2]**

```
Energy::badarg:
    The first argument must be a positive integer. The second
    argument is optional and, if specified, must be one of the
    rules: Unit -> Joule or Unit -> Erg. You entered {-2}.
```

In[59] := **Energy[2, Unit]**

```
Energy::badarg:
    The first argument must be a positive integer. The second
    argument is optional and, if specified, must be one of the
    rules: Unit -> Joule or Unit -> Erg. You entered {2, Unit}.
```

The third definition for `Energy` gives an error message if there is a wrong number of arguments:

In[60] := **Energy[2, Unit -> Erg, 3]**

> Hydrogen::badarg:
>> You called Energy with 3 argument(s)! It must have 1 or 2
>>   argument(s).

Then, without changing the context search path, which is `{"PT`Hydrogen`"`, `"Miscellaneous`Units`"`, `"System`"}`, the command `End` restores the current context to `PT`Hydrogen`:

In[61] := **End[];**

In[62] := **$Context**

Out[62] = PT`Hydrogen`

In[63] := **$ContextPath**

Out[63] = {PT`Hydrogen`, Miscellaneous`Units`, System`}

`Protect` protects the functions `Psi` and `Energy` intended for export just as it does for built-in *Mathematica* objects:

In[64] := **Protect[Psi, Energy];**

Finally, `EndPackage[]` restores the current context to the one before the preceding `BeginPackage` was executed:

In[65] := **EndPackage[]**

In[66] := **$Context**

Out[66] = Global`

While `PT`Hydrogen`Private`` has been removed from the context search path, the context `PT`Hydrogen`` is prepended to it:

In[67] := **$ContextPath**

Out[67] = {PT`Hydrogen`, Miscellaneous`Units`, Miscellaneous`SIUnits`,
        Graphics`Animation`, Global`, System`}

The package `Miscellaneous`Units`` imported the package `Miscellaneous`SIUnits``.

This package produces the side effect that the contexts `Miscellaneous`Units`` and `Miscellaneous`SIUnits`` are now on the context search path. To avoid this side effect, all we need to do is replace

```
BeginPackage["PT`Hydrogen`", "Miscellaneous`Units`"]
```

with

```
BeginPackage["PT`Hydrogen`"]
Needs["Miscellaneous`Units`"]
```

because `EndPackage[ ]` restores $ContextPath to its value before the preceding `BeginPackage` and prepends the current context to the list `$ContextPath`.

As intended, the symbols in the context `PT`Hydrogen`` are still visible, whereas those in the context `PT`Hydrogen`Private`` are not.

In[68] := **f**

Out[68] = f

In[69] := **radialFunction**

Out[69] = radialFunction

We can obtain the eigenfunctions:

In[70] := **Psi[3, 2, -2, r, theta, phi]**

$$Out[70] = \frac{E^{-2\ I\ phi\ -\ r/(3\ a)}\ r^2\ Sin[theta]^2}{162\ a^{7/2}\ Sqrt[Pi]}$$

We can determine the eigenenergies:

In[71] := **Energy[1]**

Out[71] = -13.6059 ElectronVolt

The energy can be expressed in terms of another unit:

In[72] := **Energy[1, Unit -> Joule]**

Out[72] = -2.1799 10^{-18} Joule

To use the package `PT`Hydrogen`` like a standard *Mathematica* package, we must

---

 2.2

1. Create a folder with the name PT in the Packages folder of the *Mathematica* folder
2. Enter the commands of Section 3.7.3.2 into a cell of a new notebook
3. Click the cell bracket and select **Initialization Cell** in the Attributes submenu of the Style menu
4. Name the notebook **Hydrogen.m** and put it in the PT folder

---

 2.2

1. Create a directory (or folder) with the name "pt" in the "packages" directory of the *Mathematica* directory
2. Enter the commands of Section 3.7.3.2 into a cell of a new notebook
3. Choose **Save** in the File menu
4. Select Packages as the File Type in the Save Notebook dialog box
5. Name the file **hydrogen.m**, and put it in the "pt" directory

Whereas Windows file and directory names are made up only of lowercase letters, *Mathematica* context names can comprise both lowercase and uppercase letters. To load the package, be sure to enter **Needs["PT`Hydrogen`"]** rather than **Needs["pt`hydrogen`"]**.

❀ **3.0**

1. Create a folder with the name PT in the StandardPackages folder of the AddOns folder within the *Mathematica* folder
2. Enter the commands of Section 3.7.3.2 into a cell of a new notebook
3. Click the cell bracket and choose **Initialization Cell** in the Cell Properties submenu of the Cell menu
4. Choose **Package Format** in the Save As Special submenu of the File menu
5. Name the notebook `Hydrogen.m` and save it in the PT folder
6. Close the new notebook and click Don't Save in the dialog box

---

3.7.4 Template for Packages

Here is a template for packages together with comments on its construction. It is an edited version of that suggested by Maeder [Mae91].

```
(* Skeleton.m -- a template for packages *)

(* set up the package context and import the necessary packages *)
BeginPackage["Skeleton`", {"Package1`", "Package2`"}]

(* usage messages for symbols intended for export *)
Skeleton::usage = "Skeleton.m is a package that does
nothing."

Function1::usage = "Function1[n] does nothing."

Function2::usage = "Function2[n, (m:17)] does even more nothing."

(* begin the private context *)
Begin["`Private`"]

(* definition of auxiliary functions *)
Aux[f_] := Do[something]

(* error messages for objects intended for export *)
Skeleton::badarg = "Sorry, you called `1` with
argument `2`!"

(* definition of functions intended for export *)
Function1[n_] := n

Function2[n_, m_:17] := n m /; n < 5 ||
            Message[Skeleton::badarg, Function2, n]
```

```
(* end the private context *)

End[]

(* protect symbols intended for export *)

Protect[Function1, Function2]

(* end the package context *)

EndPackage[]
```

3.7.5 Exercises

1. Write a package that exports a function for normalizing n-dimensional vectors, which are represented by n-element lists in *Mathematica*. The elements may be complex variables.

2. Write a package that exports a function for vector addition in two dimensions. The function should take as its arguments the directions and magnitudes of an arbitrary number of coplanar vectors and an option for specifying whether the angles are in degrees or radians, and it should return the direction and magnitude of the resultant.

3. Write a package for exporting two functions: one for normalizing three-dimensional, single-particle wave functions in Cartesian coordinates and another for normalizing those in spherical coordinates.

4. Write a package that provides the functions for the eigenenergies and normalized energy eigenfunctions of the simple harmonic oscillator.

$E = \frac{1}{2}\hbar\omega$

❋ 2.2 $\quad \Psi_n(x) = \left(\frac{m\omega}{\pi\hbar}\right)^{1/4} \frac{1}{\sqrt{2^n n!}} H_n(\xi) e^{-\xi^2/2}, \quad \xi = \sqrt{\frac{m\omega}{\hbar}} x$

5. First, load the package `Miscellaneous`PhysicalConstants`` with the `Needs` command. Then, load the package with the "`<<`" command. Finally, load the package again with the `Needs` command. Repeat this for several other standard packages. What can be concluded about the advantages that `Needs` has over "`<<`"?

6. Prepare the package `PT`Hydrogen`` for use like a standard *Mathematica* package in accordance with the instructions given at the end of Section 3.7.3. Load the package with the `Needs` command. Then, load the package with the "`<<`" command. Finally, load the package again with the `Needs` command. What can be concluded about the advantages that `Needs` has over "`<<`"?

❋ 3.0

7. Using the notation introduced in Example 2.5.3, rewrite the package `PT`Hydrogen``.

PART 2

Physics with *Mathematica*

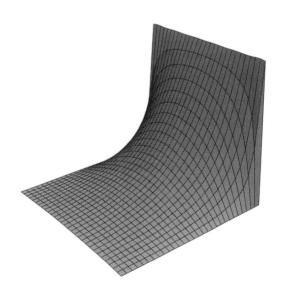

CHAPTER 4

Mechanics

4.1 Falling Bodies

4.1.1 The Problem

A small body falls downward with an initial velocity v_0 from a height h near the surface of the earth, as in Figure 4.1.1. For low velocities (less than about 24 m/s), the effect of air resistance may be approximated by a frictional force proportional to the velocity. Find the displacement and velocity of the body, and determine the terminal velocity. Plot the speed as a function of time for several initial velocities.

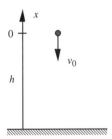

Figure 4.1.1 A small body falling downward with an initial velocity v_0 from a height h

4.1.2 Physics of the Problem

Let the x axis be directed upward, as in Figure 4.1.1. The net force on the object is

$$F = -mg - bv \tag{4.1.1}$$

349

where m is the mass, g is the magnitude of the acceleration due to gravity, and v is the velocity. The positive constant b depends on the size and shape of the object and on the viscosity of the air. From Newton's second law, the equation of motion is

$$m \frac{d^2x}{dt^2} = -mg - bv \tag{4.1.2}$$

with $x(0) = h$ and $v(0) = v_0$. We can rewrite Equation 4.1.2, a second-order equation, as two first-order equations:

$$\frac{dx}{dt} = v \tag{4.1.3}$$

$$\frac{dv}{dt} = -g - \frac{b}{m}v \tag{4.1.4}$$

(For further discussion of falling bodies, see [Sym71]; for more information on the effects of retarding forces on the motion of projectiles, refer to [MT95].)

4.1.3 Solution with Mathematica

In[1] := **ClearAll["Global`*"]**

DSolve solves Equations 4.1.3 and 4.1.4 with the initial conditions $x(0) = h$ and $v(0) = v_0$:

In[2] := **sol = DSolve[{x'[t] == v[t], v'[t] == - g - (b/m) v[t],**
                x[0] == h, v[0] == v0},
                {x[t], v[t]}, t] //ExpandAll

Out[2] = $\{\{x[t] \to h + \dfrac{g\ m^2}{b^2} - \dfrac{g\ m^2}{b^2\ E^{(b\ t)/m}} - \dfrac{g\ m\ t}{b} + \dfrac{m\ v0}{b} - \dfrac{m\ v0}{b\ E^{(b\ t)/m}},$

$v[t] \to -(\dfrac{g\ m}{b}) + \dfrac{g\ m}{b\ E^{(b\ t)/m}} + \dfrac{v0}{E^{(b\ t)/m}}\}\}$

The solution for the displacement $x(t)$ and velocity $v(t)$ is given as a list of rules.

To determine the terminal velocity, let $bt/m \gg 1$ in the rule for $v(t)$:

In[3] := **Last[Flatten[sol]] /.((b t)/m -> Infinity)**

Out[3] = $v[t] \to -(\dfrac{g\ m}{b})$

The terminal velocity v_t is $-gm/b$.

Let us determine the velocity as a function of time in a system of units where time, displacement, and therefore velocity are in units of m/b, gm^2/b^2, and gm/b, respectively:

In[4] := **v[t_] = (v[t] /. sol[[1]]) //. {g m/b -> 1, b/m -> 1}**

Out[4] = $-1 + E^{-t} + \dfrac{v0}{E^t}$

We now plot the speed as a function of time for three illustrative initial velocities:

```
In[5] := Plot[Evaluate[{Abs[v[t]] /. v0 -> -3,
                        Abs[v[t]] /. v0 -> -0.6,
                        Abs[v[t]] /. v0 -> 0,
                        1}
                      ], {t, 0, 3.5},
              PlotRange -> All,
              Frame -> True,
              FrameLabel -> {" t (m/b)", " speed (gm/b)"},
              PlotStyle -> {{RGBColor[1, 0, 0]}, {},
                            {RGBColor[0, 0, 1]}, {Dashing[{0.02}]}},
              Prolog -> {Text["v0 = 0", {0.95, 0.3}],
                         Text["|v0| < |vt|", {2.8, 0.55}],
                         Text["|v0| > |vt|", {1.5, 2.05}],
                         Text["terminal speed,|vt|", {0.9, 1.2}],
                         Line[{{0.95, 0.79}, {2.2, 0.6}}]}]
      ];
```

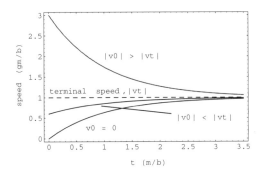

The bottom two curves indicate that the speed of the body increases, as expected, and eventually approaches the terminal speed during its descent. The top curve shows that if the initial speed exceeds the terminal speed, the falling object actually slows down and its speed approaches the terminal speed from above. (Note that though the velocity of the falling body is negative, its speed is positive.)

✤ **3.0** In the strings for the text of the plot, we can specify subscripts with the syntax \!\(... \) introduced in Section 2.5.2.1:

```
In[6] := Plot[Evaluate[{Abs[v[t]] /. v0 -> -3,
                        Abs[v[t]] /. v0 -> -0.6,
                        Abs[v[t]] /. v0 -> 0,
                        1}
                      ], {t, 0, 3.5},
              PlotRange -> All,
```

```
      Frame -> True,
      FrameLabel -> {" t (m/b)", " speed (gm/b)"},
      PlotStyle -> {{RGBColor[1, 0, 0]}, {},
                   {RGBColor[0, 0, 1]}, {Dashing[{0.02}]}},
      Prolog -> {Text["\!\(v\_0\) = 0",{0.95,0.3}],
                 Text["|\!\(v\_0\)| < |\!\(v\_t\)|",{2.8,0.55}],
                 Text["|\!\(v\_0\)| > |\!\(v\_t\)|",{1.5,2.05}],
                 Text["terminal speed,|\!\(v\_t\)|",{0.9,1.2}],
                 Line[{{0.95, 0.79}, {2.2, 0.6}}]}
   ];
```

In[7] := ClearAll[sol, v]

4.2 Projectile Motion

4.2.1 The Problem

A baseball of mass m leaves the bat with a speed v_0 at an angle θ_0 from the horizontal. The effect of air resistance may be approximated by a drag force $\mathbf{F}_D$ that depends on the square of the speed, that is,

$$\mathbf{F}_D = -mkv\mathbf{v}$$

where v and $\mathbf{v}$ are the speed and velocity of the ball, respectively, and the drag factor k is equal to 5.2×10^{-3} m^{-1}. Define a function that generates the graphs for animating the motion of the baseball. The function should have v_0 and θ_0 as arguments, and the graphs should show the paths of the baseball with and without air resistance. Evaluate the function for $v_0 = 45$ m/s and $\theta_0 = 60°$.

4.2.2 Physics of the Problem

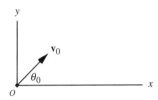

Figure 4.2.1 A baseball leaves the bat with an initial velocity $\mathbf{v_0}$ at an elevation angle θ_0

The equations of motion for the baseball are

$$\frac{d^2x}{dt^2} = -k\sqrt{\left(\frac{dx}{dt}\right)^2 + \left(\frac{dy}{dt}\right)^2}\left(\frac{dx}{dt}\right) \tag{4.2.1}$$

$$\frac{d^2y}{dt^2} = -k\sqrt{\left(\frac{dx}{dt}\right)^2 + \left(\frac{dy}{dt}\right)^2}\left(\frac{dy}{dt}\right) - g \tag{4.2.2}$$

where the x and y directions are the horizontal and vertical directions, respectively, as in Figure 4.2.1 and g is the magnitude of the acceleration due to gravity.

If air resistance is negligible (i.e., $k = 0$), these equations can be solved analytically. The path of the baseball is given by the equation

$$y = \frac{v_{y0}}{v_{x0}}x - \frac{1}{2}\frac{g}{v_{x0}^2}x^2$$

where the initial position is chosen as the origin and the initial x and y components of the velocity are

$$v_{x0} = v_0 \cos\theta_0$$

$$v_{y0} = v_0 \sin\theta_0$$

The maximum altitude occurs at

$$H = \frac{v_{y0}^2}{2g}$$

the range is

$$R = \frac{2v_{x0}v_{y0}}{g}$$

and the time of flight is

$$T = \frac{2v_{y0}}{g}$$

A realistic model for the flight of the baseball must include air resistance. Unfortunately, the equations of motion, Equations 4.2.1 and 4.2.2, can no longer be solved analytically; numerical methods must be employed. (For more information on the physics of baseball, consult [Ada95] and [Bra85].)

4.2.3   Solution with Mathematica

What follows is a function that generates the graphs for animating the motion of the baseball. We use theta, width, height, and t0 as aliases for θ_0, R, H, and T, respectively. Comments giving the details about the function are embedded in the body of the function in the form (* *comment* *). These comments are ignored by the *Mathematica* kernel. The function takes two arguments: the first is the numerical value of the initial speed in SI units, and the second is that of the initial elevation angle in degrees. The initial speed is limited to 80 m/s because higher speeds are unlikely for baseballs, and the initial angle of elevation is restricted to the range 30° to 85° in order to accommodate the sizes of most monitor screens.

```
In[1] := ClearAll["Global`*"]

In[2] := projectile[vo_ /; vo > 0 && vo <= 80,
                    theta_ /; theta >= 30 && theta <= 85]  :=

        Module[{k = 5.2 10^(-3), g = 9.81, vxo, vyo, x, y,
                width, height, tmax, t0,
                pathWithoutAirResistance, sol},

          (* initial x and y components of velocity *)
          vxo = vo Cos[theta Degree]//N;
          vyo = vo Sin[theta Degree]//N;

          (* range without air resistance *)
          width = 2 vyo vxo/g;

          (* maximum height without air resistance *)
          height = vyo^2/(2 g);

          (* Time of flight without air resistance *)
          t0 = 2 vyo/g;

          pathWithoutAirResistance =
              Plot[(vyo/vxo)x - (1/2)(g/vxo^2)x^2,
                  {x, 0, width}, PlotRange -> {0, height},
                  DisplayFunction -> Identity
                ];
```

```
(* numerical solution of the equations of motion with
   air resistance *)
sol = NDSolve[
         {x''[t] == - k Sqrt[x'[t]^2 + y'[t]^2] x'[t],
          y''[t] == - k Sqrt[x'[t]^2 + y'[t]^2] y'[t] - g,
          x[0] == 0,
          y[0] == 0,
          x'[0] == vxo,
          y'[0] == vyo
         },
         {x, y}, {t, 0, t0}
             ];


(* the x and y coordinates as a function of time *)
x[t_] = x[t] /. sol[[1]];
y[t_] = y[t] /. sol[[1]];


(* time of flight *)
tmax = t /. FindRoot[y[t] == 0, {t, t0}];


(* graphics for animation *)
Do[
   Show[
     pathWithoutAirResistance,
     Graphics[
       {AbsolutePointSize[7],
         Point[{x[(tmax/16) i], y[(tmax/16) i]}]}]
             ],
     ParametricPlot[
       {x[t], y[t]}, {t, 0, (tmax/16) i + 0.0001},
       PlotStyle -> Dashing[{0.02, 0.02}],
       DisplayFunction -> Identity
                   ],
     PlotRange -> {{-0.01, width}, {-0.01, 1.02 height}},
     Axes -> True,
     AxesLabel -> {" x (m)", " y (m)"},
     AspectRatio -> Automatic,
     DisplayFunction -> $DisplayFunction
         ],
   {i, 0, 16}
   ]
     ]
```

Let us generate the graphs for animating the motion of the baseball with $v_0 = 45$ m/s and $\theta_0 = 60°$:

In[3] := **projectile[45, 60]**

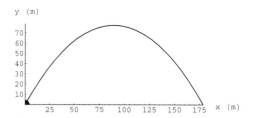

Only the head cell of the closed group of 17 graphics cells is shown here. The solid line represents the path of the baseball without air resistance. In the other graphs, the dashed line traces that with air resistance. To animate, click the group bracket and press Command-Y for the Macintosh or Control+Y for Windows.

4.3   The Pendulum

4.3.1   The Problem

Solve the equations of motion numerically and plot the phase diagrams of the plane, the damped, and the damped, driven pendulums. For the damped, driven pendulum, also determine the Poincaré section and produce the graphics for animation.

4.3.2   Physics of the Problem

4.3.2.1   THE PLANE PENDULUM

Consider a particle of mass m that is constrained by a rigid and massless rod to move in a vertical circle of radius L, as in Figure 4.3.1.

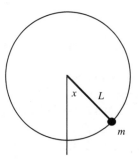

Figure 4.3.1   The plane pendulum

The equation of motion is

$$\frac{d^2x}{dt^2} + \omega_0^2 \sin x = 0 \tag{4.3.1}$$

where x is the angle that the rod makes with the vertical,

$$\omega_0 = \sqrt{\frac{g}{L}} \tag{4.3.2}$$

and g is the magnitude of the acceleration due to gravity. We can rewrite Equation 4.3.1, a second-order equation, as two first-order equations:

$$\frac{dx}{dt} = v \tag{4.3.3a}$$

$$\frac{dv}{dt} = -\omega_0^2 \sin x \tag{4.3.3b}$$

If the unit of time t is taken to be $t_0 = 1/\omega_0$, Equation 4.3.3b assumes the form

$$\frac{dv}{dt} = -\sin x \tag{4.3.3b'}$$

where x is in radians and v is in radians/t_0.

4.3.2.2 THE DAMPED PENDULUM

We now include damping that is proportional to the velocity. Equation 4.3.3b' becomes

$$\frac{dv}{dt} = -\sin x - av \tag{4.3.4}$$

where a is a parameter measuring the strength of the damping.

4.3.2.3 THE DAMPED, DRIVEN PENDULUM

With a sinusoidal driving force, Equation 4.3.4 becomes

$$\frac{dv}{dt} = -\sin x - av + f \cos wt \tag{4.3.5}$$

where f and w are the amplitude and frequency of the driving force, respectively.

The motion of the pendulum depends on the parameters $a, f,$ and w as well as on the initial conditions for x and v. Whether the motion is periodic or chaotic is determined solely by the parameters $a, f,$ and w. (For introductions to chaotic dynamics, consult [BG90] and [MT95]; for more advanced discussions, refer to [Moo87] and [Ras90].)

4.3.3 *Solution with Mathematica*

In this section we solve the equations of motion and plot the phase diagrams of the plane, the damped, and the damped, driven pendulums. For the damped, driven pendulum, we also determine the Poincaré section and produce the graphics for animation.

```
In[1] := ClearAll["Global`*"]
```

4.3.3.1 THE PLANE PENDULUM

As an example, we choose the initial conditions $x = 0$ and $v = 2 - d$, where d is a small number. (What would happen if the initial conditions are such that $x = 0$ and $v \geq 2$? See Problem 7 in Section 4.5.) NDSolve solves Equation 4.3.3 together with the initial conditions:

```
In[2] := points1 = NDSolve[{x'[t] == v[t], v'[t] == -Sin[x[t]],
                    x[0] == 0, v[0] == 2 - 0.001},
                    {x, v}, {t, 0, 19.5}];
```

With the solution points1, we can plot the phase diagram:

```
In[3] := ParametricPlot[Evaluate[{x[t], v[t]} /. points1],
                    {t, 0, 19.5},
                    AxesLabel -> {" x (rad)", " v (rad/to)"}];
```

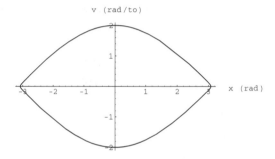

4.3.3.2 THE DAMPED PENDULUM

For an example, choose the initial conditions $x = 0$ and $v = 3$ and let $a = 0.2$. NDSolve solves Equations 4.3.3a and 4.3.4 together with the initial conditions:

```
In[4] := points2 = NDSolve[{x'[t] == v[t],
                    v'[t] == -Sin[x[t]] - 0.2 v[t],
                    x[0] == 0, v[0] == 3},
                    {x, v}, {t, 0, 30}];
```

Let us plot the phase diagram:

```
In[5] := ParametricPlot[Evaluate[{x[t], v[t]} /. points2],
                    {t, 0, 30}, PlotRange -> All,
                    AxesLabel -> {" x (rad)", " v (rad/to)"}];
```

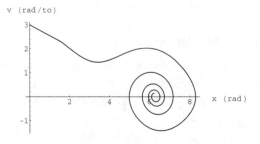

4.3.3.3 THE DAMPED, DRIVEN PENDULUM

As an example, let $a = 0.2, f = 0.52$, and $w = 0.694$:

```
In[6] := a = 0.2; f = 0.52; w = 0.694;
```

Also, choose the initial conditions $x = 0.8$ and $v = 0.8$.

For t in the range 0 to $cycles(2\pi/w)$, **NDSolve** solves Equations 4.3.3a and 4.3.5 together with the initial conditions:

```
In[7] := cycles = 50;
         sol3 = NDSolve[
                   {x'[t] == v[t],
                    v'[t] == -Sin[x[t]] - a v[t] + f Cos[w t],
                    x[0] == 0.8,
                    v[0] == 0.8},
                   {x, v}, {t, 0, cycles(2Pi/w)},
                   MaxSteps -> 20000
                      ];
```

Mathematically, the angle x can take on values from $-\infty$ to $+\infty$; physically, x can only vary from $-\pi$ to $+\pi$. In the following, the phase diagram is constructed with all the points translated back to the interval $-\pi$ to $+\pi$ and with the lines connecting adjacent points omitted for clarity:

```
In[9] := pi = N[Pi];
         reduce[x_] := Mod[x, 2pi] /; Mod[x, 2pi] <= pi;
         reduce[x_] := (Mod[x, 2pi] - 2pi) /; Mod[x, 2pi] > pi
```

```
In[12] := steps = 30;
          points3 =
             Flatten[
                Table[{x[t], v[t]}/.sol3,
                      {t, 0, cycles(2Pi/w), (1/steps)(2Pi/w)//N}
                   ], 1
                   ];
          xposition3 = Table[{i, 1}, {i, Length[points3]}];
          newpoints3 = MapAt[reduce, points3, xposition3];
          ListPlot[newpoints3, PlotRange -> {-3, 3},
                   AxesLabel -> {" x (rad)", " v (rad/to)"}];
```

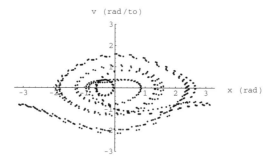

The Poincaré section provides a simplification of the phase diagram to reveal the essential features of the dynamics. For the phase diagram the coordinates $(x(t), v(t))$ were determined for the values of time $t = 0, \Delta t, 2\Delta t, 3\Delta t, \ldots$, where we have chosen $\Delta t = T/30$ in which T is the period of the driving force. For the Poincaré section we select only those points for $t = 0, T, 2T, 3T, \ldots$. To avoid transient effects, the points for the first 25 cycles are discarded in the following construction of the Poincaré section:

```
In[17] := poincare3 =
             Table[newpoints3[[n]],
                   {n, 1 + 25 steps, Length[newpoints3], steps}];
```

```
In[18] := Length[poincare3]
```

```
Out[18] = 26
```

```
In[19] := ListPlot[poincare3, PlotRange -> {{-3, 3}, {-3, 3}},
                AxesLabel -> {" x (rad)", " v (rad/to)"},
                PlotStyle -> PointSize[0.015]];
```

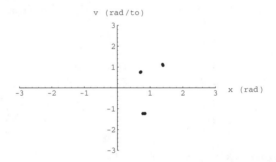

The Poincaré section has three dots and indicates periodic motion. (For an introduction to the Poincaré section, see [BG90] or [MT95].)

For another example, let $a = 1/2, f = 1.15$, and $w = 2/3$:

```
In[20] := a = 1/2; f = 1.15; w = 2/3;
```

Also, choose the initial conditions $x = 0.8$ and $y = 0.8$.

As the solution of this example with *Mathematica* requires a kernel memory partition of about 16 MB, consider beginning here a new *Mathematica* session with virtual memory, if available, turned on. Cognizant of the fact that virtual memory is too slow for animation, we can first complete the computations and produce the graphics and then show the animation in another *Mathematica* session, using only the front end and no virtual memory. If the memory demand of the kernel creates a hardship, refer to the discussion later in this section for a more frugal method of solving the equation of motion.

For t from 0 to $cycles(2\pi/w)$, NDSolve solves Equations 4.3.3a and 4.3.5 together with the initial conditions:

```
In[21] := cycles = 180;
         sol4 = NDSolve[
                    {x'[t] == v[t],
                     v'[t] == - Sin[x[t]] - a v[t] + f Cos[w t],
                     x[0] == 0.8,
                     v[0] == 0.8},
                    {x, v}, {t, 0, cycles(2Pi/w)},
                    MaxSteps -> 20000
                       ];
         steps = 30;
         points4 =
              Flatten[
                Table[{x[t], v[t]}/.sol4,
                      {t, 0, cycles(2Pi/w), (1/steps)(2Pi/w)//N}
                      ], 1
                   ];
```

where we have generated a nested list of coordinates $\{x(t), v(t)\}$, with t ranging from 0 to $cycles(2\pi/w)$ in steps of size $\Delta t = (1/steps)(2\pi/w)$, for plotting the phase diagram and the Poincaré section. We must take computer solution of differential equations in chaotic dynamics with a grain of salt, as sensitive dependence on initial conditions is a characteristic feature of chaos and round-off error is an inherent limitation of numerical methods. (For further discussion, refer to [PJS92] and [Ras90].)

❄ **3.0** Specification of the option `Method -> RungeKutta` in `NDSolve` reduces the kernel memory requirement by about 2 MB. (For an introduction to the Runge-Kutta methods of numerical solution of differential equations, see [KM90]; for a more detailed discussion, consult [Pat94].)

❄ **2.2** Users with maximum available kernel memory limited by earlier versions of *Mathematica* for Windows or by modest means can solve the equation of motion numerically with the Runge-Kutta methods since they require only a kernel memory partition of less than 10 MB. (For an introduction to the Runge-Kutta methods, see [KM90]; for a more detailed discussion, consult [Pat94].) The package `ProgrammingExamples`RungeKutta`` contains the function `RungeKutta` that implements the fourth-order Runge-Kutta method, which gives an excellent balance between accuracy and computational effort. `RungeKutta[{e1, e2, ...}, {y1, y2, ...}, {a1, a2, ...},` `{t, t0, t1, Δt}]` numerically integrates the ei as functions of time t and yi with initial values ai. The integration proceeds in steps of Δt from $t0$ to $t1$. `RungeKutta` returns a nested list of `{y1[t], y2[t], ...}` for t ranging from $t0$ to $t1$ in steps of Δt. Let us load the package:

```
Needs["ProgrammingExamples`RungeKutta`"]
```

With $t0 = 0$, $t1 = cycles(2\pi/w)$, and $\Delta t = (1/steps)(2\pi/w)$, `RungeKutta` solves Equations 4.3.3a and 4.3.5 together with the initial conditions:

```
cycles = 180; steps = 30;
points4 = RungeKutta[
                {v, -Sin[x] - a v + f Cos[w t]},
                {x, v}, {0.8, 0.8},
                {t, 0, cycles(2Pi/w), (1/steps)(2Pi/w)}
                ];
```

Let us plot the phase diagram with all the points translated back to the interval $x = -\pi$ to $+\pi$ and with the lines joining the points omitted for clarity:

```
In[25] := pi = N[Pi];
          reduce[x_] := Mod[x, 2pi] /; Mod[x, 2pi] <= pi;
          reduce[x_] := (Mod[x, 2pi] - 2pi) /; Mod[x, 2pi] > pi
```

```
In[28] := xposition4 = Table[{i, 1}, {i, Length[points4]}];
          newpoints4 = MapAt[reduce, points4, xposition4];
          ListPlot[newpoints4,
                AxesLabel -> {" x (rad)", " v (rad/to)"}];
```

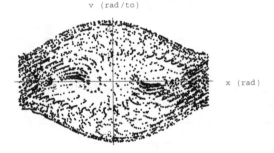

Next, we construct the Poincaré section:

```
In[31] := poincare4 =
            Table[newpoints4[[n]],
                {n, 1 + 20 steps, Length[newpoints4], steps}];
          ListPlot[poincare4, PlotRange -> {{-4, 4}, {-4, 4}},
                AxesLabel -> {" x (rad)", " v (rad/to)"}];
```

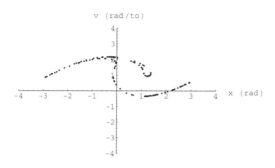

The Poincaré section has a complex geometry that is characteristic of chaotic states.

   We end this notebook with the graphics showing the chaotic motion of the pendulum. The package **PT`Pendulum`** containing the function **Motion** is included in Section 4.3.3.4. To use this package like a standard *Mathematica* package, follow the instructions at the end of Section 3.7.3.3. **Motion**[*points, n, m, j*] produces the graphics of the pendulum for animation. The first argument *points* is for the nested list of coordinates $\{x(t_i), v(t_i)\}$ from the solution of the equation of motion; the second argument *n*, for the number of graphs of the pendulum to be plotted; the third argument *m*, for the time increment; the fourth argument *j*, for the number of blank graphs. Only nonzero positive integers can be assumed by *n*, *m*, and *j*. The time interval between two successive graphs is equal to *m* times the step size Δt previously specified. For given *points* and *m*, *n* has an upper limit

   Ceiling[Length[points]/m]

Length[*points*] gives the number of elements in *points,* and **Ceiling**[*x*] is the least integer not smaller than *x*. Here is the number of elements in **points4**:

In[33] := **Length[points4]**

Out[33] = 5401

   Let us load the package **PT`Pendulum`** and call the function **Motion**:

In[34] := **Needs["PT`Pendulum`"]**
          Motion[points4, 200, 2, 20]

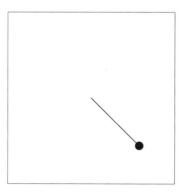

(Only the head cell of the closed group of 220 graphics cells is shown here.) To animate, click the group bracket, if it is not already selected, and press Command-Y for the Macintosh or Control+Y for Windows. When *Mathematica* comes to the end of a sequence of animated graphs, it begins again immediately with the first graph of the sequence, and so blank graphs are included to separate the two sequences. (See Section 2.3.3 for a discussion of other techniques for separating two animation sequences.) Generating 200 graphs of the pendulum and 20 blank graphs requires a great deal of memory. The actual number of graphs, the more the better, that can be produced depends, of course, on the available memory. If a color monitor is used, the color should be turned off to reduce the required memory.

4.3.3.4 PENDULUM.M: A *MATHEMATICA* PACKAGE

```
(* to use this package like a standard Mathematica
   package, follow the instructions at the end of
   Section 3.7.3.3. *)

BeginPackage["PT`Pendulum`"]

Pendulum::usage = "Pendulum.m is a package that
generates the graphs of the pendulum for animation."

Motion::usage = "Motion[points, n, m, j] plots
the graphs of the pendulum for animation."

Begin["`Private`"]

Motion::badarg = "For the given nested list of
coordinates and your choice of time increment,
the maximum number of graphs that can be plotted
is `1`. You requested `2` graphs."

Motion[points_List, n_Integer?Positive,
       m_Integer?Positive, j_Integer?Positive] :=
  (list = Table[points[[i, 1]], {i, 1, Length[points], m}];
   x = Sin[list];
   y = -Cos[list];
   Do[Show[Graphics[{
                 Line[{{0, 0}, {x[[k]], y[[k]]}}],
                 PointSize[0.05],
                 Point[{x[[k]], y[[k]]}]
                    }
                    ],
          PlotRange -> {{-1.25, 1.25}, {-1.25, 1.25}},
          AspectRatio -> Automatic,
          Frame -> True, FrameTicks -> None
         ], {k, n}
     ];

Do[Show[Graphics[{GrayLevel[1],
```

```
                    Line[{{0, 0}, {1.25, 1.25}}]]}
                ],
            PlotRange -> {{-1.25, 1.25}, {-1.25, 1.25}},
            AspectRatio -> Automatic,
            Frame -> True, FrameTicks -> None
          ], {i, j}
      ]
  ) /; n <= Ceiling[Length[points]/m];

Motion[points_List, n_Integer?Positive,
      m_Integer?Positive, j_Integer?Positive] :=
  (Message[Motion::badarg, Ceiling[Length[points]/m], n]
  ) /; n > Ceiling[Length[points]/m]

End[]

Protect[Motion]

EndPackage[]
```

4.4   The Spherical Pendulum

4.4.1   The Problem

The spherical pendulum consists of a particle of mass m constrained by a rigid and massless rod to move on a sphere of radius R. With several choices of initial conditions for the pendulum, explore its motion.

4.4.2   Physics of the Problem

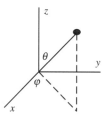

Figure 4.4.1   The spherical pendulum

We adopt a system of units in which $m = 1$, $R = 1$, and $g = 1$, where g is the magnitude of the acceleration due to gravity. Let θ and φ be the spherical coordinates of the particle, and let the fixed end of the rod be the origin, as in Figure 4.4.1. In terms of the spherical coordinates, the kinetic energy is

$$T = \frac{1}{2}(\dot{\theta}^2 + \sin^2\theta\,\dot{\varphi}^2) \tag{4.4.1}$$

The gravitational potential energy relative to the horizontal plane is

$$V = \cos\theta \tag{4.4.2}$$

Hence, the Lagrangian, $T - V$, can be expressed as

$$L = \frac{1}{2}(\dot{\theta}^2 + \sin^2\theta\,\dot{\varphi}^2) - \cos\theta \tag{4.4.3}$$

The Lagrange equations of motion give

$$\frac{d\dot{\theta}}{dt} - \sin\theta\cos\theta\,\dot{\varphi}^2 - \sin\theta = 0 \tag{4.4.4}$$

$$\frac{d}{dt}(\sin^2\theta\,\dot{\varphi}) = 0 \tag{4.4.5}$$

Equations 4.4.4 and 4.4.5 together with the initial conditions θ_0, $\dot{\theta}_0$, φ_0, and $\dot{\varphi}_0$ completely specify the motion of the system. (For an introduction to Lagrangian mechanics, see [Sym71].)

We gain insight into the motion of the pendulum by recognizing that there are two constants of the motion: the angular momentum about the z axis and the total energy. Since the coordinate φ is ignorable (i.e., it does not appear explicitly in the Lagrangian), the corresponding generalized momentum (also known as canonical momentum or conjugate momentum)

$$p_\varphi = \frac{\partial L}{\partial\dot{\varphi}} = \sin^2\theta\,\dot{\varphi} \tag{4.4.6}$$

is a constant of the motion, as can be seen from Equation 4.4.5. The total energy

$$E = T + V = \frac{1}{2}(\dot{\theta}^2 + \sin^2\theta\,\dot{\varphi}^2) + \cos\theta \tag{4.4.7}$$

can also be shown to be a constant of the motion because the spherical pendulum is a natural system (i.e., the Lagrangian has the form $T_2 - V$, where the kinetic energy T_2 contains only terms quadratic in the generalized velocities and the potential energy V is velocity-independent), the Lagrangian of which does not depend explicitly on time. (For a detailed discussion of the energy integral of the motion, see [CS60].) In terms of the angular momentum p_φ, Equations 4.4.4 and 4.4.5 can be written as

$$\ddot{\theta} - \frac{\cos\theta}{\sin^3\theta}p_\varphi^2 - \sin\theta = 0 \tag{4.4.8}$$

$$\dot{\varphi} = \frac{p_\varphi}{\sin^2\theta} \tag{4.4.9}$$

and Equation 4.4.7 can be put in the form

$$E = \frac{1}{2}\left(\dot{\theta}^2 + \frac{p_\varphi^2}{\sin^2\theta}\right) + \cos\theta \tag{4.4.10}$$

Let us define an effective potential

$$U(\theta) = \frac{1}{2}\frac{p_\varphi^2}{\sin^2\theta} + \cos\theta \tag{4.4.11}$$

Equation 4.4.10 becomes

$$\frac{1}{2}\dot{\theta}^2 = E - U(\theta) \tag{4.4.12}$$

If $p_\varphi = 0$, Equations 4.4.8 and 4.4.9 are the equations of motion of a simple pendulum oscillating in a plane perpendicular to the xy plane. If $p_\varphi \neq 0$, Equation 4.4.9 indicates that, except for the two singularities at θ equals $0°$ and $180°$, the sign of $\dot{\varphi}$ remains constant. Thus, the particle rotates about the z axis. We obtain further insight into the motion of the pendulum by examining the effective potential curve in Figure 4.4.2.

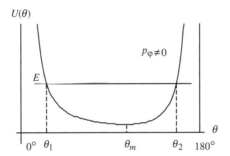

Figure 4.4.2 Effective potential for the spherical pendulum

The effective potential rises to infinity at $\theta = 0°$ and $\theta = 180°$. Therefore, these angles are inaccessible to the system unless the total energy is infinite because $\dot{\theta}^2$ in Equation 4.4.12 cannot be negative. In general, the angle θ oscillates between two turning points θ_1 and θ_2, which are the roots of the equation

$$E - U(\theta) = 0 \qquad (4.4.13)$$

If $E = U(\theta_m)$ where θ_m is the angle for the minimum value of $U(\theta)$, these roots coincide and the particle rotates about the z axis at the fixed angle θ_m and, from Equation 4.4.9, with constant $\dot{\varphi}$.

4.4.3 Solution with Mathematica

In[1] := **ClearAll["Global`*"]**

We begin with the definition of a function for plotting the effective potential together with the total energy. The first argument of the function is θ_0 in degrees from $0°$ to $180°$; the second argument, $\dot{\theta}_0$; the third argument, $\dot{\varphi}_0$. Options, except **AxesLabel**, **PlotStyle**, and **Ticks**, for passing to the **Plot** function may be specified after the third argument.

```
(* Mathematica Version 2.2 Specifics *)

effectivePotential[theta0_ /; theta0 >= 0 && theta0 <= 180,
               dtheta0_, dphi0_, opts___Rule] :=
 Module[{p},
    p = (Sin[theta0 Degree]^2) dphi0;
    Plot[{(1/2)(p^2/Sin[theta Degree]^2) +
          Cos[theta Degree],
          (1/2)(dtheta0^2 + p^2/Sin[theta0 Degree]^2) +
```

```
          Cos[theta0 Degree]},
      {theta, 0 + 10^(-3), 180 - 10^(-3)},
      Ticks -> {{0, 90, 180}, Automatic},
      AxesLabel -> {FontForm["q (∞)", {"Symbol", 9}],
                   " U (mgR)"},
      PlotStyle ->
        {{Thickness[0.0075]}, {Dashing[{0.05, 0.05}]}},
      opts
      ];
   ]
```

In[2] := (* 🏵 **3.0** *)

```
    effectivePotential[theta0_ /; theta0 >= 0 && theta0 <= 180,
                   dtheta0_, dphi0_, opts___Rule] :=
         Module[{p},
             p = (Sin[theta0°]^2) dphi0;
             Plot[{(1/2)(p^2/Sin[theta°]^2) +
                   Cos[theta°],
                   (1/2)(dtheta0^2 + p^2/Sin[theta0°]^2) +
                   Cos[theta0°]},
                 {theta, 0 + 10^(-3), 180 - 10^(-3)},
                 Ticks -> {{0, 90, 180}, Automatic},
                 AxesLabel -> {"θ (°)", " U (mgR)"},
                 PlotStyle ->
                   {{Thickness[0.0075]}, {Dashing[{0.05, 0.05}]}},
                 opts
                 ];
             ]
```

We now define the function that generates the graphs for animating the spherical pendulum. The first argument of the function is θ_0 in degrees from $0°$ to $180°$; the second argument, $\dot{\theta}_0$ limited between -10 and 10; the third argument, φ_0 in degrees from $0°$ to $360°$; the fourth argument, $\dot{\varphi}_0$ confined between -10 and 10; the fifth argument, the interval of time *tmax* restricted from 1 to 40 for the animation sequence; the sixth argument, the number of graphs *nv* generated with a minimum value of 2 and a computed default value of the integer closest to 3*tmax*. Options, with the exception of PlotRange, BoxRatios, and DisplayFunction, for passing to the Show function may be specified after these arguments. The construct of the function is explained in the body of the function with statements in the form (* *comment* *). The maximum possible values for *tmax* and *nv* depend on the available memory. Be forewarned that this function, which generates the three-dimensional graphics for an animation sequence, is memory- and time-intensive. If a color monitor is used, the color should be turned off to reduce the required memory. If there is insufficient memory for animation, we can still use the function to plot the path of the pendulum. With *nv* = 2, the function generates two graphs: one for time $t = 0$ and another for $t = tmax$.

```
In[3] := sphericalPendulum[theta0_ /; theta0 >= 0 && theta0 <= 180,
                           dtheta0_ /; Abs[dtheta0 // N] < 10,
                           phi0_ /; phi0 >= 0 && phi0 <= 360,
                           dphi0_ /; Abs[dphi0 // N] < 10,
                           tmax_ /; tmax >= 1 && tmax <= 40,
                           nv_Integer:999, opts___Rule] :=

      Module[{n = nv, p, sol, theta, phi, x, y, z, sphere},

       (* number of graphs must be an integer greater than one *)
       If[n > 1,

        (* computed default for number of graphs *)
        If[n == 999, n = Round[3 tmax]];

        (* angular momentum *)
        p = (((Sin[theta0 Degree]^2) dphi0)//N);

        (* numerical solution of equations of motion *)
        sol = NDSolve[
                {theta''[t]
                   - p^2 (Cos[theta[t]]/Sin[theta[t]]^3)
                   - Sin[theta[t]] == 0,
                 phi'[t] == p/(Sin[theta[t]]^2),
                 theta[0] == theta0 Degree,
                 theta'[0] == dtheta0,
                 phi[0] == phi0 Degree
                },
                {theta, phi}, {t, 0, tmax + 0.01},
                MaxSteps -> 6000
                  ];
        (* spherical coordinates *)
        theta = theta /. First[sol];
        phi = phi /. First[sol];

        (* rectangular coordinates *)
        x[t_] := Sin[theta[t]] Cos[phi[t]];
        y[t_] := Sin[theta[t]] Sin[phi[t]];
        z[t_] := Cos[theta[t]];

        (* graphics *)
        sphere = ParametricPlot3D[
                    {{0, Sin[t], Cos[t]},
                     {Sin[t], 0, Cos[t]}},
                    {t, 0, 2 Pi},
                    DisplayFunction -> Identity
                       ];
```

```
Do[
 Show[

  (* background sphere *)
  sphere,
  Graphics3D[Line[{{0, 0, -1}, {0, 0, 1}}]],

  (* trace *)
  ParametricPlot3D[{x[t], y[t], z[t]},
                   {t, 0, (tmax/(n-1)) i + 0.0001},
                   PlotPoints -> (25 + Round[12 tmax/(n-1) i]),
                   DisplayFunction -> Identity
                   ],

  (* pendulum *)
  Graphics3D[{Thickness[0.0125],
              Line[{{0, 0, 0},
                    {x[(tmax/(n - 1)) i],
                     y[(tmax/(n - 1)) i],
                     z[(tmax/(n - 1)) i]}}
                  ],
              PointSize[0.04],
              Point[{x[(tmax/(n - 1)) i],
                     y[(tmax/(n - 1)) i],
                     z[(tmax/(n - 1)) i]}]}]
          ],

  (* options *)
  PlotRange ->
     {{-1.25, 1.25}, {-1.25, 1.25}, {-1.25, 1.25}},
  BoxRatios -> {1, 1, 1},
  DisplayFunction -> $DisplayFunction,
  opts,
  AxesLabel -> {"x (R)", "y (R)", "z (R)"}
     ],
 {i, 0, n - 1}
 ],

(* error message *)
sphericalPendulum::badarg =
 "Number of graphs must be an integer greater than one!
 You requested `1` graph(s).";
Message[sphericalPendulum::badarg, nv]
 ]
    ]
```

   With the functions `effectivePotential` and `sphericalPendulum`, we are ready to explore the motion of the spherical pendulum with several choices of initial conditions.

4.4.3.1 $\theta_0 = 120,\ \dot{\theta}_0 = 0,\ \varphi_0 = 120,\ \dot{\varphi}_0 = 0$

Since $p_\varphi = 0$, the spherical pendulum behaves as a simple pendulum. The function effectivePotential plots the effective potential together with the total energy:

In[4] := **effectivePotential[120, 0, 0]**

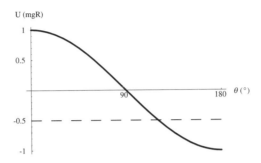

In the graph, the solid curve and the dashed line represent the effective potential and the total energy, respectively. The effective potential has a minimum at $\theta = 180°$. The pendulum oscillates about this angle, and the turning point $\theta = 120°$ occurs at the intersection of the effective potential curve and the total energy line. The function sphericalPendulum generates the graphs for animation:

In[5] := **sphericalPendulum[120, 0, 45, 0, 14]**

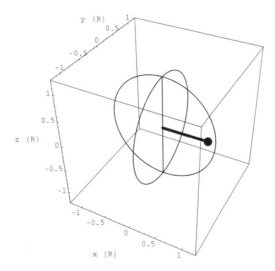

Only the head cell of the closed group of 42 graphics cells is shown here. In the other graphs, the curve traces the path of the particle from its initial position. To animate, click the cell bracket and press Command-Y for the Macintosh or Control+Y for Windows. In this case, the spherical pendulum behaves like a simple pendulum, as expected.

4.4.3.2 $\theta_0 = 135, \dot{\theta}_0 = 0, \varphi_0 = 90, \dot{\varphi}_0 = 2^{1/4}$

Let us plot the effective potential:

```
In[6] := effectivePotential[135, 0, 2^(1/4), PlotRange -> {-0.5, 2.0}]
```

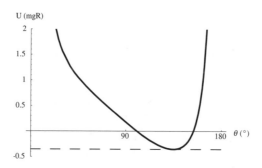

Since $p_\varphi \neq 0$ and $E = U(\theta_m)$, the particle rotates uniformly in a horizontal circle about the z axis. We now generate the graphs for animation:

```
In[7] := sphericalPendulum[135, 0, 90, 2^(1/4), 15]
```

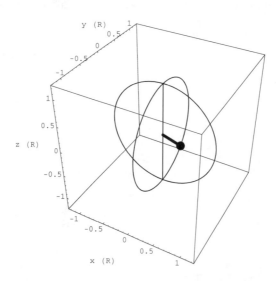

Only the head cell of the closed group of 45 graphics cells is shown here. To animate, click the cell bracket and press Command-Y for the Macintosh or Control+Y for Windows. The pendulum indeed rotates uniformly about the z axis.

4.4.3.3 $\theta_0 = 135, \dot{\theta}_0 = 2.5, \varphi_0 = 90, \dot{\varphi}_0 = 1.5 \times 2^{1/4}$

Again, we plot the effective potential:

```
In[8] := effectivePotential[135, 2.5, 1.5 2^(1/4),
                         PlotRange -> {-0.25, 4.0}]
```

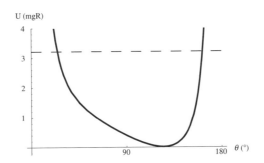

The particle rotates about the z axis while the angle θ oscillates between two turning points, $\theta \approx 24°$ and $\theta \approx 162°$, which occur at the intersections of the effective potential curve and the energy line. Let us generate the graphs for animation:

```
In[9] := sphericalPendulum[135, 2.5, 90, 1.5  2^(1/4), 20]
```

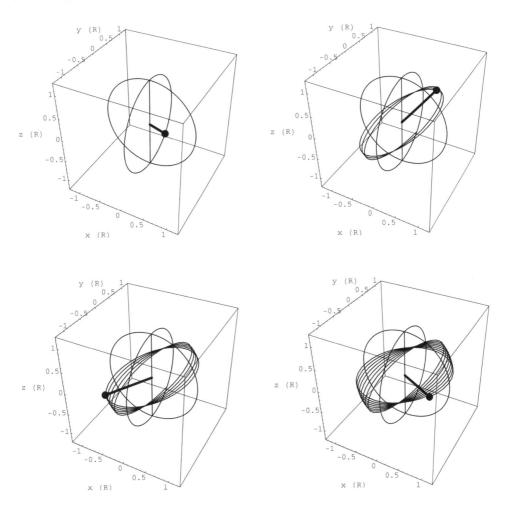

To animate, select all the graphics cells and press Command-Y for the Macintosh or Control+Y for Windows. (Only 4 of the 60 graphs are shown here.) We see that the orbit of the pendulum rotates about an axis.

4.4.3.4 $\theta_0 = 120, \dot{\theta}_0 = 0.75, \varphi_0 = 90, \dot{\varphi}_0 = 2.0 \times 2^{1/4}$

Once more, we plot the effective potential:

```
In[10] := effectivePotential[120, 0.75, 2.0  2^(1/4), PlotRange -> {0, 4}]
```

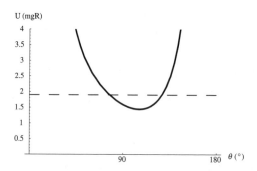

Here the turning points for the angle θ are about 77° and 127°. We now generate the graphs for animation:

```
In[11] := sphericalPendulum[120, 0.75, 90, 2.0 2^(1/4), 20]
```

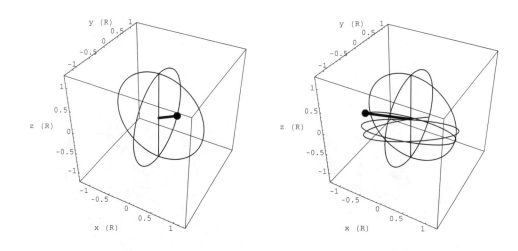

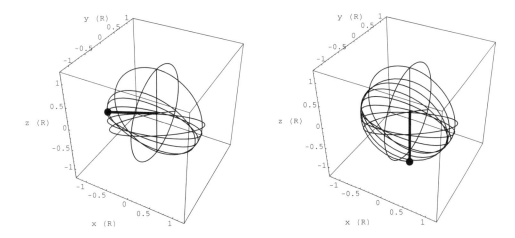

To animate, select all the graphics cells and press Command-Y for the Macintosh or Control+Y for Windows. (Only 4 of the 60 graphs are shown here.) We see that the motion of the pendulum is more complex than those of the preceding cases. As discussed before, the angular momentum about the z axis must remain constant during the motion.

It is instructive to observe the motion of the pendulum from another viewpoint.

```
In[12] := sphericalPendulum[120, 0.75, 90, 2.0 2^(1/4), 20,
                   ViewPoint -> {3.225, 1.018, 0.109}]
```

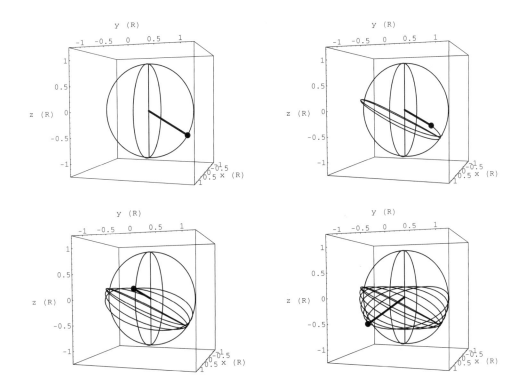

(Only 4 of the 60 graphs are shown here.) The path of the particle is clearly confined between two planes parallel to the xy plane; that is, the angle θ oscillates between two turning points.

```
In[13] := ClearAll[effectivePotential, sphericalPendulum]
```

4.5   Problems

Problems of intermediate difficulty are marked with one asterisk; those of considerable challenge, with two asterisks.

1. In an experiment, the motion of a block down a smooth incline is observed and the following data are obtained:

$\sin \theta$	$a\ (\text{m/s}^2)$
0.271	2.78
0.282	2.66
0.302	2.65
0.319	3.11
0.336	3.27
0.350	3.31
0.360	3.55
0.376	3.60
0.389	3.68
0.400	3.92
0.418	3.90

   where θ is the angle the incline makes with the horizontal and a is the acceleration of the block down the incline.
   (a) Find the equation of the straight line that best fits the data.
   (b) Determine the gravitational acceleration g.
   (c) Calculate the percentage error of the experimental value for g from the accepted value of 9.80 m/s^2.
   (d) Superpose the best fit curve on the data to examine the fit.

2. A physicist of mass m_p (including paint) is about to paint her house. She places a uniform ladder of length L and mass m_L against the house at an angle θ with respect to the ground. Being cautious, she decides to do a quick calculation to see how far up the ladder she can safely climb. Suppose there are rollers at the top of the ladder, that is, let the coefficient of (static) friction between the ladder and the house be zero, let the coefficient of (static) friction between the ladder and the ground be μ, and express the position of the physicist measured up the ladder from its base as fL. What fraction of

the total length of the ladder does she calculate she can climb before the ladder slips, and what are the forces that the ground and the house exert on the ladder? Obtain numerical answers for the values $m_L = 9.50$ kg, $m_p = 88.5$ kg, $\mu = 0.55$, $\theta = 50°$, and $g = 9.80$ m/s^2. (For solutions of this problem with MACSYMA, Maple, Mathcad, *Mathematica*, and Theorist, see [CDSTD92a].)

3. Consider a body falling from rest near the surface of the earth. For a small heavy body with large terminal velocity, the effect of air resistance may be approximated by a frictional force proportional to the square of the speed. Determine the motion of the body.

4. For an acoustic membrane clamped at radius a, the $n = 2$ normal mode of vibration is

$$z(r, \theta, t) = J_2(\omega r/v) \sin(2\theta) \cos(\omega t)$$

where z, (r, θ), and t denote the membrane displacement, polar coordinates, and time, respectively. J_2 is the Bessel function of order 2; v, the acoustic speed; ω, the frequency. The boundary condition requires that $\omega a/v = 5.13562$. Animate the vibration of the membrane for a set of parameters.

*5. A projectile of mass m leaves the origin with a speed v_0 at an angle θ_0 from the horizontal. Assume that the effect of air resistance may be approximated by a frictional force $-b\mathbf{v}$, where $\mathbf{v}$ is the velocity.
   (a) Obtain an equation for the trajectory, that is, $y = f(x)$, where the x and y axes are along the horizontal and vertical directions, respectively. *Hint:* Use `DSolve` and `Solve`.
   (b) Then let $m = 0.14$ kg, $v_0 = 45$ m/s, $\theta_0 = 60°$, and $b = 0.033$ kg/s in the equation obtained in (a). On the same graph, plot the path of the projectile with and without air resistance.

**6. Consider the Duffing-Holmes equation for the forced vibrations of a buckled elastic beam or the forced oscillations of a particle in a two-well potential:

$$\frac{d^2x}{dt^2} + \gamma \frac{dx}{dt} - \frac{1}{2}x(1 - x^2) = f \cos \omega t$$

Solve the equation numerically, plot the phase diagram, and determine the Poincaré section for the parameters: (a) $\gamma = 0.15$, $\omega = 0.8$, $f = 0.32$; (b) $\gamma = 0.15$, $\omega = 0.8$, $f = 0.2$; (c) $\gamma = 0.15$, $\omega = 0.8$, $f = 0.15$.

7. Section 4.3.3.1 specified, for the plane pendulum, the initial conditions $x = 0$ and $v = 2 - d$, where d is a small number.
   (a) What is the significance of this choice of initial conditions?
   (b) Solve Equation 4.3.3 and plot the phase diagram for the initial conditions $x = 0$ and $v = 2 + d$, where d is again a small number. Explain the resulting phase diagram.

(c) Repeat (b) for the initial conditions $x = 0$ and $v = 0$. Is the resulting phase diagram what you expected? If not, why not?

8. In Section 4.4.3, the first five arguments of the function `sphericalPendulum` are restricted. Are these restrictions necessary? Justify your answer.

❀ **3.0**

9. In Section 4.1.3, can we replace `v0` by the two-dimensional form v_0 in `DSolve`? If not, why not?

CHAPTER 5

Electricity and Magnetism

5.1 Electric Field Lines and Equipotentials

5.1.1 The Problem

Plot the electric field lines and equipotentials for two point charges q_1 and q_2 when (a) $q_1 = q_2 = +q$ and (b) $q_1 = +2q$ and $q_2 = -q$.

5.1.2 Physics of the Problem

5.1.2.1 ELECTRIC FIELD LINES

At position $\mathbf{r}$, the electric field of a charge distribution consisting of a discrete distribution of N point charges $q_1, q_2, \ldots, q_N$ at positions $\mathbf{r}_1, \mathbf{r}_2, \ldots, \mathbf{r}_N$, respectively, and a continuous distribution of volume charge density $\rho(\mathbf{r}')$ in the volume V and surface charge density $\sigma(\mathbf{r}')$ on the surface S that bounds V is

$$\mathbf{E}(\mathbf{r}) = \frac{1}{4\pi\varepsilon_0} \sum_{i=1}^{N} q_i \frac{\mathbf{r} - \mathbf{r}_i}{|\mathbf{r} - \mathbf{r}_i|^3} + \frac{1}{4\pi\varepsilon_0} \int_V \frac{\mathbf{r} - \mathbf{r}'}{|\mathbf{r} - \mathbf{r}'|^3} \rho(\mathbf{r}') \, dv' + \frac{1}{4\pi\varepsilon_0} \int_S \frac{\mathbf{r} - \mathbf{r}'}{|\mathbf{r} - \mathbf{r}'|^3} \sigma(\mathbf{r}') \, da' \quad (5.1.1)$$

where ε_0 is the permittivity of free space. The electric field is a vector point function or a vector field.

To represent the electric field graphically, we can draw vectors at points on a grid such that at each point the direction of the vector is in the direction of the field and the length of the vector is proportional to the magnitude of the field. `Graphics`PlotField`` and `Graphics`PlotField3D``, two standard *Mathematica* packages, contain functions for generating such field maps. When the lengths of vectors vary over wide ranges, these plots do not provide clear representations of the fields. As an example, let us map on the *xy* plane the electric field of two equal positive charges:

```
In[1] := Needs["Graphics`PlotField`"]
         PlotGradientField[-1/Sqrt[(x + 1)^2 + y^2] -
                 1/Sqrt[(x - 1)^2 + y^2],
                 {x, -2, 2}, {y, -2, 2}];
```

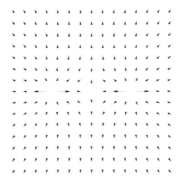

Most vectors in this picture are too short to be visible.

Electric field lines offer a better means for visualization. These directed curves are drawn such that at each point the tangent to the curve is in the direction of the electric field and the number of lines per unit area perpendicular to the lines (the density of the lines) is proportional to the magnitude of the electric field. Electric field lines have the following properties:

1. The lines must begin on positive charges and end on negative charges or at infinity. It is possible for lines to come from infinity and terminate on negative charges.

2. The number of lines leaving or entering a charge is proportional to the magnitude of the charge. If n lines are at charge q, then $n' = |q'/q|n$ lines must be at charge q'. The value of the proportionality constant is chosen so that the graph shows the essential features of the electric field.

3. Electric field lines cannot cross each other except at points of equilibrium where the electric field vanishes.

4. The field lines near a charge are symmetrical about the charge and radially directed.

(For an introduction to electric field lines, see [FGT93]; for an advanced treatment, refer to [Jea66].)

5.1.2.2   EQUIPOTENTIALS

The electric field in Equation 5.1.1 can be written as

$$\mathbf{E}(\mathbf{r}) = -\nabla\varphi(\mathbf{r}) \tag{5.1.2}$$

where the scalar function $\varphi(\mathbf{r})$, called the electric potential, is

$$\varphi(\mathbf{r}) = \frac{1}{4\pi\varepsilon_0}\sum_{i=1}^{N}\frac{q_i}{|\mathbf{r}-\mathbf{r}_i|} + \frac{1}{4\pi\varepsilon_0}\int_V\frac{\rho(\mathbf{r}')\,dv'}{|\mathbf{r}-\mathbf{r}'|} + \frac{1}{4\pi\varepsilon_0}\int_S\frac{\sigma(\mathbf{r}')\,da'}{|\mathbf{r}-\mathbf{r}'|} \tag{5.1.3}$$

Equipotentials are regions where the electric potential has constant values. For a system of point charges, the equipotentials are surfaces (lines in two dimensions). Except at points of equilibrium where the electric field vanishes, the equipotential surfaces do not intersect and the electric field lines are perpendicular to these surfaces everywhere.

5.1.2.3   ELECTRIC FIELD LINES AND EQUIPOTENTIALS FOR TWO POINT CHARGES

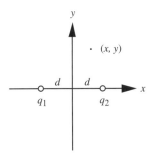

Figure 5.1.1   A coordinate system for two point charges

Let the midpoint between the two point charges be the origin of the coordinate system and the x axis be directed from q_1 to q_2, as in Figure 5.1.1. For these charges, it is necessary to plot only the electric field lines and equipotentials on the upper xy plane because of the rotational symmetry about the x axis. However, for clarity, we show the lines on the entire xy plane. To obtain a three-dimensional picture, rotate the plot about the x axis.

For these charges, Equation 5.1.3 gives

$$\varphi(x, y) = \frac{1}{4\pi\varepsilon_0}\left\{ \frac{q_1}{\sqrt{(x + d)^2 + y^2}} + \frac{q_2}{\sqrt{(x - d)^2 + y^2}}\right\} \tag{5.1.4}$$

If q_1 and q_2, x and y, and $\varphi(x, y)$ are expressed in units of q, d, and $q/4\pi\varepsilon_0 d$, respectively, Equation 5.1.4 becomes

$$\varphi(x, y) = \frac{q_1}{\sqrt{(x + 1)^2 + y^2}} + \frac{q_2}{\sqrt{(x - 1)^2 + y^2}} \tag{5.1.5}$$

Equation 5.1.2 for the electric field implies

$$\mathbf{E}(x, y) = -\frac{\partial\varphi(x, y)}{\partial x}\mathbf{i} - \frac{\partial\varphi(x, y)}{\partial y}\mathbf{j} \tag{5.1.6}$$

where $\mathbf{i}$ and $\mathbf{j}$ are unit vectors in the x and y directions, respectively.

5.1.3   Solution with Mathematica

There is a simple method for generating the electric field lines of collinear charges (see Problem 3 in Section 5.4). Here we adopt another algorithm in order to illustrate list manipulation and offer easy generalization for noncollinear charges.

5.1.3.1 $q_1 = q_2 = +q$

In[3] := **ClearAll["Global`*"]**

For the electric field lines, let us define a set of starting points symmetrically distributed around and near q_1:

In[4] := **x0[n_] := -1 + 0.2 Cos[n Pi/12]//N**
           y0[n_] := 0.2 Sin[n Pi/12]//N

For each point with coordinates (x, y) on an electric field line, the coordinates of the adjacent point a distance ds away down the line is

$$\left(x + \frac{E_x}{E} \, ds, y + \frac{E_y}{E} \, ds \right)$$

A function for such mapping from one point to another is

In[6] := **f[{x_, y_}] := {x + 0.02 (efx[x, y]/ef[x, y]),**
                          y + 0.02 (efy[x, y]/ef[x, y])}

where ds is set equal to 0.02 and **efx**, **efy**, and **ef** are the aliases of the x component, y component, and magnitude of the electric field, respectively. The components of the electric field are given by Equation 5.1.6 in terms of the electric potential in Equation 5.1.5:

In[7] := **phi[x_, y_] := 1/Sqrt[(x + 1)^2 + y^2] +**
                         1/Sqrt[(x - 1)^2 + y^2]
           efx[x_, y_] = D[-phi[x, y], x];
           efy[x_, y_] = D[-phi[x, y], y];

The magnitude of the electric field can be determined from its components:

In[10] := **ef[x_, y_] = Sqrt[efx[x, y]^2 + efy[x, y]^2];**

Starting with a point that has coordinates (x, y), a function for generating a list of coordinates of consecutive points down an electric field line in the plotting region is

In[11] := **g[{x_, y_}] := FixedPointList[f, {x, y},**
                           SameTest ->
                           ((Sqrt[#2[[1]]^2 + #2[[2]]^2] > 3.0)
                           || (#2[[1]] > 0) &)]

FixedPointList[*f*, *x*, **SameTest ->** *test***]** generates the list $\{x, f[x], f[f[x]], f[f[f[x]]], \ldots\}$, stopping when the function *test* yields **True**. The function *test* is normally used to compare, in the form **(Abs[#1 - #2] <** *condition* **&)**, two consecutive elements of the list; we use it here to test whether the second element of this pair reaches outside the plotting region, which is a quarter-circle located in the second quadrant and centered at the origin.

Now apply **g** to each element in a list of starting points to obtain a nested list of coordinates for the electric field lines in the second quadrant:

In[12] := **g /@ Table[{x0[n], y0[n]}, {n, 1, 11, 2}];**

The coordinates in the other quadrants can easily be determined from the symmetries across the x and y axes:

```
In[13] := coordinates1 = Join[%,
                   % /. {x_, y_} -> {x, -y},
                   % /. {x_, y_} -> {-x, y},
                   % /. {x_, y_} -> {-x, -y}
                   ];
```

ListPlot generates the electric field lines, and ContourPlot maps the equipotentials on the same graph:

```
In[14] := SetOptions[ListPlot, PlotJoined -> True,
                        PlotStyle -> RGBColor[1, 0, 0],
                        DisplayFunction -> Identity];
     Show[ListPlot /@ coordinates1,
          ContourPlot[phi[x, y], {x, -2, 2}, {y, -2, 2},
                      ContourShading -> False,
                      ContourSmoothing -> Automatic,
                      PlotRange -> {1.1, 6.0},
                      Contours -> 16,
                      PlotPoints -> 50,
                      Frame -> False,
                      ContourStyle -> {RGBColor[0, 0, 1],
                                       Dashing[{0.01, 0.01}]},
                      DisplayFunction -> Identity
                     ],
          Graphics[{Text["+q", {-1, 0}], Text["+q", {1, 0}]}],
          AspectRatio -> Automatic,
          PlotRange -> {{-2, 2}, {-2, 2}},
          Axes -> False,
          DisplayFunction -> $DisplayFunction
         ];
```

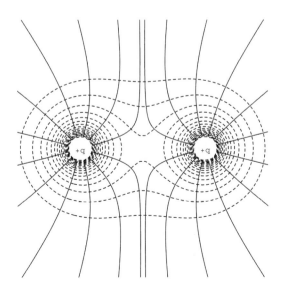

The solid red lines directed from the charges to infinity are the electric field lines, and the dashed blue closed contours are the equipotentials. The electric field lines and equipotentials are perpendicular to each other everywhere. Since the equipotentials are plotted with a sequence of equally spaced values for the potential, the electric field is strong in regions where the equipotentials are close together, in accordance with Equation 5.1.6. The origin is a point of equilibrium where the electric field vanishes. The uniqueness of this point is not obvious from the graph. Let us plot the electric field lines and the equipotential through this point:

```
In[16] := (* generate coordinates for electric field lines *)
          coordinates2 = g /@ {{-1 + 0.2, 0}, {0, 0.01}};


          (* enter coordinates of vertices of arrowheads for
             labeling the senses of electric field lines *)
          arrowheads = {{{0.5, 0}, {0.6, 0.1}, {0.6, -0.1}},
                        {{0, 1}, {0.1, 0.9}, {-0.1, 0.9}}};



      Show[ListPlot /@
          Join[coordinates2,
              coordinates2 /. {x_, y_} -> {x, -y},
              coordinates2 /. {x_, y_} -> {-x, y},
              coordinates2 /. {x_, y_} -> {-x, -y}
             ],
          ContourPlot[phi[x, y], {x, -2, 2}, {y, -2, 2},
                  ContourShading -> False,
                  ContourSmoothing -> Automatic,
                  Contours -> {2},
                  PlotPoints -> 100,
                  Frame -> False,
                  ContourStyle -> {RGBColor[0, 0, 1],
                                   Dashing[{0.01, 0.01}]},
                  DisplayFunction -> Identity
                 ],
          Graphics[Polygon /@
                  arrowheads ~ Join ~
                  (arrowheads /. {x_, y_} -> {-x, -y})
                 ],
          Graphics[{Text["+q", {-1, 0}], Text["+q", {1, 0}]}],
          AspectRatio -> Automatic,
          PlotRange -> {{-2, 2}, {-2, 2}},
          Axes -> False,
          DisplayFunction -> $DisplayFunction
         ];
```

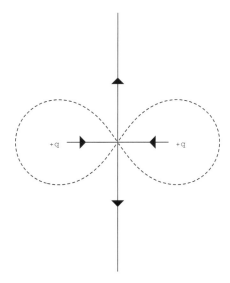

At the origin, an equipotential intersects itself and the two electric field lines, one from each charge, coalesce before splitting in two lines, one going to $+\infty$ and the other to $-\infty$. Also, the electric field lines are not perpendicular to the equipotential there.

5.1.3.2 $q_1 = +2q$ and $q_2 = -q$

```
In[21] := ClearAll["Global`*"]
```

For these charges, it is not sufficient to calculate the coordinates for the electric field lines in only one quadrant because the reflection symmetry across the y axis no longer exists. Here, we determine the coordinates within a semicircle centered at the origin, less two small semicircles centered at the charges. The procedure outlined in the preceding section for mapping the electric field can easily be modified:

```
In[22] := x0[n_] := ((-1 + 0.2 Cos[n Pi/16])//N)
         y0[n_] := ((0.2 Sin[n Pi/16])//N)
         f[{x_, y_}] := {x + 0.02 (efx[x, y]/ef[x, y]),
                         y + 0.02 (efy[x, y]/ef[x, y])}
         phi[x_, y_] := (2/Sqrt[(x + 1)^2 + y^2] -
                         1/Sqrt[(x - 1)^2 + y^2])
         efx[x_, y_] = (-D[phi[x, y], x]);
         efy[x_, y_] = (-D[phi[x, y], y]);
         ef[x_, y_] = Sqrt[efx[x, y]^2 + efy[x, y]^2];
         g[{x_, y_}] := FixedPointList[f, {x, y},
                           SameTest ->
                     ((Sqrt[#2[[1]]^2 + #2[[2]]^2] > 3.5)
                      || (Sqrt[(#2[[1]] - 1)^2 +
                              #2[[2]]^2] < 0.2) &)
                     ]
```

```
coordinates3 =
      (g /@ Table[{x0[n], y0[n]}, {n, 1, 15, 2}]);
SetOptions[ListPlot, PlotJoined -> True,
                     PlotStyle -> RGBColor[1, 0, 0],
                     DisplayFunction -> Identity];
Show[ListPlot /@
      Join[coordinates3,
            coordinates3 /. {x_, y_} -> {x, -y}],
    ContourPlot[phi[x, y], {x, -2.7, 2.7}, {y, -2.7, 2.7},
                ContourShading -> False,
                ContourSmoothing -> Automatic,
                PlotRange -> {-5.0, 10},
                Contours -> 24,
                PlotPoints -> 50,
                Frame -> False,
                ContourStyle -> {RGBColor[0, 0, 1],
                                 Dashing[{0.01, 0.01}]},
                DisplayFunction -> Identity
              ],
    Graphics[{Text["+2q", {-1, 0}], Text["-q", {1, 0}]}],
    AspectRatio -> Automatic,
    PlotRange -> {{-2.7, 2.7}, {-2.7, 2.7}},
    Axes -> False,
    DisplayFunction -> $DisplayFunction
    ];
```

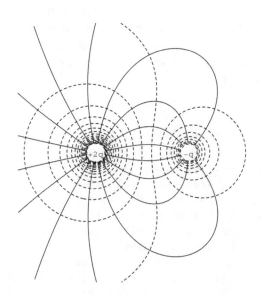

The red solid lines are the electric field lines and the blue dashed lines are the equipotentials. Of the 16 lines emanating from $q_1 = +2q$, half of them terminate on $q_2 = -q$, as they should because $|q_1/q_2| = 2$, and the other half go to infinity. The point of equilibrium at $\{5.83, 0\}$ is not shown because it lies outside the plotting region.

5.1.3.3   *MATHEMATICA* VERSION 3.0 SPECIFICS

Utilizing special characters and two-dimensional forms, we upgrade here the *Mathematica* input in Section 5.1.3.2. To proceed efficiently and correctly, paste below a copy of the *Mathematica* input, click the bracket of the second new cell, choose **StandardForm** in the Convert To submenu of the Cell menu, and edit the result:

```
ClearAll["Global`*"]

x₀[n_] := N[-1 + 0.2 Cos[ nπ/16 ]]

y₀[n_] := N[0.2 Sin[ nπ/16 ]]

f[{x_, y_}] := {x + 0.02 Eₓ[x, y]/ℰ[x, y], y + 0.02 Eᵧ[x, y]/ℰ[x, y]}

φ[x_, y_] := 2/√((x + 1)² + y²) - 1/√((x - 1)² + y²)

Eₓ[x_, y_] = -∂ₓφ[x, y];
Eᵧ[x_, y_] = -∂ᵧφ[x, y];

ℰ[x_, y_] = √(Eₓ[x, y]² + Eᵧ[x, y]²);

g[{x_, y_}] := FixedPointList[f, {x, y}, SameTest →

(√((#2[[1]])² + (#2[[2]])²) > 3.5 || √(((#2[[1]]) - 1)² + (#2[[2]])²) < 0.2&)]

coordinates3 = g/@Table[{x₀[n], y₀[n]}, {n, 1, 15, 2}];
SetOptions[ListPlot, PlotJoined → True,
                    PlotStyle → RGBColor[1, 0, 0],
                    DisplayFunction → Identity
          ];
Show[
   ListPlot/@
     Join[coordinates3, coordinates3 /. {x_, y_} → {x, -y}],
   ContourPlot[
     φ[x, y], {x, -2.7, 2.7}, {y, -2.7, 2.7},
     ContourShading → False,
     ContourSmoothing → Automatic,
     PlotRange → {-5.0, 10},
     Contours → 24,
     PlotPoints → 50,
     Frame → False,
     ContourStyle → {RGBColor[0, 0, 1], Dashing[{0.01, 0.01}]},
     DisplayFunction → Identity
            ],
   Graphics[{Text["+2q", {-1, 0}], Text["-q", {1, 0}]}],
   AspectRatio → Automatic,
```

```
    PlotRange → {{-2.7, 2.7}, {-2.7, 2.7}},
    Axes → False,
    DisplayFunction → $DisplayFunction
     ];
```

where x_0, y_0, E_x, E_y, ε, and φ have replaced, respectively, x0, y0, efx, efy, ef, and phi, before the conversion to StandardForm. The Greek letter E used here is different from the similar-looking keyboard letter E, which represents the exponential constant in *Mathematica*; the subscripts of E are the script letters x and y rather than the ordinary letters x and y. We have used the script letter ε for the magnitude of the electric field.

5.2   Laplace's Equation

5.2.1   The Problem

Figure 5.2.1 shows two grounded, semi-infinite, parallel electrodes separated by a distance b. A third electrode located at $x = 0$ is maintained at potential V_0. For the points between the plates, solve Laplace's equation analytically and numerically and plot the electric potential.

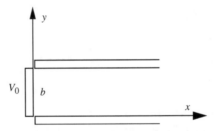

Figure 5.2.1   Two grounded, plane electrodes, one at $y = 0$ and the other at $y = b$, lie parallel to the xz plane and extend along the positive x direction from zero to infinity. At $x = 0$, a third electrode, in the form of a plane strip, is maintained at potential V_0. These electrodes are infinite along the z direction perpendicular to the page.

5.2.2   Physics of the Problem

The electric potential V between the plates satisfies Laplace's equation

$$\nabla^2 V = 0 \tag{5.2.1}$$

with the boundary conditions

$$
\begin{array}{lll}
V = 0 & y = 0 & \text{(5.2.2)} \\
V = 0 & y = b & \text{(5.2.3)} \\
V = V_0 & x = 0 & \text{(5.2.4)} \\
V \to 0 & x \to \infty & \text{(5.2.5)}
\end{array}
$$

Because of the translational symmetry along the z direction, the potential must be independent of the z coordinate, and Equation 5.2.1 becomes

$$\frac{\partial^2 V}{\partial x^2} + \frac{\partial^2 V}{\partial y^2} = 0 \qquad (5.2.6)$$

In what follows, let x as well as y be expressed in units of b, and V in units of V_0.

5.2.2.1 ANALYTICAL SOLUTION

With the method of separation of variables and Fourier analysis, we can determine the solution to the two-dimensional Laplace's equation with the boundary conditions. The potential V at any point (x, y, z) between the plates is

$$V(x, y) = \frac{4}{\pi} \sum_{m}^{\infty} \frac{1}{(2m-1)} \sin(2m-1)\pi y \, e^{-(2m-1)\pi x} \qquad (5.2.7)$$

where m varies over all positive integers and V depends only on x and y. (For a derivation of Equation 5.2.7, see [LC70].)

5.2.2.2 NUMERICAL SOLUTION

For a numerical solution of the two-dimensional Laplace's equation, we approximate the boundary condition Equation 5.2.5 by

$$V \to 0 \qquad x \to x_f \qquad (5.2.8)$$

where x_f is finite but sufficiently large so that the solution has the desired accuracy. Thus, we are to determine the potential at points inside a rectangular region bounded by the lines $x = 0$, $x = x_f$, $y = 0$, and $y = 1$.

Let us solve Laplace's equation with the finite difference method that imposes a grid of points (x_i, y_i) on the rectangular region such that

$$x_i = i\Delta \qquad i = 0, 1, 2, \ldots, imax \qquad (5.2.9)$$

$$y_j = j\Delta \qquad j = 0, 1, 2, \ldots, jmax \qquad (5.2.10)$$

and approximates the second derivatives by

$$\frac{\partial^2 V}{\partial x^2} \to \frac{V_{i+1,j} - 2V_{i,j} + V_{i-1,j}}{\Delta^2} \qquad (5.2.11)$$

$$\frac{\partial^2 V}{\partial y^2} \to \frac{V_{i,j+1} - 2V_{i,j} + V_{i,j-1}}{\Delta^2} \qquad (5.2.12)$$

where Δ is the separation between adjacent points and we have introduced the notation

$$V_{i,j} = V(x_i, y_j)$$

(For more information on the finite difference method, see [DeV94] and [KM90].) With Equations 5.2.11 and 5.2.12, Equation 5.2.6 becomes a finite difference equation

$$\frac{V_{i+1,j} - 2V_{i,j} + V_{i-1,j}}{\Delta^2} + \frac{V_{i,j+1} - 2V_{i,j} + V_{i,j-1}}{\Delta^2} = 0 \tag{5.2.13}$$

Solving for $V_{i,j}$, we find

$$V_{i,j} = \frac{1}{4}\left[V_{i+1,j} + V_{i-1,j} + V_{i,j+1} + V_{i,j-1}\right] \tag{5.2.14}$$

which states that the potential at any interior grid point inside the boundary is the average of the four nearest neighbors. This equation is a discrete approximation of the two-dimensional Laplace's equation (Equation 5.2.6) and approaches the latter as Δ goes to zero. The potential at the boundary grid points is given by Equations 5.2.2, 5.2.3, 5.2.4, and 5.2.8, which imply

$$\begin{aligned} V_{i,0} &= 0 \\ V_{i,jmax} &= 0 \\ V_{0,j} &= 1 \\ V_{imax,j} &= 0 \end{aligned} \tag{5.2.15}$$

for $i = 0, 1, 2, \ldots, imax$ and $j = 1, \ldots, jmax$.

Equation 5.2.14 yields a system of $(imax - 1) \times (jmax - 1)$ linear equations in the unknowns $V_{i,j}$ with $i = 1, 2, \ldots, imax - 1$ and $j = 1, 2, \ldots, jmax - 1$. These unknowns can, in principle, be determined exactly with the methods for solving systems of linear equations such as the Gaussian elimination or factorization methods (see [Pat94]). When the number of equations is large, these direct methods require such an enormous amount of computation that iterative methods, called relaxation methods, are often used. In one relaxation method, the Jacobi method, Equation 5.2.14 is written as

$$V_{i,j}^{n+1} = \frac{1}{4}\left[V_{i+1,j}^n + V_{i-1,j}^n + V_{i,j+1}^n + V_{i,j-1}^n\right] \tag{5.2.16}$$

where the superscripts denote the iteration numbers. We start with an initial guess $V_{i,j}^0$ for the potential at the interior grid points and use the set of equations given by Equation 5.2.16 to compute the first iterate $V_{i,j}^1$, which is then substituted back into the equations to determine the second iterate $V_{i,j}^2$, and so on. In this manner, the initial guess eventually converges to the correct solution of Equation 5.2.14. In practice, the iteration terminates when the condition

$$\left|V_{i,j}^{n+1} - V_{i,j}^n\right| < tol \tag{5.2.17}$$

is satisfied for all interior grid points. The parameter tol is chosen to be small enough to ensure that the solution attains a specified level of accuracy.

The rate of convergence of the Jacobi method depends on the initial guess. The closer the initial guess is to the correct solution, the faster is the convergence. The rate of convergence can be substantially improved by using the successive overrelaxation (SOR) method in which Equation 5.2.16 is replaced by

$$V_{i,j}^{n+1} = (1 - w)V_{i,j}^n + \frac{w}{4}\left[V_{i+1,j}^n + V_{i-1,j}^{n+1} + V_{i,j+1}^n + V_{i,j-1}^{n+1}\right] \tag{5.2.18}$$

with $1 < w < 2$. The relaxation parameter w affects the rate of convergence, and the optimal value for w depends on the size of the rectangle, the grid spacing Δ, and the iteration parameter *tol*. (For further discussion of this method, see [Pat94] and [Won92].)

Aside from the usual round-off error, there are three kinds of errors in the numerical solution: boundary error in replacing Equation 5.2.5 with Equation 5.2.8, discretization error in approximating Equation 5.2.6 by Equation 5.2.14, and iteration error in using relaxation methods to determine the correct solution to Equation 5.2.14. We can reduce these errors by increasing x_f and decreasing Δ as well as *tol*. Of course, the sizes of these parameters are limited by the required computational resources.

5.2.3 *Solution with Mathematica*

```
In[1] := ClearAll["Global`*"]
```

5.2.3.1 ANALYTICAL SOLUTION

The solution given in Equation 5.2.7 is exact. Before plotting V as a function of x and y, we must examine the convergence of the series. The series converges very rapidly because of the factor $1/(2m-1)$ and the exponential, except for $x \approx 0$. At $x = 0$, the exponential equals one. Therefore, we must determine how many terms are needed to approximate the boundary condition $V(0, y) = 1$. The electric potential is

```
In[2] := v[x_, y_, nmax_] := (
                (4/Pi) Sum[
            (1/(2 m - 1)) Sin[(2 m - 1) Pi y] Exp[-(2 m - 1) Pi x],
            {m, 1, nmax}
                    ]
                )
```

where **nmax** is the number of terms in the series and **v** is the alias of the potential V. Here are the plots of $V(0, y)$ with 10, 200, and 400 terms:

```
In[3] := Plot[Evaluate[v[0, y, 10]], {y, 0, 1},
            PlotRange -> {0, 1.3},
            AxesLabel -> {" y (b)", " V (Vo)"}];
```

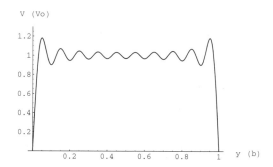

```
In[4] := Plot[Evaluate[v[0, y, 200]], {y, 0, 1},
            PlotRange -> {0, 1.3},
            AxesLabel -> {" y (b)", " V (Vo)"}];
```

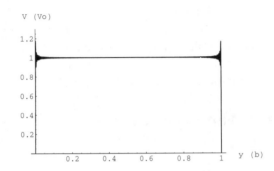

```
In[5] := Plot[Evaluate[v[0, y, 400]], {y, 0, 1},
            PlotRange -> {0, 1.3},
            AxesLabel -> {" y (b)", " V (Vo)"}];
```

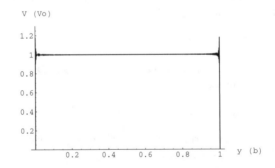

With 400 terms, the series gives an excellent approximation of the boundary condition $V(0, y) = 1$.

Let us plot the potential $V(x, y)$ with 400 terms:

```
In[6] := Plot3D[Evaluate[v[x, y, 400]], {x, 0, 1.3}, {y, 0, 1},
               PlotRange -> {0, 1}, Boxed -> False,
               ViewPoint -> {1.741, -2.565, 1.356},
               PlotPoints -> 27, BoxRatios -> {1, 1, 1},
               AxesLabel -> {"x (b)", "y (b)", "V (Vo)"}
             ]//Timing
```

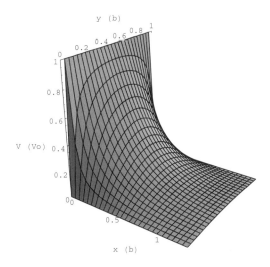

```
{140.433 Second, -SurfaceGraphics-}
```

Plotting the electric potential with 400 terms takes 140 seconds on a Macintosh IIci and 36 seconds on a Gateway 2000 4DX2-66V. (In this book, all timings have been performed with *Mathematica* version 2.2.)

5.2.3.2 NUMERICAL SOLUTION

As discussed in Section 5.2.2.2, we must make an initial guess for the potential at the interior grid points and choose the values for the parameters w, x_f, Δ, and *tol*. While the initial guess and all the parameters affect the rate of convergence, the boundary limit x_f, the grid spacing Δ, and the tolerance *tol* also influence the error in the solution.

Though any initial guess for the potential at the interior grid points converges toward the correct solution of Equation 5.2.14, a good guess substantially improves the rate of convergence, that is, the number of necessary iterations. (See Problem 6 of Section 5.4.) From the boundary conditions (Equation 5.2.15), we surmise that the initial potential is a product of an exponentially decreasing function of the coordinate x and a sinusoidal function of the coordinate y such that the potential vanishes at $y = 0$ and $y = 1$. Let us just take the first term of the analytical series solution in Equation 5.2.7. With Equations 5.2.9 and 5.2.10 for $i = 1, 2, \ldots, imax - 1$ and $j = 1, 2, \ldots, jmax - 1$, we have

```
In[7] := V[0][i_Integer?Positive /; i < imax,
            j_Integer?Positive /; j < jmax] :=
        V[0][i, j] = ((4/Pi) Sin[Pi j del] Exp[- Pi i del])//N
```

where **del** is the alias of Δ. Note that dynamic programming, introduced in Section 3.3.3, is used in the definition in order to reduce the amount of computation.

We set $w = 1.0$ for the moment and discuss its appropriate value later. This choice of w reduces the successive overrelaxation method to what is known as the Gauss-Seidel method.

```
In[8] := w = 1.0;
```

The smaller the Δ and *tol*, the smaller the error. However, their sizes are limited by the necessary computation effort. We choose the reasonable values

```
In[9] := del = 1/20;
        tol = 0.00025;
```

Equation 5.2.10 implies $1 = jmax\,\Delta$. Thus,

```
In[11] := jmax = 20;
```

The choice of x_f is somewhat arbitrary. We may start with a reasonable guess and determine the solution to Equation 5.2.14. Then, increase x_f by a fixed amount and determine another solution, and so on, until the solutions of the last two iterations differ by at most a specified small constant, resulting in an acceptable accuracy in the final solution. We may also seek guidance from the available analytical solution. It can be shown that the rate of convergence and the error in the solution are rather insensitive to choices for x_f as long as they are reasonable. (See Problem 7 of Section 5.4.) Let $x_f = 5/2$. Equation 5.2.9 implies $x_f = imax\,\Delta$. Hence,

```
In[12] := imax = 50;
```

The iteration equation (Equation 5.2.18) becomes

```
In[13] := V[n_Integer?Positive] [i_Integer?Positive /; i < imax,
                          j_Integer?Positive /; j < jmax] :=
            V[n] [i, j] =
                ((1 - w)V[n - 1] [i, j] +
                (w/4)(V[n - 1] [i + 1, j] + V[n] [i - 1, j] +
                    V[n - 1] [i, j + 1] + V[n] [i, j - 1]))
```

Again, it is important to use dynamic programming in the definition.

Equation 5.2.15 for the boundary conditions requires

```
In[14] := V[n_Integer?NonNegative] [i_Integer?NonNegative /; i <= imax,
                          0
                          ] := V[n] [i, 0] = 0
        V[n_Integer?NonNegative] [i_Integer?NonNegative /; i <= imax,
                          jmax
                          ] := V[n] [i, jmax] = 0
        V[n_Integer?NonNegative] [0,
                          j_Integer?Positive /; j <= jmax
                          ] := V[n] [0, j] = 1
        V[n_Integer?NonNegative] [imax,
                          j_Integer?Positive /; j <= jmax
                          ] := V[n] [imax, j] = 0
```

To implement the condition in Equation 5.2.17, we use the `For` loop discussed in Section 3.4.3.2:

```
In[18] := For[m = 1,
            Max[Table[Abs[V[m] [i, j] - V[m - 1] [i, j]],
```

```
                        {i, 1, imax - 1}, {j, 1, jmax - 1}
                        ]
                    ] >= tol,
            m++,
            If[m > 40,
                Print["Iteration limit of 40 exceeded."];
                Break[]
                ];
            Print["Iteration Number = " <> ToString[m]]
            ];
Iteration Number = 1
Iteration Number = 2
Iteration Number = 3
Iteration Number = 4
Iteration Number = 5
Iteration Number = 6
Iteration Number = 7
Iteration Number = 8
Iteration Number = 9
Iteration Number = 10
Iteration Number = 11
Iteration Number = 12
Iteration Number = 13
Iteration Number = 14
Iteration Number = 15
Iteration Number = 16
Iteration Number = 17
Iteration Number = 18
Iteration Number = 19
Iteration Number = 20
Iteration Number = 21
Iteration Number = 22
Iteration Number = 23
Iteration Number = 24
Iteration Number = 25
Iteration Number = 26
```

For each iteration, the command **Print** prints the iteration number. The operator **<>** with the full name **StringJoin** concatenates two strings and **ToString**[*expr*] gives a string corresponding to the printed form of *expr*. The loop terminates when Equation 5.2.17 or the first argument of the **If** function is true. The function **If** provides a graceful exit and prints a message when the iteration number exceeds the specified limit.

An important result of the **For** loop is the final iteration number at which the loop terminates:

```
In[19] := Print["Final Iteration Number = " <> ToString[m]]

        Final Iteration Number = 27
```

With this number already assigned to **m**, we can plot the electric potential:

```
In[20] := Show[
            SurfaceGraphics[
               Table[V[m][i, j], {j, 0, jmax}, {i, 0, 26}],
               MeshRange -> {{0, 26 del}, {0, jmax del}}
                 ],
            BoxRatios -> {1, 1, 1}, PlotRange -> {0, 1},
            Boxed -> False, Axes -> True,
            AxesLabel -> {"x (b)", "y (b)", "V (Vo)"},
            ViewPoint -> {1.741, -2.565, 1.356}
          ];
```

The graphics object **SurfaceGraphics** [*array*], introduced in Section 3.5.1, represents a three-dimensional plot of a surface, with heights of each point on a grid specified by values in *array*. We have suppressed the plot of the electric potential because it looks identical to that in Section 5.2.3.1 for the analytical solution.

For comparison, let us tabulate the values of the electric potential from the analytical and numerical solutions:

```
In[21] := Table[Evaluate[{v[i del, j del, 400]//N, V[m][i, j]}],
              {i, 0, 50, 5}, {j, 2, 10, 2}];

In[22] := NumberForm[
            Prepend[
              Flatten[#, 1]& /@
              Transpose @ {
                Table["x = "<>ToString[i del //N], {i, 0, 50, 5}],
                %
                      },
              Prepend[
              Table["y = "<>ToString[j del //N], {j, 2, 10, 2}], {}]
                 ]
              ] // TableForm, {4, 3}
              ]
```

Out[22]//NumberForm=

	y = 0.1	y = 0.2	y = 0.3	y = 0.4	y = 0.5
	0.997	0.999	0.999	0.999	0.999
x = 0	1.	1.	1.	1.	1.
	0.218	0.379	0.477	0.529	0.545
x = 0.25	0.219	0.38	0.478	0.53	0.545
	0.085	0.159	0.215	0.249	0.261
x = 0.5	0.085	0.16	0.216	0.25	0.261
	0.038	0.071	0.098	0.115	0.12
x = 0.75	0.038	0.071	0.098	0.115	0.121

	0.017	0.032	0.045	0.052	0.055
x = 1.	0.017	0.032	0.045	0.052	0.055
	0.008	0.015	0.02	0.024	0.025
x = 1.25	0.008	0.015	0.02	0.024	0.025
	0.004	0.007	0.009	0.011	0.011
x = 1.5	0.004	0.007	0.009	0.011	0.011
	0.002	0.003	0.004	0.005	0.005
x = 1.75	0.002	0.003	0.004	0.005	0.005
	0.001	0.001	0.002	0.002	0.002
x = 2.	0.001	0.001	0.002	0.002	0.002
	0.	0.001	0.001	0.001	0.001
x = 2.25	0.	0.001	0.001	0.001	0.001
	0.	0.	0.	0.	0.
x = 2.5	0.	0.	0.	0.	0.

The upper number in each entry is from the analytical solution, and the lower number is from the numerical solution. The agreement between the two solutions is excellent to three digits after the decimal point. Including the values for $y > 0.5$ in the table is unnecessary because the potential is symmetric about $y = 0.5$. (With *Mathematica* version 3.0, make sure that Default Output FormatType in the Cell menu is set at **OutputForm** before evaluating the previous input. Also, either choose, in the Screen Style Environment submenu of the Format menu, **Printout** before and **Working** after the evaluation of the input, or replace *expr*`//TableForm` by `TableForm[`*expr*`, TableSpacing -> {1, 1}]` in the input.)

In our computations, we have set $w = 1.0$. What is the optimal value for the relaxation parameter w? The following table shows the final iteration numbers for *jmax* $= 20$, *tol* $= 0.00025$, and w ranging from 1.0 to 1.9 in steps of 0.1. (See Problem 5 of Section 5.4.)

Relaxation Parameter w	1.0	1.1	1.2	1.3	1.4	1.5	1.6	1.7	1.8	1.9
Final Iteration Number	27	23	21	19	17	16	15	21	28	>40

The relaxation parameter w indeed controls the rate of convergence, and its optimal value is 1.6.

In this section, we have used *Mathematica* as an interactive system rather than as a programming language in order to explain and justify our solution of Laplace's equation. For symbol localization, clarity, convenience, and elegance, we ought to assemble the commands into procedures in function definitions if we are to use them for more than one set of parameter values or if they compose only a subset of the commands in a *Mathematica* session. For example, we can write a procedural program in the form of a function that prints the preliminary as well as final iteration numbers and plots the electric potential. The function takes four arguments: the first must be an even integer and is for *jmax*, which equals $1/\Delta$; the

second for the relaxation parameter w, which must be a real number between 1 and 2; the
third for the tolerance *tol*; and the fourth for the graphics options.

```
In[23] := laplace2D[jmax_Integer?EvenQ, w_Real /; w > 1 && w < 2,
             tol_, opts___] :=
          Module[{imax = (5/2) jmax, del = 1/jmax, m, V},

            (* initial guess *)
            V[0][i_Integer?Positive /; i < imax,
                j_Integer?Positive /; j < jmax] :=
             V[0][i, j] = ((4/Pi) Sin[Pi j del] Exp[- Pi i del])//N;

            (* boundary conditions *)
            V[n_Integer?NonNegative][i_Integer?NonNegative /; i <= imax,
                                     0] := V[n][i, 0] = 0;
            V[n_Integer?NonNegative][i_Integer?NonNegative /; i <= imax,
                                     jmax] := V[n][i, jmax] = 0;
            V[n_Integer?NonNegative][0,
                                     j_Integer?Positive /; j <= jmax
                                     ] := V[n][0, j] = 1;
            V[n_Integer?NonNegative][imax,
                                     j_Integer?Positive /; j <= jmax
                                     ] := V[n][imax, j] = 0;

            (* iteration equation *)
            V[n_Integer?Positive][i_Integer?Positive /; i < imax,
                                  j_Integer?Positive /; j < jmax] :=
             V[n][i, j] =
                 ((1 - w)V[n - 1][i, j] +
                  (w/4)(V[n - 1][i + 1, j] + V[n][i - 1, j] +
                        V[n - 1][i, j + 1] + V[n][i, j - 1]));

            (* the For loop *)
            For[m = 1,
             Max[Table[Abs[V[m][i, j] - V[m - 1][i, j]],
                       {i, 1, imax - 1}, {j, 1, jmax - 1}
                       ]
                 ] >= tol,
             m++,
             If[m > 40,
                Print["Iteration limit of 40 exceeded."];
                Break[]
                ];
             Print["Iteration Number = " <> ToString[m]]
                ];
            Print["Final Iteration Number = " <> ToString[m]];

            (* graphics *)
```

```
Show[
  SurfaceGraphics[
    Table[V[m][i, j], {j, 0, jmax}, {i, 0, 26}],
    MeshRange -> {{0, 26 del}, {0, jmax del}}
                  ],
  opts,
  BoxRatios -> {1, 1, 1}, PlotRange -> {0, 1},
  Boxed -> False, Axes -> True,
  AxesLabel -> {"x (b)", "y (b)", "V (Vo)"},
  ViewPoint -> {1.741, -2.565, 1.356}
    ];
    ]
```

5.3 Charged Particle in Crossed Electric and Magnetic Fields

5.3.1 The Problem

For a charge moving in uniform electric and magnetic fields that are perpendicular to each other, solve the equation of motion, determine the nature of the motion, and plot the trajectories for several choices of field strengths and initial conditions. (For a comparison of solutions of this problem with MACSYMA, Maple, Mathcad, *Mathematica*, and Theorist, see [CDSTD92b].)

5.3.2 Physics of the Problem

On a particle of charge q and mass m moving in electric and magnetic fields $\mathbf{E}$ and $\mathbf{B}$, the Lorentz force is

$$\mathbf{F} = q(\mathbf{E} + \mathbf{v} \times \mathbf{B}) \tag{5.3.1}$$

where $\mathbf{v}$ is the particle velocity. Thus, the equation of motion can be written as

$$m\frac{d^2\mathbf{r}}{dt^2} = q\mathbf{E} + q\frac{d\mathbf{r}}{dt} \times \mathbf{B} \tag{5.3.2}$$

in which $\mathbf{r}$ is the position vector of the particle. We orient the coordinate axes so that the uniform crossed fields are given by

$$\mathbf{E} = E_0\hat{\mathbf{y}} \qquad \mathbf{B} = B_0\hat{\mathbf{z}} \tag{5.3.3}$$

and locate the origin so that the initial position of the particle is specified by

$$\mathbf{r}(0) = 0 \tag{5.3.4}$$

If $\mathbf{v}_0$ is the initial velocity,

$$\frac{d\mathbf{r}(0)}{dt} = \mathbf{v}_0 \tag{5.3.5}$$

5.3.3 Solution with Mathematica

In[1] := **ClearAll["Global`*"]**

❈ **2.2**   To use the functions **CrossProduct** and **ComplexToTrig** later, we load the packages **Calculus`VectorAnalysis`** and **Algebra`Trigonometry`**:

> **Needs["Calculus`VectorAnalysis`"]**

> **Needs["Algebra`Trigonometry`"]**

CrossProduct[*v1*, *v2*] computes the cross product of the vectors *v1* and *v2* given in terms of their Cartesian components, and **ComplexToTrig[*expr*]** writes complex exponentials in *expr* using trigonometric functions.

Mathematica represents vectors by lists. In Cartesian coordinates, the position vector **r** is defined as

In[2] := **r[t_] := {x[t], y[t], z[t]}**

With **EF** and **BF** as, respectively, aliases of **E** and **B**, Equation 5.3.3 takes the form

In[3] := **EF := {0, E0, 0}**
           BF := {0, 0, B0}

To solve Equation 5.3.2 subject to the initial conditions (Equations 5.3.4 and 5.3.5), first obtain a list of their Cartesian components:

❈ **2.2**

```
Thread /@ {m r''[t] == q EF + q CrossProduct[r'[t], BF],
          r[0] == {0, 0, 0},
          r'[0] == {vx0, vy0, vz0}
          } // Flatten
```

❈ **3.0**

```
In[5] := Thread /@ {m r''[t] == q EF + q r'[t] × BF,
                   r[0] == {0, 0, 0},
                   r'[0] == {vx0, vy0, vz0}
                   } // Flatten
```

Out[5] = {m x''[t] == B0 q y'[t], m y''[t] == E0 q - B0 q x'[t],
       m z''[t] == 0,
       x[0] == 0, y[0] == 0, z[0] == 0, x'[0] == vx0,
       y'[0] == vy0, z'[0] == vz0}

Then **DSolve** solves this system of equations:

```
❄ 2.2

   sol = DSolve[%, {x[t], y[t], z[t]}, t] //.
          {B0 -> m w/q, E0 -> vd B0} // ComplexToTrig // ExpandAll
```

```
❄ 3.0

   In[6] := sol = DSolve[%, {x[t], y[t], z[t]}, t] //.
          {B0 -> m w/q, E0 -> vd B0} // ExpandAll
```

$Out[6]$ = {{x[t]->t vd + $\frac{vy0}{w}$ - $\frac{vy0\ Cos[t\ w]}{w}$ - $\frac{vd\ Sin[t\ w]}{w}$ + $\frac{vx0\ Sin[t\ w]}{w}$,

y[t]->$\frac{vd}{w}$ - $\frac{vx0}{w}$ - $\frac{vd\ Cos[t\ w]}{w}$ + $\frac{vx0\ Cos[t\ w]}{w}$ + $\frac{vy0\ Sin[t\ w]}{w}$,

z[t]->t vz0}}

where we have introduced the definitions

$$w = \frac{q}{m}B_0 \tag{5.3.6}$$

$$v_d = \frac{E_0}{B_0} \tag{5.3.7}$$

To determine the nature of the particle's motion, convert the rules returned by DSolve into equations:

```
In[7] := eqn = sol[[1]] /. Rule -> Equal
```

$Out[7]$ = {x[t] == t vd + $\frac{vy0}{w}$ - $\frac{vy0\ Cos[t\ w]}{w}$ - $\frac{vd\ Sin[t\ w]}{w}$ + $\frac{vx0\ Sin[t\ w]}{w}$,

y[t] == $\frac{vd}{w}$ - $\frac{vx0}{w}$ - $\frac{vd\ Cos[t\ w]}{w}$ + $\frac{vx0\ Cos[t\ w]}{w}$ + $\frac{vy0\ Sin[t\ w]}{w}$,

z[t] == t vz0}

Subtract the term (t vd + vy0/w) from both sides of the first equation and then square both sides of the resultant equation:

```
In[8] := Map[((# - (t vd + vy0/w))^2)&, eqn[[1]]]
```

$Out[8]$ = (-(t vd) - $\frac{vy0}{w}$ + x[t])^2 == (-($\frac{vy0\ Cos[t\ w]}{w}$) - $\frac{vd\ Sin[t\ w]}{w}$ + $\frac{vx0\ Sin[t\ w]}{w}$)^2

Repeat with the second equation but this time subtract the term (vd/w - vx0/w) from both sides of the equation:

```
In[9] := Map[((# - (vd/w - vx0/w))^2)&, eqn[[2]]]
```

$Out[9]$ = (-($\frac{vd}{w}$) + $\frac{vx0}{w}$ + y[t])^2 == Power[-($\frac{vd\ Cos[t\ w]}{w}$) + $\frac{vx0\ Cos[t\ w]}{w}$ + $\frac{vy0\ Sin[t\ w]}{w}$, 2]

Add the two equations:

In[10] := **Thread[Plus[%%, %], Equal]**

Out[10] = $(-(t\ vd)\ -\ \frac{vy0}{w}\ +\ x[t])^2\ +\ (-(\frac{vd}{w})\ +\ \frac{vx0}{w}\ +\ y[t])^2$ ==

$(-(\frac{vy0\ Cos[t\ w]}{w})\ -\ \frac{vd\ Sin[t\ w]}{w}\ +\ \frac{vx0\ Sin[t\ w]}{w})^2\ +$
$(-(\frac{vd\ Cos[t\ w]}{w})\ +\ \frac{vx0\ Cos[t\ w]}{w}\ +\ \frac{vy0\ Sin[t\ w]}{w})^2$

Finally, simplify the *rhs* of the equation:

In[11] := **MapAt[Simplify, %, {{2}}]**

Out[11] = $(-(t\ vd)\ -\ \frac{vy0}{w}\ +\ x[t])^2\ +\ (-(\frac{vd}{w})\ +\ \frac{vx0}{w}\ +\ y[t])^2$ ==
$\frac{vd^2\ -\ 2\ vd\ vx0\ +\ vx0^2\ +\ vy0^2}{w^2}$

This is the equation of a circle on the *xy* plane, with radius

$$R = \left(\frac{v_d^2 - 2v_d v_{x0} + v_{x0}^2 + v_{y0}^2}{w^2}\right)^{1/2} \tag{5.3.8}$$

and center at coordinates

$$x_c = \frac{v_{y0}}{w} + v_d t \tag{5.3.9}$$

$$y_c = \frac{v_d}{w} - \frac{v_{x0}}{w} \tag{5.3.10}$$

The motion along the *z* direction is described by

In[12] := **eqn[[3]]**

Out[12] = $z[t]$ == $t\ vz0$

Hence, the general motion of the particle consists of three components: (a) circular motion perpendicular to **B** with angular frequency w, called the cyclotron frequency, (b) uniform translational motion parallel to **B**, and (c) constant drifting motion in the direction of $\mathbf{E} \times \mathbf{B}$ with velocity v_d, called the drift velocity.

To plot the trajectories of the particle for several choices of numerical values of field strengths and initial conditions, we first obtain the Cartesian coordinates of the particle as a function of time from the solution returned by **DSolve**:

In[13] := **{x[t_], y[t_], z[t_]} = {x[t], y[t], z[t]} /. sol[[1]];**

For a given charge, specifying w and v_d is equivalent to specifying E_0 and B_0. Let us express time in units of $1/w$ and velocity in units of v_d, that is, $w = 1$ and $v_d = 1$. We now plot the paths for $v_{z0} = 0.75, v_{y0} = 0$, and $v_{x0} = -1, 0, \ldots, 4$:

2.2

```
g[i_] :=
  (vx0 = i;
   Show[ParametricPlot3D[Evaluate[{x[t], y[t], 0}],
                          {t, 0, 22},
                          DisplayFunction -> Identity
                         ],
        ParametricPlot3D[Evaluate[r[t]],
                          {t, 0, 22},
                          DisplayFunction -> Identity
                         ],
        Graphics3D[
             {Thickness[0.008],
              Line[{{0, 0, 0}, {0, 0, 5}}],
              Line[{{0, 0, 0}, {0, 5, 0}}],
              Line[{{0, 0, 0}, {22, 0, 0}}],
              Text[FontForm["E", {"Times-Bold", 10}],
                   {1.3, 5.5, 0}],
              Text[FontForm["B", {"Times-Bold", 10}],
                   {0, 0, 7}],
              Text[FontForm[
                     "vx0 = "<>ToString[vx0], {"Courier", 9}
                           ],
                   Scaled[{0.4, 0.9, 0.85}]]
             }
                  ],
        Axes -> False,
        PlotRange -> {{-2, 24}, {-7, 7}, {0, 22}}
       ]
  );
```

3.0

```
In[14] := g[i_] :=
  (vx0 = i;
   Show[ParametricPlot3D[Evaluate[{x[t], y[t], 0}],
                          {t, 0, 22},
                          DisplayFunction -> Identity
                         ],
        ParametricPlot3D[Evaluate[r[t]],
                          {t, 0, 22},
                          DisplayFunction -> Identity
                         ],
```

```
            Graphics3D[
                {Thickness[0.008],
                 Line[{{0, 0, 0}, {0, 0, 5}}],
                 Line[{{0, 0, 0}, {0, 5, 0}}],
                 Line[{{0, 0, 0}, {22, 0, 0}}],
                 Text[StyleForm["E",
                                FontFamily -> "Times",
                                FontWeight -> "Bold",
                                FontSize -> 10],
                       {1.3, 5.5, 0}],
                 Text[StyleForm["B",
                                FontFamily -> "Times",
                                FontWeight -> "Bold",
                                FontSize -> 10],
                       {0, 0, 7}],
                 Text[StyleForm["vx0 = "<>ToString[vx0],
                                FontFamily -> "Courier",
                                FontSize -> 9],
                       Scaled[{0.4, 0.9, 0.85}]]
                }
                     ],
            Axes -> False,
            PlotRange -> {{-2, 24}, {-7, 7}, {0, 22}}
            ]
        );
```

```
In[15] := w = 1; vd = 1; vz0 = 0.75; vy0 = 0;
          Print["w = ", w, "    vd = ", vd,
                "    vz0 = ", vz0, "    vy0 = ", vy0
               ];
          Show[GraphicsArray[Partition[Table[g[i], {i, -1, 4}], 3]]];

          w = 1    vd = 1    vz0 = 0.75    vy0 = 0
```

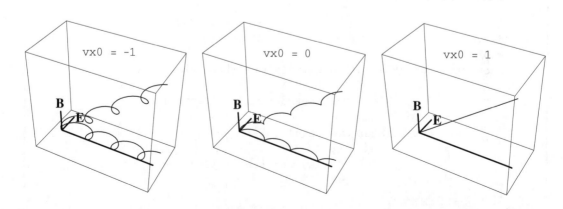

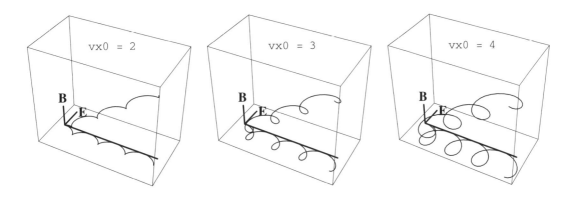

The trajectories of the particle with $v_{y0} = 0$ are shown together with their projections on the xy plane. If $v_{y0} \neq 0$, the paths are essentially the same except that the x_c coordinate of the center at $t = 0$ is shifted and the radius R is enlarged in accordance with Equations 5.3.8 and 5.3.9. (The command Show[GraphicsArray[{{*g11*, *g12*, ...}, ...}]] displays a two-dimensional array of graphics objects. See Exercise 9 of Section 2.3.4.)

Let us generate the graphs for animating a representative motion of the particle:

```
In[18] := w = 1; vd = 1; vx0 = 4; vy0 = 0; vz0 = 0.75;
```

☀ **2.2**

```
Do[Show[
        ParametricPlot3D[Evaluate[r[t]],
                        {t, 0, i + 0.0001},
                        DisplayFunction -> Identity
                        ],
        Graphics3D[{
                    {Thickness[0.008],
                     Line[{{0, 0, 0}, {0, 0, 5}}],
                     Line[{{0, 0, 0}, {0, 5, 0}}],
                     Line[{{0, 0, 0}, {22, 0, 0}}]},
                    Text[FontForm["E", {"Times-Bold", 10}],
                         {1.3, 5.5, 0}],
                    Text[FontForm["B", {"Times-Bold", 10}],
                         {0, 0, 7}],
                    {Dashing[{0.01, 0.01}],
                     Line[{r[i], {x[i], y[i], 0}}]},
                    {PointSize[0.03], Point[r[i]],
                     GrayLevel[0.55], Point[{x[i], y[i], 0}]}
                  }
                ],
```

```
        DisplayFunction -> $DisplayFunction,
        Axes -> False,
        PlotRange -> {{-2, 24}, {-7, 7}, {0, 22}}];,
   {i, 0, 24}
 ]
```

✿ **3.0**

```
In[19] := Do[Show[
              ParametricPlot3D[Evaluate[r[t]],
                       {t, 0, i + 0.0001},
                       DisplayFunction -> Identity
                       ],
              Graphics3D[{
                      {Thickness[0.008],
                       Line[{{0, 0, 0}, {0, 0, 5}}],
                       Line[{{0, 0, 0}, {0, 5, 0}}],
                       Line[{{0, 0, 0}, {22, 0, 0}}]},
                       Text[StyleForm["E",
                                   FontFamily -> "Times",
                                   FontWeight -> "Bold",
                                   FontSize -> 10],
                           {1.3, 5.5, 0}],
                       Text[StyleForm["B",
                                   FontFamily -> "Times",
                                   FontWeight -> "Bold",
                                   FontSize -> 10],
                           {0, 0, 7}],
                      {Dashing[{0.01, 0.01}],
                       Line[{r[i], {x[i], y[i], 0}}]},
                      {PointSize[0.03], Point[r[i]],
                       GrayLevel[0.55], Point[{x[i], y[i], 0}]}
                      }
                      ],
              DisplayFunction -> $DisplayFunction,
              Axes -> False,
              PlotRange -> {{-2, 24}, {-7, 7}, {0, 22}}];,
         {i, 0, 24}
            ]
```

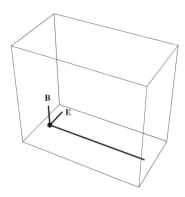

Only the head cell of the closed group of 25 graphics cells is shown here. The vertical dashed line in each of the other cells connects the particle with its projection on the xy plane. To animate, click the group bracket, if it is not already selected, and press Command-Y for the Macintosh or Control+Y for Windows.

```
In[20] := ClearAll[r, EF, BF, sol, eqn, x, y, z, w, vd, vx0, vy0, vz0, g]
```

5.4   *Problems*

1. Determine the electric field for points on the axis of a uniformly charged disk of radius a and charge density b. Verify that at large distances from the disk, the electric field approaches that produced by a point charge.

2. Plot the electric field lines and equipotentials for (a) two equal but opposite point charges and (b) three equal positive point charges at the corners of an equilateral triangle.

3. Consider a distribution of collinear point charges $q_1, q_2, q_3, \ldots, q_N$, at positions $x_1, x_2, x_3, \ldots, x_N$, respectively. Show that the differential equation for the electric field lines on the xy plane is

$$\frac{dy}{dx} = \frac{E_y}{E_x}$$

Verify that the solution to this equation is

$$\sum_{i=1}^{N} \frac{q_i(x - x_i)}{\sqrt{(x - x_i)^2 + y^2}} = constant$$

With `ContourPlot`, generate the electric field lines when there are (a) 2 equal charges, (b) 2 equal but opposite charges, (c) 2 charges, $+2q$ and $-q$, and (d) 21 equal and evenly spaced charges. What can be inferred from the plot in (d)?

*4. For a distribution of two point charges, $+2q$ and $-q$, plot the electric field lines and equipotential that pass through the point of equilibrium.

*5. Modify the function `laplace2D` in Section 5.2.3.2 to omit printing of the preliminary iteration numbers and plotting of the electric potential. The new function, named `newLaplace2D`, prints the final iteration number and the table for comparing values of the electric potential from the analytical and numerical solutions. The function `newLaplace2D` takes only three arguments because the fourth argument for the graphics options is no longer necessary. Evaluate `newLaplace2D[jmax, w, tol]` for

$jmax = 20, 10$

$w = 1.2, 1.3, 1.4, 1.5, 1.6, 1.7$

$tol = 0.0025, 0.00025.$

Does the optimal value for w depend on $jmax$ and tol? How do the rate of convergence and the error in the numerical solution depend on $jmax$, w, and tol? Note that $jmax$ equals $1/\Delta$, where Δ is the grid spacing. (The legitimate values $w = 1.1$, 1.8, and 1.9 have been dropped to make this problem a bit less time consuming.)

*6. Replace the initial guess for the electric potential of the interior grid points in the function `newLaplace2D` of Problem 5 with

$$V_{i,j}^0 = \frac{4}{\pi} \sum_{m=1}^{30} \frac{1}{(2m-1)} \sin(2m-1)\pi y \, e^{-(2m-1)\pi x}$$

Call the new function `newLaplace2Dg1`. Repeat with the initial guess

$$V_{i,j}^0 = f(y) e^{-\pi x}$$

where

$$f(y) = \begin{cases} 2y & 0 \le y \le 1/2 \\ -2(y-1) & 1/2 < y \le 1 \end{cases}$$

Call this function `newlaplace2Dg2`. Evaluate `newLaplace2D`, `newLaplace2Dg1`, and `newlaplace2Dg2` for $jmax = 20$, $tol = 0.00025$, and $w = 1.2, 1.3, 1.4, 1.5, 1.6, 1.7$. What can be concluded about the dependence of the optimal value for w, the rate of convergence, and the error in the numerical solution on the initial guess?

*7. Modify the function `newLaplace2D` in Problem 5 to include a fourth argument for the boundary limit x_f. Name this function `xfnewLaplace`. Evaluate this function for $jmax = 20$, $tol = 0.00025$, $w = 1.6$, and $x_f = 5/2$, 3, and 7/2. Are the rate of convergence and the error in the numerical solution sensitive to the choice for x_f?

8. The following figure shows two sets of parallel conducting plates of finite width but infinite length along the z direction perpendicular to the page. The plates separated by a distance b are grounded, and those at a distance a apart are maintained at a potential

V_0. For the points enclosed by the plates, solve analytically and numerically Laplace's equation and plot the electric potential.

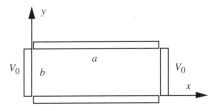

✿ **3.0**

9. In Section 5.1.3.3, can the keyboard letters **x** and **y** rather than the script letters x and y be the subscripts of the Greek letter **ε**? Explain.

CHAPTER 6

Quantum Physics

6.1 Blackbody Radiation

6.1.1 The Problem

The Planck formula for blackbody radiation is

$$u(\nu, T) = \frac{8\pi h}{c^3} \frac{\nu^3}{e^{h\nu/kT} - 1} \tag{6.1.1}$$

where $u(\nu, T)$ is the energy density as a function of frequency ν and temperature T, k is Boltzmann's constant, c is the speed of light, and h is Planck's constant. Obtain an expression for the energy density as a function of wavelength λ and temperature T. Also, plot this energy density for several temperatures. Then, derive Wien's displacement law relating T and $\lambda_{\max}$, the wavelength for which the energy density is maximal. Finally, use 5700 K as an estimate of the Sun's surface temperature to determine $\lambda_{\max}$ for solar radiation.

6.1.2 Physics of the Problem

The frequency ν, the wavelength λ, and the speed of light c are related by

$$\nu = \frac{c}{\lambda} \tag{6.1.2}$$

Thus,

$$d\nu = -\left(\frac{c}{\lambda^2}\right) d\lambda \tag{6.1.3}$$

The energy density as a function of wavelength λ, $u(\lambda, T)$, is given in terms of the energy density as a function of frequency ν, $u(\nu, T)$, by

$$u(\lambda, T)\, d\lambda = -u(\nu, T)\, d\nu \tag{6.1.4}$$

(For an elaboration of this equation, see [Eis61].) With Equation 6.1.3, Equation 6.1.4 takes the form

$$u(\lambda, T) = u(\nu, T)\frac{c}{\lambda^2} \tag{6.1.5}$$

From Equations 6.1.1, 6.1.2, and 6.1.5, we have

$$u(\lambda, T) = \frac{8\pi hc}{\lambda^5} \frac{1}{e^{hc/\lambda kT} - 1} \tag{6.1.6}$$

Wien's displacement law can be written as

$$\lambda_{\text{max}} = \frac{0.201405hc}{k}\frac{1}{T} \tag{6.1.7}$$

Therefore, λ_{max} is displaced toward shorter wavelengths as the temperature increases. (For further discussions of blackbody radiation, see [ER85], [Gas74], and [Eis61].)

6.1.3 Solution with Mathematica

In[1] := **ClearAll["Global`*"]**

6.1.3.1 $u(\lambda, T)$ AT SEVERAL TEMPERATURES

Let x be the alias of λ. Equation 6.1.6 for the energy density can be entered as

In[2] := **u[x_, T_] := ((8 Pi h c)/((x^5)(Exp[(h c)/(x k T)] - 1)))**

To plot $u(x, T)$ (i.e., $u(\lambda, T)$) at several temperatures, we need the values for c, h, and k, which can be obtained from the package **Miscellaneous`PhysicalConstants`**. Let us load the package:

In[3] := **Needs["Miscellaneous`PhysicalConstants`"]**

In plotting graphs, we do not want to carry the units along. To discard the units of c, h, and k and to express u in units of 10^6 joule/m^3-m and x in units of 10^4 A, define a function **f**:

```
In[4] := f[x_, T_] :=  ((u[x 10^-6, T]/10^6) /.
                {c -> SpeedOfLight (Second/Meter),
                 h -> PlanckConstant/(Joule Second),
                 k -> BoltzmannConstant (Kelvin/Joule)}
              )
```

Let us plot the energy density at four selected temperatures:

```
In[5] := Plot[Evaluate[Table[f[x, n 1000], {n, 4, 7}]],
            {x, 0.1, 1.5},
            PlotRange -> {0, 3.15},
            Frame -> True,
            PlotLabel -> "Energy Density vs. Wavelength",
            FrameTicks -> {{0, 0.5, 1.0, 1.5}, {0, 1, 2, 3},
                        None, None},
            FrameLabel -> {"x (10^4 A)", "u (10^6 joule/m^3-m)"}
          ];
```

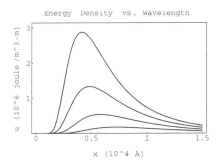

The curve at the top is for $T = 7000$ K, the next one down is for $T = 6000$ K, the one below that is for $T = 5000$ K, and the bottom curve is for $T = 4000$ K. Consistent with Wien's displacement law, $\lambda_{\max}$ decreases as T increases.

6.1.3.2   WIEN'S DISPLACEMENT LAW

Let *xmax* be the alias of $\lambda_{\max}$. To determine the wavelength *xmax* at which $u(x, T)$ is a maximum for a given T, we set the derivative of $u(x, T)$ with respect to x to zero and solve for x.

```
In[6] := dudx = D[u[x, T], x]//Together
```

$Out[6] = (-8 \text{ c h Pi } (-(c \text{ E}^{(c \text{ h})/(k \text{ T x})} \text{ h}) - 5 \text{ k T x} + 5 \text{ E}^{(c \text{ h})/(k \text{ T x})} \text{ k T x})) / ((-1 + \text{E}^{(c \text{ h})/(k \text{ T x})})^2 \text{ k T x}^7)$

This derivative is zero if its numerator is zero unless, of course, the denominator is also zero. The function **Numerator** picks out the numerator:

```
In[7] := Numerator[dudx]
```

$Out[7] = -8 \text{ c h Pi } (-(c \text{ E}^{(c \text{ h})/(k \text{ T x})} \text{ h}) - 5 \text{ k T x} + 5 \text{ E}^{(c \text{ h})/(k \text{ T x})} \text{ k T x})$

Let us make a change of variable from x to y such that

$$x = \frac{(ch/ky)}{T} \tag{6.1.8}$$

```
In[8] := (% /. x -> c h/(T k y))//Simplify
```

$$Out[8] = \frac{8 \text{ c}^2 \text{ h}^2 \text{ Pi } (5 - 5 \text{ E}^y + \text{E}^y \text{ y})}{y}$$

In terms of y, the condition $du(x, T)/dx = 0$ becomes

$$5 - 5e^y + e^y y = 0 \tag{6.1.9}$$

Dividing Equation 6.1.9 by $\exp(y)$, we have the equation for y:

```
In[9] := Simplify[%[[6]]/Exp[y]] == 0
```

$$Out[9] = -5 + \frac{5}{\text{E}^y} + y == 0$$

Observe that the solution is $y \approx 5$, since $5/\exp(y) << 5$ for $y = 5$. FindRoot yields a more accurate solution:

```
In[10] := FindRoot[Evaluate[%], {y, 5}]
```

```
Out[10] = {y -> 4.96511}
```

(We can also determine graphically the starting value for FindRoot with the procedure outlined in Section 2.3.1.4.) Wien's displacement law follows immediately from Equation 6.1.8 together with the value of y just calculated:

```
In[11] := xmax = (((c h)/(k y))/T) /. %
```

$$Out[11] = \frac{0.201405 \; c \; h}{k \; T}$$

What remains is the verification of our assumption that the denominator of the derivative $du(x, T)/dx$ is not zero at $x = xmax$:

```
In[12] := Denominator[dudx] /. x -> xmax
```

$$Out[12] = \frac{0.272306 \; c^7 \; h^7}{k^6 \; T^6}$$

It is indeed not zero.

6.1.3.3 λ_{max} FOR SOLAR RADIATION

With 5700 K as the estimated surface temperature of the Sun, Wien's displacement law gives λ_{max} for solar radiation:

```
In[13] := xmax /. {T -> 5700 Kelvin,
                   c -> SpeedOfLight,
                   h -> PlanckConstant,
                   k -> BoltzmannConstant}
```

```
Out[13] = 5.08378 10^-7 Meter
```

Measured in angstroms, λ_{max} for solar radiation is

```
In[14] := N[% 10^10 A/Meter, 4]
```

```
Out[14] = 5084. A
```

The graphical capability of the Macintosh or Windows front end provides another method for determining λ_{max} (see Section 2.3.1.4).

```
In[15] := Plot[Evaluate[f[x, 5700]], {x, 0.1, 1.5},
              Frame -> True,
              PlotRange -> {0, 1.1},
              PlotLabel -> "Energy Density vs. Wavelength",
              FrameTicks -> {{0, 0.5, 1.0, 1.5}, {0, 0.5, 1},
                             None, None},
              FrameLabel -> {"x (10^4 A)", "u (10^6 joule/m^3-m)"}
              ];
```

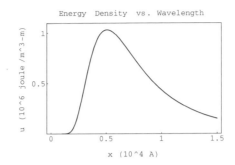

$$\{0.50384, 1.0317\}$$

Graphically, we have found $\lambda_{\max} = 5038$ A, which is within 1% of the value 5084 A determined earlier with Wien's displacement law.

```
In[16] := ClearAll[u, f, dudx, xmax]
```

6.2 Wave Packets

6.2.1 The Problem

Demonstrate graphically the spreading of the one-dimensional, free-particle, Gaussian wave packet.

6.2.2 Physics of the Problem

The one-dimensional, free-particle, Gaussian wave packet can be written as

$$\Psi(x,t) = \left[\frac{1}{2\pi L^2 \left(1 + i\dfrac{\hbar}{2mL^2}t\right)^2}\right]^{1/4} e^{i(k_0 x - \omega_0 t)} \exp\left[-\frac{1}{4L^2}\frac{(x - v_{gr}t)^2}{\left(1 + i\dfrac{\hbar}{2mL^2}t\right)}\right]$$

where m is the mass of the particle and L is the position uncertainty at time $t = 0$. Also,

$$v_{gr} = \frac{\hbar k_0}{m}$$

$$\omega_0 = \frac{\hbar k_0^2}{2m}$$

and k_0 is the average wave number.

The position probability density, $\Psi^*(x,t)\,\Psi(x,t)$, can be expressed as

$$P(x,t) = \left[\frac{1}{2\pi L^2 \left(1 + \dfrac{\hbar^2}{4m^2 L^4}t^2\right)}\right]^{1/2} \exp\left[-\frac{1}{2L^2}\frac{(x - v_{gr}t)^2}{\left(1 + \dfrac{\hbar^2}{4m^2 L^4}t^2\right)}\right] \qquad (6.2.1)$$

Let x, t, and $P(x, t)$ be measured, respectively, in units of L, T, and A, where

$$T = \frac{2mL^2}{\hbar}$$

$$A = \frac{1}{\sqrt{2\pi L^2}}$$

Equation 6.2.1 becomes

$$P(x, t) = \left[\frac{1}{1 + t^2} \right]^{1/2} \exp\left[-\frac{(x - vt)^2}{2(1 + t^2)} \right] \tag{6.2.2}$$

in which

$$v = \frac{2mL}{\hbar} v_{gr}$$

In units of L, the position uncertainty or the width of the wave packet at time t is

$$\Delta x(t) = \sqrt{1 + t^2} \tag{6.2.3}$$

(For a discussion that "plumbs the depths of a Gaussian wave packet," see [Mor90].)

6.2.3   Solution with Mathematica

```
In[1] := ClearAll["Global`*"]
```

There are several ways to demonstrate the spreading of the wave packet. The simplest is to plot $\Delta x(t)$ given in Equation 6.2.3:

```
In[2] := Plot[{Sqrt[1 + t^2], t}, {t, 0, 6},
            PlotRange -> {0, 6},
            PlotStyle -> {{}, {Dashing[{0.02, 0.02}]}},
            AxesLabel -> {" t (T)", " Δx (L)"}];
```

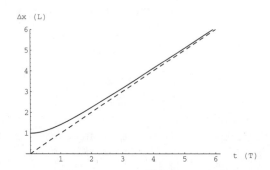

A dashed straight line is added to the graph to show that $\Delta x(t) \approx t$ for $t \gg 1$.

Another way to show the spreading of the wave packet is to plot $P(x, t)$ in Equation 6.2.2 for several values of t. In what follows, we set $\nu = 1$.

```
In[3] := p[x_, t_] :=  ((1/Sqrt[1 + t^2]) Exp[-((x - t)^2)/(2(1 + t^2))])
```

```
In[4] := Plot[Evaluate[Table[p[x, i], {i, 0, 3}]], {x, -4, 12},
            PlotStyle ->
                  {{Thickness[0.01]}, {},
                   {Dashing[{0.05, 0.05}]},
                   {Dashing[{0.01, 0.05, 0.01, 0.05}]}},
            PlotRange -> {0, 1},
            AxesLabel -> {" x (L)", " P(x,t) (A)"},
            Epilog ->
                  {Text["SPREADING OF A\nGAUSSIAN WAVE PACKET",
                        {6.8, 0.8}],
                   Text["t = 0", {2, 0.95}],
                   Text["t = T", {3.5, 0.6}],
                   Text["t = 2T", {5.2, 0.4}],
                   Text["t = 3T", {8, 0.24}]}
          ];
```

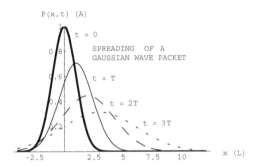

There is another way to plot $P(x, t)$ for several values of t:

```
In[5] := g[n_] := ParametricPlot3D[{x, n, p[x, n]}, {x, -5.6, 21},
                            PlotPoints -> 100,
                            DisplayFunction -> Identity]
```

```
In[6] := Show[Table[{g[i],
                Graphics3D[{Dashing[{0.01, 0.01}],
                            Line[{{-5.6, i, 0}, {21, i, 0}}]}]},
                {i, 0, 7}],
            PlotRange -> {{-5.6, 21}, {0, 7}, {0, 1}},
            ViewPoint -> {2.230, 2.323, 1.040},
            BoxRatios -> {1, 2, 0.5},
            Boxed -> False,
            AxesLabel -> {"x (L)", " t (T)", "P(x,t) (A)"},
```

```
Ticks -> {Automatic, Automatic, {0, 1}},
DisplayFunction -> $DisplayFunction
];
```

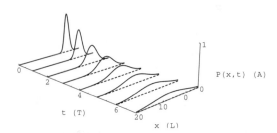

Here the added dimension is time. Let us observe the spreading of the wave packet from another viewpoint:

```
In[7] := Show[%, BoxRatios -> {1, 1, 1}, Boxed -> True,
              ViewPoint -> {0.017, 3.375, 0.245}];
```

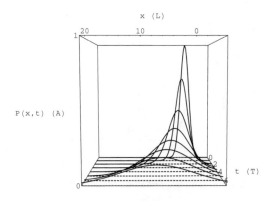

Finally, we generate a three-dimensional surface plot of $P(x, t)$ in Equation 6.2.2 as a function of x and t:

```
In[8] := Plot3D[p[x, t], {x, -5, 14}, {t, 0, 4},
              PlotRange -> {0, 1},
              ViewPoint -> {0.017, 3.375, 0.245},
              BoxRatios -> {1, 1, 1},
              PlotPoints -> 45,
              AxesLabel -> {"x (L)", " t (T)", "P(x,t) (A)"},
              Ticks -> {Automatic, Automatic, {0, 1}}
          ];
```

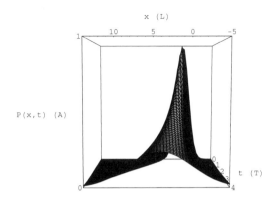

To animate the spreading of the wave packet, click the graphics cell bracket and select the **Render PostScript** command in the Graph menu for *Mathematica* version 2.2, or the **Rerender Graphics** command in the Cell menu for version 3.0.

```
In[9] := ClearAll[p, g]
```

6.3   *Particle in a One-Dimensional Box*

6.3.1   *The Problem*

Animate the time evolution of a wave packet in an infinite one-dimensional potential well.

6.3.2   *Physics of the Problem*

Consider a particle of mass m moving in a one-dimensional potential

$$V(x) = \begin{cases} 0 & |x| < a \\ \infty & \text{elsewhere} \end{cases} \tag{6.3.1}$$

In a system of units with $m = 1$, $a = 1$, and $\hbar = 1$, the solutions to the time-independent Schrödinger equation are

$$\phi_n^{(-)}(x) = \sin n\pi x \tag{6.3.2}$$

$$\phi_n^{(+)}(x) = \cos\left[n - \frac{1}{2}\right]\pi x \tag{6.3.3}$$

and the corresponding energy eigenvalues are

$$E_n^{(-)} = n^2\pi^2/2 \tag{6.3.4}$$

$$E_n^{(+)} = \left[n - \frac{1}{2}\right]^2 \pi^2/2 \tag{6.3.5}$$

for all positive integers n.

The wave function $\Psi(x,t)$ can be expressed as an infinite superposition

$$\Psi(x,t) = \sum_{n=1}^{\infty}[c_n^{(+)}\phi_n^{(+)}(x)e^{-iE_n^{(+)}t} + c_n^{(-)}\phi_n^{(-)}(x)e^{-iE_n^{(-)}t}] \tag{6.3.6}$$

with

$$c_n^{(\pm)} = \int_{-1}^{1} dx\,\phi_n^{(\pm)^*}(x)\,\Psi(x,0) \tag{6.3.7}$$

Let us consider the case in which $\Psi(x,0)$ is a Gaussian function, that is,

$$\Psi(x,0) = \left(\frac{1}{2\pi L^2}\right)^{1/4} e^{ik_0x}e^{-x^2/4L^2} \tag{6.3.8}$$

where the width L and the average wave number k_0 of the wave packet are in units of a and $1/a$, respectively. The Gaussian function cannot actually be an initial wave function, since it does not satisfy the boundary conditions that all wave functions must vanish at $x = \pm1$. In what follows, we compute the coefficients $c_n^{(\pm)}$ that give an initial wave function $\Psi(x,0)$ that approximates a Gaussian function and satisfies the boundary conditions. The approximation is reasonable if the width of the Gaussian function is a fraction of that of the potential well. Equation 6.3.8 can be written as

$$\Psi(x,0) = \Psi^{(+)}(x,0) + i\Psi^{(-)}(x,0) \tag{6.3.9}$$

with

$$\Psi^{(+)}(x,0) = \left(\frac{1}{2\pi L^2}\right)^{1/4} \cos k_0x\, e^{-x^2/4L^2} \tag{6.3.10}$$

$$\Psi^{(-)}(x,0) = \left(\frac{1}{2\pi L^2}\right)^{1/4} \sin k_0x\, e^{-x^2/4L^2} \tag{6.3.11}$$

Note that $\phi_n^{(+)}(x)$ and $\Psi^{(+)}(x,0)$ are even functions of x, whereas $\phi_n^{(-)}(x)$ and $\Psi^{(-)}(x,0)$ are odd functions. Equations 6.3.7 and 6.3.9 imply

$$c_n^{(+)} = 2\int_{0}^{1} dx\,\phi_n^{(+)^*}(x)\,\Psi^{(+)}(x,0) \tag{6.3.12}$$

$$c_n^{(-)} = 2i\int_{0}^{1} dx\,\phi_n^{(-)^*}(x)\,\Psi^{(-)}(x,0) \tag{6.3.13}$$

The probability density is

$$P(x,t) = \Psi^*(x,t)\Psi(x,t) \tag{6.3.14}$$

Our task is to explore the time evolution of the probability density.

6.3.3   Solution with Mathematica

```
In[1] := ClearAll["Global`*"]
```

6.3.3.1 FUNCTION DEFINITIONS

In this section, we define a function that generates the graphics for animating the time evolution of the probability density. The first argument of the function is the Gaussian width L restricted from $1/8$ to $1/4$; the second argument, the average wave number k_0 limited from 15 to 30; the third argument, the number m of graphs generated, which must be a positive integer. (Why should L and k_0 be restricted? See Problem 7 in Section 6.7.) Options for passing to the `Plot` function may be specified after the third argument. The time interval for animation is equal to $(m - 1)/5k_0$. Statements explaining the construct of the function are embedded in the body of the function in the form (* *comment* *) ignored by the kernel.

■ *Mathematica* Version 2.2 Specifics

```
particle[width_ /; width >= 1/8 && width <= 1/4,
        k0_ /; k0 >= 15 && k0 <= 30,
        m_Integer /; m > 0, opts___] :=
Module[{cplus, cminus, psiplus, psiminus,
        phiplus, phiminus, eplus, eminus,
        psi, psistar, p},

        (* evaluate the coefficients with Equations
           6.3.12 and 6.3.13 and set those less than
           10^(-2) identically to zero, i.e., determine
           a set of coefficients that gives a reasonable
           approximation of the Gaussian function for
           the initial wave function that satisfies the
           boundary conditions *)
        cplus[n_] := cplus[n] =
            Chop[2 NIntegrate[Evaluate[
                                phiplus[n, x] psiplus[x] // N
                                    ],
                            {x, 0, 1}
                            ],
                    10^(-2)
                ];
        cminus[n_] := cminus[n] =
            Chop[2 I NIntegrate[Evaluate[
                                phiminus[n, x] psiminus[x] // N
                                    ],
                            {x, 0, 1}
                            ],
                    10^(-2)
                ];
        psiplus[x_] := (
            (1/(2 Pi width^2))^(1/4) Cos[k0 x] Exp[-x^2/(4 width^2)]
                    );
        psiminus[x_] := (
```

```
                    (1/(2 Pi width^2))^(1/4) Sin[k0 x] Exp[-x^2/(4 width^2)]
                        );
        phiplus[n_, x_] :=  Cos[(n - 1/2) Pi x];
        phiminus[n_, x_] :=  Sin[n Pi x];

        (* define the wave function and its complex
           conjugate *)
        psi[x_, t_] :=
           Sum[(cplus[n] phiplus[n, x] Exp[-I eplus[n] t]) +
              (cminus[n] phiminus[n, x] Exp[-I eminus[n] t]),
              {n, 1, 14}
              ];
        eplus[n_] :=  ((n - 1/2)^2 Pi^2) / 2;
        eminus[n_] :=  (n^2 Pi^2)/2;
        psistar[x_, t_] :=
           (psi[x, t] /. Complex[p_, q_] -> Complex[p, -q]);

        (* define the probability density *)
        p[x_, t_] :=  psi[x, t] psistar[x, t] // N //Expand;

        (* generate the graphics for animation *)
        Do[Plot[Evaluate[p[x, (1/(5 k0)) i]], {x, -1, 1},
                opts,
                Ticks -> {{-1, 0, 1}, None},
                AxesLabel -> {" x (a)", "P(x, t)"},
                PlotRange -> {0, 6},
                Prolog -> {Line[{{1, 0}, {1, 6}}],
                           Line[{{-1, 0}, {-1, 6}}]}
                ],
           {i, 0, m - 1}
           ]
           ]
```

■ *Mathematica* Version 3.0 Specifics

$In[2] :=$ $\mathtt{particle}\left[\mathtt{L\_} \ /; \mathtt{L} \geq \dfrac{1}{8} \bigwedge \mathtt{L} \leq \dfrac{1}{4},\right.$

            k0_ /; k0 ≥ 15 ∧ k0 ≤ 30,
            m_Integer?Positive,
            opts___] :=
      Module[{SuperPlus, SuperMinus, Φ, SuperStar, P},
      (* evaluate the coefficients with Equations 6.3.12
         and 6.3.13 and set those less than 10^{-2} identically
         to zero, i.e., determine a set of coefficients that
         gives a reasonable approximation of the Gaussian function
         for the initial wave function that satisfies the
         boundary conditions *)
      c_n^+_ := c_n^+ =

```
      Chop[2 NIntegrate[Evaluate[ϕₙ⁺[x] Ψ⁺[x] // N], {x, 0, 1}], 10⁻²];
   cₙ⁻ := cₙ⁻ =
      Chop[2 I NIntegrate[Evaluate[ϕₙ⁻[x] Ψ⁻[x] // N], {x, 0, 1}], 10⁻²];
```

$$\Psi^+[x\_] := \left(\frac{1}{2\,\pi\,L^2}\right)^{1/4}\,\mathrm{Cos}[k0\,x]\,\mathrm{Exp}\!\left[-\frac{x^2}{4\,L^2}\right];$$

$$\Psi^-[x\_] := \left(\frac{1}{2\,\pi\,L^2}\right)^{1/4}\,\mathrm{Sin}[k0\,x]\,\mathrm{Exp}\!\left[-\frac{x^2}{4\,L^2}\right];$$

$$\phi_n^+[x\_] := \mathrm{Cos}[(n-1/2)\,\pi\,x];$$

$$\phi_n^-[x\_] := \mathrm{Sin}[n\,\pi\,x];$$

```
(* define the wave function and its complex conjugate *)
```

$$\Psi[x\_,\ t\_] := \sum_{n=1}^{14}(c_n^+\ \phi_n^+[x]\,\mathrm{Exp}[-I\,e_n^+\,t] + c_n^-\ \phi_n^-[x]\,\mathrm{Exp}[-I\,e_n^-\,t]);$$

$$e_n^+ := \frac{(n-1/2)^2\,\pi^2}{2};$$

$$e_n^- := \frac{n^2\,\pi^2}{2};$$

$$\Psi^*[x\_,\ t\_] := (\Psi[x,\ t]\ /.\ \mathrm{Complex}[p\_,\ q\_] \rightarrow \mathrm{Complex}[p,\ -q]);$$

```
(* define the probability density *)
P[x_, t_] := Ψ[x, t] Ψ*[x, t] // N // Expand // Chop;

(* generate the graphics for animation *)
```

$$Do\!\left[\mathrm{Plot}\!\left[\mathrm{Evaluate}\!\left[P\!\left[x,\ \frac{i}{5\,k0}\right]\right],\ \{x,\ -1,\ 1\},\right.\right.$$

```
            opts,
            Ticks -> {{-1, 0, 1}, None},
            AxesLabel -> {" x (a)", "P(x, t)"},
            PlotRange -> {0, 6},
            Prolog -> {Line[{{1, 0}, {1, 6}}],
                       Line[{{-1, 0}, {-1, 6}}]}
         ],
      {i, 0, m - 1}
    ]
  ]
```

6.3.3.2 GRAPHICS

As an example, consider the case with $L = 1/4$, $k_0 = 20$, and $m = 100$. The function Timing allows us to determine the time required for generating the graphs. Timing [*expr*] evaluates *expr*, returns the result, and gives the time used:

In[3] := **Timing[particle[1/4, 20, 100]]**

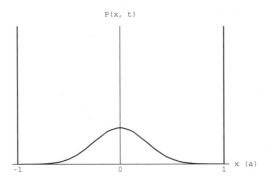

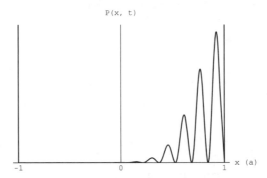

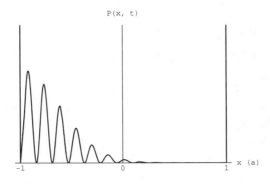

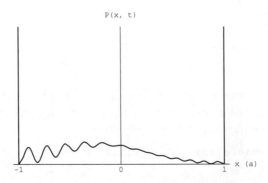

```
{1379.43 Second, Null}
```

It takes 1379 seconds, or 23 minutes, to generate the 100 graphs on a Macintosh IIci and 212 seconds, or 3.5 minutes, on a Gateway 2000 4DX2-66V. (In this book, all timings have been performed with *Mathematica* version 2.2.) To animate, select all the graphics cells and press Command-Y for the Macintosh or Control+Y for Windows. (Only 4 of the 100 graphs are shown here.) We observe the interference among the component waves as the wave packet sloshes between the two walls. It is a mistake to assume that the probability density will eventually become uniform within the well. On the contrary, the probability density restores its original shape in time $T = 2\pi/E_1^{(+)} = 16/\pi$, in our system of units. The 100 graphs cover only the time interval $(m-1)/5k_0 = 0.99 \approx T/5$. (To generate 510 graphs spanning the entire time interval T, a PowerMac 7500/100 requires memory partitions of 20 MB for the front end and 10 MB for the kernel and takes about 650 seconds, or 11 minutes.)

```
In[4] := ClearAll[particle]
```

6.4 The Square Well Potential

6.4.1 The Problem

Determine analytically and numerically the bound states of a particle in a one-dimensional square potential well.

6.4.2 Physics of the Problem

Consider a particle of mass m moving in a one-dimensional potential

$$V(x) = \begin{cases} 0 & |x| > a \\ -V_0 & |x| < a \end{cases} \tag{6.4.1}$$

with $V_0 > 0$. The time-independent Schrödinger equation is

$$-\frac{\hbar^2}{2m}\frac{d^2u(x)}{dx^2} + V(x)u(x) = Eu(x) \tag{6.4.2}$$

where $u(x)$ is the eigenfunction corresponding to the energy eigenvalue E. For bound states, $E < 0$.

6.4.2.1 ANALYTICAL SOLUTION

Let

$$k = \frac{\sqrt{2m|E|}}{\hbar} \tag{6.4.3}$$

$$q = \frac{\sqrt{2m(V_0 - |E|)}}{\hbar} \tag{6.4.4}$$

$$\lambda = \frac{2mV_0a^2}{\hbar^2} \tag{6.4.5}$$

$$y = qa \tag{6.4.6}$$

Equations 6.4.4, 6.4.5, and 6.4.6 imply that the energy eigenvalue E can be expressed as

$$E = -\left(1 - \frac{y^2}{\lambda}\right)V_0 \tag{6.4.7}$$

and Equations 6.4.3, 6.4.5, and 6.4.7 suggest that k can be written as

$$k = \frac{\sqrt{\lambda - y^2}}{a} \tag{6.4.8}$$

As a consequence of the symmetry of the potential, the time-independent Schrödinger equation (Equation 6.4.2) has two classes of solutions that are bounded at infinity. The even solutions are

$$u(x) = \begin{cases} Be^{kx} & x < -a \\ A\cos qx & -a < x < a \\ Be^{-kx} & x > a \end{cases} \tag{6.4.9}$$

The continuity of the eigenfunctions and their derivatives at $x = +a$ (or $x = -a$) imposes the conditions

$$B = A\cos qa\, e^{ka} \tag{6.4.10}$$

$$k = q\tan qa \tag{6.4.11}$$

In terms of λ and y defined in Equations 6.4.5 and 6.4.6, Equation 6.4.11 can be written as

$$\frac{\sqrt{\lambda - y^2}}{y} = \tan y \tag{6.4.12}$$

The corresponding equations for the odd solutions are

$$u(x) = \begin{cases} De^{kx} & x < -a \\ C\sin qx & -a < x < a \\ -De^{-kx} & x > a \end{cases} \tag{6.4.13}$$

$$D = -C\sin qa\, e^{ka} \tag{6.4.14}$$

$$k = -q\cot qa \tag{6.4.15}$$

$$\frac{\sqrt{\lambda - y^2}}{y} = -\cot y \tag{6.4.16}$$

(For more information on the bound states in a square potential well, see [Gas95].)

6.4.2.2   NUMERICAL SOLUTION

Though specifying values for the potential at $x = \pm a$ is unnecessary for the analytical solution of the time-independent Schrödinger equation (Equation 6.4.2), it is prudent, for the numerical solution, to modify Equation 6.4.1 to include the reasonable value $-V_0$ for $V(x)$ at these points. Thus,

$$V(x) = \begin{cases} 0 & |x| > a \\ -V_0 & |x| \leq a \end{cases} \tag{6.4.17}$$

with $V_0 > 0$. Equation 6.4.2 can be written as

$$\frac{d^2 u(x)}{dx^2} = F(x)u(x) \tag{6.4.18}$$

where lengths are measured in units of a, that is, $a = 1$, and

$$F(x) = \begin{cases} -\lambda e & |x| > 1 \\ -\lambda(e + 1) & |x| \leq 1 \end{cases} \tag{6.4.19}$$

with the energy parameter

$$e = \frac{E}{V_0} \tag{6.4.20}$$

The parameter λ measuring the strength of the potential well is defined in Equation 6.4.5. For bound states, $0 > e > -1$.

As mentioned in Section 6.4.2.1, the symmetry of the potential implies that there are two classes of solutions: even solutions and odd solutions (see [Sch55]). For even solutions, $u'(0) = 0$. Without loss of generality, we let $u(0) = 1$ and determine its appropriate value later with the normalization condition

$$\int_{-\infty}^{\infty} u^2(x)\,dx = 1 \tag{6.4.21}$$

For the odd solutions, $u(0) = 0$. We may set $u'(0) = 1$ and impose the normalization condition on the eigenfunctions later. For both even and odd solutions, it is only necessary to numerically solve Equation 6.4.18 for $x \geq 0$.

6.4.3 *Solution with Mathematica*

```
In[1] := ClearAll["Global`*"]
```

As an example, we take, rather arbitrarily, $\lambda = 16$:

```
In[2] := lambda = 16;
```

where `lambda` is the alias of λ. We consider first the analytical solution and then the numerical solution for the bound states.

6.4.3.1 ANALYTICAL SOLUTION

Equation 6.4.12 is a transcendental equation, and its roots can be determined only by graphical or numerical methods. Following the procedure outlined in Section 2.3.1.4, "FindRoot," we plot the *lhs* and *rhs* of the equation on the same graph and look for their intersections:

In[3] := `Plot[{Sqrt[lambda - y^2]/y, Tan[y]}, {y, 0, 4},`
          `PlotRange -> {0, 5}];`

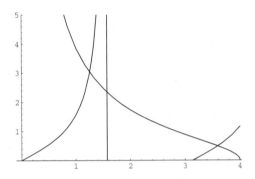

        `{{1.25524, 3.04686}, {3.60764, 0.494768}}`

There are two intersections, and the approximate roots are obtained graphically: $y = 1.25524$ and $y = 3.60764$. (The vertical line represents an asymptote of the tangent function.) The accuracy of these roots can be improved with `FindRoot`:

In[4] := `y[1] = y/.FindRoot[Sqrt[lambda - y^2]/y == Tan[y], {y, 1.25524}]`

Out[4] = `1.25235`

In[5] := `y[3] = y/.FindRoot[Sqrt[lambda - y^2]/y == Tan[y], {y, 3.60764}]`

Out[5] = `3.5953`

The roots are `y[1]` $= 1.25235$ and `y[3]` $= 3.5953$.

Similarly, we can determine the roots of Equation 6.4.16:

In[6] := `Plot[{Sqrt[lambda - y^2]/y, -Cot[y]}, {y, 0, 4},`
          `PlotRange -> {0, 5}];`

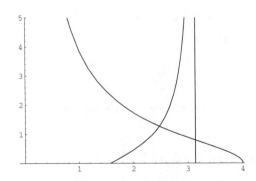

     `{2.44496, 1.27602}`

There is only one intersection and, therefore, one root. FindRoot improves the accuracy of the root:

In[7] := y[2] = y/.FindRoot[Sqrt[lambda - y^2]/y == -Cot[y], {y, 2.44496}]

Out[7] = 2.47458

The root is y[2] = 2.47458.

When a root of Equation 6.4.12 or 6.4.16 is known, Equation 6.4.7 gives the corresponding energy eigenvalue in terms of V_0:

In[8] := energy[n_] := -(1 - (y[n]^2)/lambda)V0

where energy is the alias of E. The potential and the energy eigenvalues are illustrated to scale in the following:

In[9] := V[x_ /; Abs[x] > 1] := 0
        V[x_ /; Abs[x] < 1] := -1
        Plot[V[x], {x, -2, 2},
            PlotStyle -> Thickness[0.0125],
            PlotRange -> {0.05, -1.05},
            Axes -> False,
            Ticks -> {None, Automatic},
            DefaultFont -> {"Courier", 9},
            Epilog ->
             ({Line[{{1, energy[1]}, {-1, energy[1]}}],
               Text["E1", {1.25, energy[1]}],
               Text[ToString[energy[1]], {0, energy[1] + 0.055}],
               Line[{{1, energy[2]}, {-1, energy[2]}}],
               Text["E2", {1.25, energy[2]}],
               Text[ToString[energy[2]], {0, energy[2] + 0.055}],
               Line[{{1, energy[3]}, {-1, energy[3]}}],
               Text[ToString[energy[3]], {0, energy[3] + 0.055}],
               Text["E3", {1.25, energy[3]}],
               Text["-V0", {-1.25, -1}],
               Text["0", {-0.85, 0}]} /. V0 -> 1)
            ];

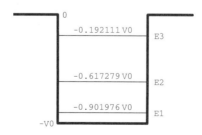

DefaultFont remains a valid option for the graphics functions in *Mathematica* version 3.0. We have used the function **ToString** in the **Text** primitives; **ToString**[*expr*] gives a string corresponding to the printed form of *expr*.

If y is known, Equations 6.4.6 and 6.4.8 yield q and k, respectively. Equations 6.4.9 and 6.4.10 or Equations 6.4.13 and 6.4.14 then give the yet to be normalized $u(x)$. To plot the eigenfunctions, we define a plotting function in order to streamline the input. The first argument of the function accepts a root of Equation 6.4.12 or 6.4.16, namely, **y[1]**, **y[2]**, or **y[3]**. The second argument takes on either the unnormalized (i.e., $A = 1$) even eigenfunction given in Equations 6.4.9 and 6.4.10 or the odd one specified in Equations 6.4.13 and 6.4.14. Explanatory comments are embedded into the body of the function, and lengths are measured in units of a, that is, $a = 1$.

```
In[12] := eigenfunctionPlot[y_, u_] :=
              Module[{normalizationConstant, normalizedu, energy},

                 (* Equation 6.4.6 *)
                 q = y;

                 (* Equation 6.4.8 *)
                 k = Sqrt[lambda - y^2];

                 (* normalize the eigenfunction *)
                 normalizationConstant = 1/Sqrt[
                     NIntegrate[(u[x])^2, {x, -Infinity, Infinity}]
                                            ];
                 normalizedu[x_] = normalizationConstant u[x];

                 (* Equation 6.4.7 *)
                 energy = -(1 - (y^2)/lambda)V0;

                 (* plot the eigenfunction *)
                 Plot[normalizedu[x], {x, -3, 3},
                     Epilog -> {Line[{{1, -1}, {1, 1}}],
                                 Line[{{-1, -1}, {-1, 1}}]},
                     Frame -> True,
                     FrameLabel -> {"x (a)", "u(x) (1/Sqrt[a])"},
                     PlotLabel -> "E = "<>ToString[energy]
                     ];
                 ]
```

The operator **<>** with the full name **StringJoin** concatenates two strings.

Let **u[1][x]**, **u[2][x]**, and **u[3][x]** represent the unnormalized eigenfunctions corresponding to the energy eigenvalues *E1*, *E2*, and *E3*, respectively. Note that *Mathematica* expressions can have complicated heads.

```
In[13] := uEven[x_ /; x < -1] :=  (Cos[q] Exp[k]) Exp[k x]
          uEven[x_ /; Abs[x] <= 1] :=  Cos[q x]
          uEven[x_ /; x > 1] :=  (Cos[q] Exp[k]) Exp[-k x]
          uOdd[x_ /; x < -1] :=  (-Sin[q] Exp[k]) Exp[k x]
          uOdd[x_ /; Abs[x] <= 1] :=  Sin[q x]
          uOdd[x_ /; x > 1] :=  (Sin[q] Exp[k]) Exp[-k x]
          u[1] = u[3] = uEven;
          u[2] = uOdd;
```

We now plot the eigenfunctions:

```
In[21] := Table[eigenfunctionPlot[y[n], u[n]], {n, 1, 3}];
```

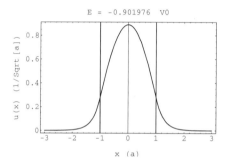

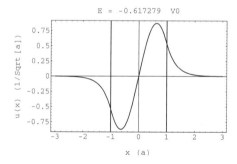

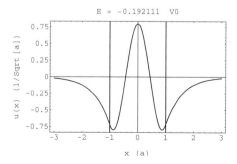

The vertical lines depict the boundaries of the potential well.

```
In[22] := ClearAll[V, q, k, y, uEven, uOdd, u,
                   energy, eigenfunctionPlot]
```

6.4.3.2 NUMERICAL SOLUTION

Following Equation 6.4.19, we define

```
In[23] := F[x_ /; Abs[x] > 1]  := -lambda e
          F[x_ /; Abs[x] <= 1] := -lambda (e + 1)
```

Given a value for the energy parameter e, we can solve Equation 6.4.18 for the corresponding eigenfunction $u(x)$. However, only for certain discrete values of the energy parameter are the eigenfunctions bounded, that is,

$$u(\infty) = 0 \tag{6.4.22}$$

For numerical analysis, we choose a point $x = x_f$ at a finite distance from the potential well but sufficiently far away from it so that Equation 6.4.22 can be replaced by the condition of $u(x)$ approaching zero at $x = x_f$. The choice of x_f depends on λ. For $\lambda = 16$, we take, somewhat arbitrarily, $x_f = 4$. (To ascertain the validity of the choice for x_f, see Problem 8 in Section 6.7.)

To find the values of e for which the eigenfunctions approach zero at $x = 4$, we define the function

```
In[25] := eqnEven[r_] :=
            (e = r;
             sol = NDSolve[{u''[x] == F[x] u[x], u'[0] == 0,
                           u[0] == 1}, u, {x, 0, 4}
                          ];
             u[4] /. sol[[1]]
            )
```

for the even solutions, and the function

```
In[26] := eqnOdd[r_] :=
            (e = r;
             sol = NDSolve[{u''[x] == F[x] u[x], u'[0] == 1,
                           u[0] == 0}, u, {x, 0, 4}
                          ];
             u[4] /. sol[[1]]
            )
```

for the odd solutions. With a value for e as their argument, these functions return the value of $u(4)$ and name sol as the solution to Equation 6.4.18. Thus, finding the values of e for which the eigenfunctions equal zero at $x = 4$ is the same as finding the roots of the equation *eqnEven*(*r*) = 0 for the even solutions and of the equation *eqnOdd*(*r*) = 0 for the odd solutions. Furthermore, the solutions **sol** for these values of *e* yield the corresponding bound state eigenfunctions. (Of course, we should verify graphically that these functions indeed approach zero at *x* = 4 to eliminate the remote possibility that they happen to cross the *x* axis at that point.) **FindRoot**[*eqn*[*r*] == 0, {*r*, *r*0}] introduced in Section 2.3.1.4 searches for a root using the Newton-Raphson method. Using this method requires that the function *eqn*(*r*) is differentiable. (See Problem 10 of Section 3.3.7.) Well, *eqnEven*(*r*) and *eqnOdd*(*r*) are not:

In[27] := **D[eqnEven[r], r]**

> NDSolve::ndnum:
> > Differential equation does not evaluate to a number at x = 0..
>
> InterpolatingFunction::dmval:
> > Input value lies outside domain of the interpolating function.
>
> InterpolatingFunction::dmval:
> > Input value lies outside domain of the interpolating function.

Out[27] = 0

Fortunately, **FindRoot**[*eqn*[*r*] == 0, {*r*, {*r*0, *r*1}}] uses a variant of the secant method that does not require *eqn*(*r*) being differentiable. Yet it requires two starting values *r*0 and *r*1. While it is not necessary that *eqn*(*r*0) and *eqn*(*r*1) have opposite signs, it is important that *r*0 and *r*1 are sufficiently close to a root.

Consider the even solutions. To determine the appropriate values for *r*0 and *r*1, we generate a list of {*e_i*, *eqnEven*(*e_i*)} with *e_i* ranging from −1 to 0 in steps of 0.1 as −1 < *e* < 0 for bound states:

In[28] := **Table[{i, eqnEven[i]}, {i, -1, 0, 0.1}]**

Out[28] = {{-1, 81378.6}, {-0.9, -735.207}, {-0.8, -16150.5},
            {-0.7, -12767.2}, {-0.6, -7007.73}, {-0.5, -3049.71},
            {-0.4, -1040.02}, {-0.3, -238.913}, {-0.2, -6.59458},
            {-0.1, 22.8456}, {0., 8.42791}}

We look for the pairs (*e_i*, *e_{i+1}*), that is, two consecutive values of *e*, for which (*eqnEven*(*e_i*) *eqnEven*(*e_{i+1}*)) ≤ 0, because each of these pairs can serve as a set of starting values (*r*0, *r*1). As mentioned before, though it is not necessary that *eqn*(*r*0) and *eqn*(*r*1) have opposite signs, the fact that they do ensures the existence of at least one root between *r*0 and *r*1. Then, if *r*0 and *r*1 are also close together, they must be close to a root. To automate the selection of starting values, partition the list above into nonoverlapping sublists of two pairs with an offset of 1:

In[29] := **Partition[%, 2, 1]**

```
Out[29] = {{{-1, 81378.6}, {-0.9, -735.207}},
            {{-0.9, -735.207}, {-0.8, -16150.5}},
            {{-0.8, -16150.5}, {-0.7, -12767.2}},
            {{-0.7, -12767.2}, {-0.6, -7007.73}},
            {{-0.6, -7007.73}, {-0.5, -3049.71}},
            {{-0.5, -3049.71}, {-0.4, -1040.02}},
            {{-0.4, -1040.02}, {-0.3, -238.913}},
            {{-0.3, -238.913}, {-0.2, -6.59458}},
            {{-0.2, -6.59458}, {-0.1, 22.8456}},
            {{-0.1, 22.8456}, {0., 8.42791}}}
```

Then choose those sublists in which the second elements of the pairs have opposite signs:

```
In[30] := Select[%, (Sign[#[[1, 2]]] Sign[#[[2, 2]]] <= 0)&]//N

Out[30] = {{{-1., 81378.6}, {-0.9, -735.207}},
            {{-0.2, -6.59458}, {-0.1, 22.8456}}}
```

Finally, pick out the pairs (e_i), (e_{i+1}) from the nested list:

```
In[31] := {#[[1, 1]], #[[2, 1]]}& /@ %

Out[31] = {{-1., -0.9}, {-0.2, -0.1}}
```

The two sets of starting values for `FindRoot[eqn[r] == 0, {r, {r0, r1}}]` are $\{-1, -0.9\}$ and $\{-0.2, -0.1\}$. We can define a function to carry out all the foregoing operations for locating the starting values. This function has four arguments. The first argument is for the minimum value of e; the second argument, the maximum value of e. The third argument plus one equals the number of sampling values for e, and the last argument accepts either *eqnEven* or *eqnOdd*.

```
In[32] := initialValues[emin_, emax_, n_, eqn_] :=
             ({#[[1, 1]], #[[2, 1]]}& /@ (Select[
                Partition[
                   Table[{i, eqn[i]}, {i, emin, emax, (emax - emin)/n}
                       ], 2, 1
                       ],
                   (Sign[#[[1, 2]]] Sign[#[[2, 2]]] <= 0)&
                                                 ]//N
                       )
                )
```

For the even solutions, we have

```
In[33] := evenvalues = initialValues[-1, 0, 10, eqnEven]

Out[33] = {{-1., -0.9}, {-0.2, -0.1}}
```

For the odd solutions, we have

```
In[34] := oddvalues = initialValues[-1, 0, 10, eqnOdd]

Out[34] = {{-0.7, -0.6}}
```

For each pair of starting values, `FindRoot[`*eqn*`[r] == 0, {r, {r0, r1}}]` returns the energy parameter *e* and names `sol`, as a result of the call to *eqn(e)*, the corresponding bound state solution to Equation 6.4.18. For the starting values $\{-1., -0.9\}$, we have

```
In[35] := FindRoot[eqnEven[r] == 0, {r, evenvalues[[1]]}]

Out[35] = {r -> -0.901976}
```

With Equation 6.4.20, the energy eigenvalue is

```
In[36] := energy[1] = V0 r /. %

Out[36] = -0.901976 V0
```

and the associated eigenfunction is

```
In[37] := u[1][x_ /; x >= 0 && x <= 4] = u[x] /. sol[[1]];
          u[1][x_ /; x < 0 && x >= -4] := u[1][-x]
```

Similarly, we have, for the starting values $\{-0.7, -0.6\}$,

```
In[39] := FindRoot[eqnOdd[r] == 0, {r, oddvalues[[1]]}]

Out[39] = {r -> -0.61728}


In[40] := energy[2] = V0 r /. %

Out[40] = -0.61728 V0


In[41] := u[2][x_ /; x >= 0 && x <= 4] = u[x] /. sol[[1]];
          u[2][x_ /; x < 0 && x >= -4] := -u[2][-x]
```

and, for the starting values $\{-0.2, -0.1\}$,

```
In[43] := FindRoot[eqnEven[r] == 0, {r, evenvalues[[2]]}]

Out[43] = {r -> -0.192105}


In[44] := energy[3] = V0 r /. %

Out[44] = -0.192105 V0


In[45] := u[3][x_ /; x >= 0 && x <= 4] = u[x] /. sol[[1]];
          u[3][x_ /; x < 0 && x >= -4] := u[3][-x]
```

The energy eigenvalues agree to at least four significant figures with those determined analytically in Section 6.4.3.1. We can produce the diagram, drawn to scale, of the potential and the energy eigenvalues:

```
In[47] := V[x_ /; Abs[x] > 1] := 0
          V[x_ /; Abs[x] <= 1] := -1
          Plot[V[x], {x, -2, 2},
               PlotStyle -> Thickness[0.0125],
               Axes -> False,
```

```
               Ticks -> {None, Automatic},
               DefaultFont -> {"Courier", 9},
               Epilog ->
                ({Line[{{1, energy[1]}, {-1, energy[1]}}],
                  Text["E1", {1.25, energy[1]}],
                  Text[ToString[energy[1]], {0, energy[1] + 0.055}],
                  Line[{{1, energy[2]}, {-1, energy[2]}}],
                  Text["E2", {1.25, energy[2]}],
                  Text[ToString[energy[2]], {0, energy[2] + 0.055}],
                  Line[{{1, energy[3]}, {-1, energy[3]}}],
                  Text[ToString[energy[3]], {0, energy[3] + 0.055}],
                  Text["E3", {1.25, energy[3]}],
                  Text["-V0", {-1.25, -1}],
                  Text["0", {-0.85, 0}]} /. V0 -> 1)
            ];
```

The diagram is not shown because it is essentially the same as that in Section 6.4.3.1.
   We can plot the normalized eigenfunctions:

```
In[50] := plotFunc[u_] :=
            Module[
               {normalizationConstant, normalizedu},
               normalizationConstant = 1/Sqrt[
                            2 NIntegrate[(u[x])^2, {x, 0, 4}]
                                              ];
               normalizedu[x_] := normalizationConstant u[x];
               Plot[normalizedu[x], {x, -3, 3},
                     Epilog -> {Line[{{1, -1}, {1, 1}}],
                                   Line[{{-1, -1}, {-1, 1}}]
                                   },
                     Frame -> True,
                     FrameLabel -> {"x (a)", "u(x) (1/Sqrt[a])"},
                     PlotLabel -> "E = "<>ToString[energy[n]]
                     ];
               ]
In[51] := Table[plotFunc[u[n]], {n, 1, 3}];
```

We have also omitted the graphs of the eigenfunctions because they appear identical to those
from the analytical solution in Section 6.4.3.1.
   Our numerical method, a form of what is known as the shooting method, can be adapted
for other one-dimensional potentials, and it can still be used when the time-independent
Schrödinger equation cannot be solved by even the most powerful analytical tools. However, a
few words of caution are in order. With this numerical method, it is quite possible to miss some
energy eigenvalues, but this pitfall can be avoided. A general property of one-dimensional
bound states can serve as a safeguard. If the bound states are arranged in the order of
increasing energies $E1, E2, \ldots, En, \ldots$, the nth eigenfunction has $n - 1$ nodes (see [Mes62]).

Thus if, for example, the eigenfunctions corresponding to two consecutive energies in our spectrum have, respectively, n and $n + 2$ nodes, we must have missed an energy eigenvalue. In determining the appropriate starting values $r0$ and $r1$ with `initialValues[ei, ef, n, eqn]`, consider making n bigger or reducing $ef - ei$ to scan regions that are likely to have roots. (For a discussion of the shooting method, see [Pat94].)

```
In[52] := ClearAll[V, eqnEven, eqnOdd, f, F, energy, lambda, u,
                   plotFunc, initialValues, evenvalues, oddvalues]
```

6.5 *Angular Momentum*

6.5.1 *The Problem*

The Hamiltonian of a force-free rigid rotator is

$$H = \frac{1}{2I_1}L_x^2 + \frac{1}{2I_2}L_y^2 + \frac{1}{2I_3}L_z^2 \tag{6.5.1}$$

where I_1, I_2, and I_3 are the principal moments of inertia and L_x, L_y, and L_z are the components of the total angular momentum operator along the principal axes. Find the eigenvalues of H when the total angular momentum quantum number equals one, two, and three.

6.5.2 *Physics of the Problem*

6.5.2.1 ANGULAR MOMENTUM IN QUANTUM MECHANICS

In classical mechanics, angular momentum of a particle is a vector defined in the coordinate space in terms of the position and momentum vectors of the particle. Corresponding to this classical dynamical variable is a quantum mechanical observable, called the orbital angular momentum. It is a vector operator obtained according to the general correspondence rule. The components of the orbital angular momentum, as we will show in the next section, obey a set of commutation relations. A particle can also have an intrinsic angular momentum or spin, which has no classical analog and cannot be defined in the coordinate space. The commutation relations for the orbital angular momentum are so fundamental that spin is defined in terms of them together with a supplementary condition on the spectrum of the square of the spin. These commutation relations provide the general definition of all angular momenta even including those of a system of particles.

In general, angular momentum is a vector operator the components of which are observables satisfying the commutation relations

$$[J_x, J_y] = i\hbar J_z \tag{6.5.2a}$$

$$[J_y, J_z] = i\hbar J_x \tag{6.5.2b}$$

$$[J_z, J_x] = i\hbar J_y \tag{6.5.2c}$$

An observable is a Hermitian operator possessing a complete, orthonormal set of eigenvectors.

Not all Hermitian operators possess such a set of eigenvectors; those capable of representing physical quantities do.

It follows from Equation 6.5.2 that

$$[\mathbf{J}^2, J_z] = 0 \tag{6.5.3}$$

where the square of the angular momentum is

$$\mathbf{J}^2 = J_x^2 + J_y^2 + J_z^2 \tag{6.5.4}$$

Since $\mathbf{J}^2$ and J_z commute with each other they possess a common set of eigenvectors. The eigenvalue equations can be written as

$$\mathbf{J}^2|jm> = \hbar^2 j(j+1)|jm> \tag{6.5.5}$$

$$J_z|jm> = \hbar m|jm> \tag{6.5.6}$$

From the general definition of angular momentum, we can show the following properties for the spectrum of $\mathbf{J}^2$ and J_z, and for the matrix elements of J_+, J_-, and J_z:

(A)   The only possible values of j are

$$j = 0, \frac{1}{2}, 1, \frac{3}{2}, 2, \frac{5}{2}, \ldots, \infty \tag{6.5.7}$$

(B)   For a given j, the only possible values of m are

$$m = -j, -j+1, \ldots, +j \tag{6.5.8}$$

(C)   The matrix elements of J_+, J_-, and J_z are

$$<jm|J_z|j'm'> = m\hbar\,\delta_{jj'}\,\delta_{mm'} \tag{6.5.9}$$

$$<jm|J_\pm|j'm'> = \hbar\,\sqrt{j(j+1) - mm'}\,\delta_{jj'}\delta_{mm'\pm1} \tag{6.5.10}$$

where we have introduced two Hermitian conjugate operators

$$J_\pm = J_x \pm iJ_y \tag{6.5.11}$$

(For an introduction to angular momentum in quantum mechanics, see [Lib92]; for an advanced treatment, see [Mes62].)

6.5.2.2   ORBITAL ANGULAR MOMENTUM

In classical mechanics, the angular momentum **l** of a particle about a point O is defined as

$$\mathbf{l} = \mathbf{r} \times \mathbf{p} \tag{6.5.12}$$

where **r** and **p** are respectively the position vector from O and the linear momentum of the particle. Corresponding to this classical dynamical variable is a quantum mechanical observable, called the orbital angular momentum l. This vector operator is obtained by replacing the vectors **r** and **p** with their corresponding observables r and p in Equation 6.5.12. The orbital angular momentum l is given by

$$\mathbf{l} = \mathbf{r} \times \mathbf{p} \tag{6.5.13}$$

The components of **r** and **p** satisfy the commutation relations

$$[r_i, r_j] = 0, \qquad [p_i, p_j] = 0 \tag{6.5.14a}$$

$$[r_i, p_j] = i\hbar\delta_{ij} \tag{6.5.14b}$$

where i and j can be x, y, or z. From Equations 6.5.13 and 6.5.14, we can show

$$[l_x, l_y] = i\hbar l_z \tag{6.5.15a}$$

$$[l_y, l_z] = i\hbar l_x \tag{6.5.15b}$$

$$[l_z, l_x] = i\hbar l_y \tag{6.5.15c}$$

In wave mechanics, Equation 6.5.13 takes the form

$$\mathbf{l} = -i\hbar \mathbf{r} \times \nabla \tag{6.5.16}$$

Consider a system of N particles. The orbital angular momentum of the nth particle about a point O is

$$\mathbf{l}^{(n)} = \mathbf{r}^{(n)} \times \mathbf{p}^{(n)} \tag{6.5.17}$$

The total orbital angular momentum **L** is defined as

$$\mathbf{L} = \sum_{n=1}^{N} \mathbf{l}^{(n)} \tag{6.5.18}$$

From Equation 6.5.15 for each $\mathbf{l}^{(n)}$, Equations 6.5.17 and 6.5.18, and the fact that the components of the orbital angular momenta for different particles commute with each other, we can show

$$[L_x, L_y] = i\hbar L_z \tag{6.5.19a}$$

$$[L_y, L_z] = i\hbar L_x \tag{6.5.19b}$$

$$[L_z, L_x] = i\hbar L_y \tag{6.5.19c}$$

Thus, we have verified that the total orbital angular momentum **L** is an angular momentum according to the general definition in Section 6.5.2.1. Therefore, **L** assumes all the properties of **J**. There is, however, one important difference. The fact that **L** is defined in the coordinate space puts a further restriction on the spectrum of $\mathbf{L}^2$ and L_z. The eigenvalue equations can be written as

$$\mathbf{L}^2|lm> = \hbar^2 l(l+1)|lm> \tag{6.5.20}$$

$$L_z|lm> = \hbar m|lm> \tag{6.5.21}$$

Whereas j in Equation 6.5.7 can take on integral and half-odd integral values, l must be an integer. To show l is an integer, we first show that the quantum number l_i for each particle must be an integer. This can be done by considering the eigenvalue equation of l_z for each particle in wave mechanics where the orbital angular momentum of a particle is defined in Equation 6.5.16. The single-valuedness of the wave function in coordinate space requires m_i and, consequently, l_i to be integers. (For discussions of how the single-valuedness of the wave function leads to integral values for m_i, see [Lib92], [Mes62], and [Tow92]; for dissenting opinions, see [Gas95] and [Oha90].) Since l_i are integers for all particles, the theorem for addition of angular momenta requires l must be an integer. (For an introduction to the addition of angular momenta, see Sections 9.4 and 9.5 of [Lib92].) Properties (A), (B), and (C) in Section 6.5.2.1 now, for **L**, become

(A')   The only possible values of l are

$$l = 0, 1, 2, \ldots, \infty \tag{6.5.22}$$

(B')   For a given l, the only possible values of m are

$$m = -l, -l + 1, \ldots, +l \tag{6.5.23}$$

(C')   The matrix elements of L_+, L_- and L_z are

$$<lm|L_z|l'm'> = m\hbar\, \delta_{ll'}\, \delta_{mm'} \tag{6.5.24}$$

$$<lm|L_z|l'm'> = \hbar\, \sqrt{l(l+1) - mm'}\, \delta_{ll'}\, \delta_{mm'\pm 1} \tag{6.5.25}$$

where

$$L_\pm = L_x \pm iL_y \tag{6.5.26}$$

Equation 6.5.26 implies

$$L_x = \frac{1}{2}(L_+ + L_-) \tag{6.5.27}$$

$$L_y = \frac{1}{2i}(L_+ - L_-) \tag{6.5.28}$$

6.5.2.3   THE EIGENVALUE PROBLEM

If $I_1 = I_2 = I_3$, the problem of finding the eigenvalues of H in Equation 6.5.1 is trivial because, in this case,

$$H = \frac{\mathbf{L}^2}{2I} \tag{6.5.29}$$

The eigenvalues are simply

$$\frac{\hbar^2 l(l+1)}{2I} \tag{6.5.30}$$

where the values of l are given in Equation 6.5.22.

If $I_1 = I_2 \neq I_3$, the Hamiltonian becomes

$$H = \frac{\mathbf{L}^2}{2I_1} + \left(\frac{1}{2I_3} - \frac{1}{2I_1} \right) L_z^2 \tag{6.5.31}$$

The eigenvalues are

$$\frac{\hbar^2 l(l+1)}{2I_1} + \left(\frac{1}{2I_3} - \frac{1}{2I_1} \right) \hbar^2 m^2 \tag{6.5.32}$$

The values for l and m are given in Equations 6.5.22 and 6.5.23.

If $I_1 \neq I_2 \neq I_3$, we diagonalize the matrix of H in the $\{|lm>\}$ representation to find the eigenvalues. Let

$$a = \frac{1}{2I_1} \qquad b = \frac{1}{2I_2} \qquad c = \frac{1}{2I_3} \tag{6.5.33}$$

The Hamiltonian can be written as

$$H = aL_x^2 + bL_y^2 + cL_z^2 \tag{6.5.34}$$

The matrix of H can be determined with Equations 6.5.22, 6.5.23, 6.5.24, 6.5.25, 6.5.27, 6.5.28, and 6.5.34. Since there are an infinite number of eigenvectors $|lm>$, we are faced with a horrendous task of diagonalizing an infinite dimensional matrix. Fortunately, from Equations 6.5.24, 6.5.25, 6.5.27, 6.5.28, and 6.5.34, we observe that $<lm|H|l'm'>$ vanishes unless $l = l'$. In other words, we can diagonalize the submatrix for each l one at a time. Since the order of the submatrix is $2l + 1$, it is still not a simple matter unless l is small.

6.5.3 Solution with Mathematica

```
In[1] := ClearAll["Global`*"]
```

For simplicity, set

$$\hbar = 1 \tag{6.5.35}$$

We begin by defining the Kronecker delta:

```
In[2] := kdel[n_, m_] := 1 /; n == m; kdel[n_, m_] := 0 /; n != m
```

Equation 6.5.24 gives us the matrix elements of L_z:

```
In[3] := elementLz[n_, m_] := m kdel[n, m]
```

Equation 6.5.25 gives us the matrix elements of L_+:

```
In[4] := elementLplus[l_, n_, m_] := (Sqrt[l(1 + 1) - n m] kdel[n, m + 1])
```

Since L_- is the Hermitian conjugate of L_+ and the matrix elements of L_+ are real, the matrix elements of L_- are

```
In[5] := elementLminus[l_, n_, m_] := elementLplus[l, m, n]
```

Equations 6.5.27 and 6.5.28 give us the matrix elements of L_x and L_y:

```
In[6] := elementLx[l_, n_, m_] := (elementLplus[l, n, m] +
                                    elementLminus[l, n, m])/2
         elementLy[l_, n_, m_] := (elementLplus[l, n, m] -
                                    elementLminus[l, n, m])/(2I)
```

For a given l, the matrices of L_x, L_y, L_z, and H are

```
In[9] := matrixLx[l_] := Table[elementLx[l, n, m],
                              {n, 1, -1, -1}, {m, 1, -1, -1}]
         matrixLy[l_] := Table[elementLy[l, n, m],
                              {n, 1, -1, -1}, {m, 1, -1, -1}]
```

```
matrixLz[l_] := Table[elementLz[n, m],
                    {n, 1, -1, -1}, {m, 1, -1, -1}]
matrixH[l_] := (a matrixLx[l].matrixLx[l] +
                b matrixLy[l].matrixLy[l] +
                c matrixLz[l].matrixLz[l])
```

The eigenvalues of H for a given l are given by

In[16] := **eigenvaluesH[l_] := Eigenvalues[matrixH[l]]**

We can now determine the eigenvalues of H when the total angular momentum quantum number equals one, two, and three:

In[17] := **one = eigenvaluesH[1]**

Out[17] = {a + b, a + c, b + c}

In[18] := **two = eigenvaluesH[2]**

Out[18] = {4 a + b + c, a + 4 b + c, a + b + 4 c,
           2 (a + b + c - Sqrt[a^2 - a b + b^2 - a c - b c + c^2]),
           2 (a + b + c + Sqrt[a^2 - a b + b^2 - a c - b c + c^2])}

In[19] := **three = eigenvaluesH[3]**

Out[19] = {4 (a + b + c), 5 a + 5 b + 2 c -
           2 Sqrt[4 a^2 - 7 a b + 4 b^2 - a c - b c + c^2],
           5 a + 5 b + 2 c + 2 Sqrt[4 a^2 - 7 a b + 4 b^2 - a c - b c + c^2],
           2 a + 5 b + 5 c - 2 Sqrt[a^2 - a b + 4 b^2 - a c - 7 b c + 4 c^2],
           2 a + 5 b + 5 c + 2 Sqrt[a^2 - a b + 4 b^2 - a c - 7 b c + 4 c^2],
           5 a + 2 b + 5 c - 2 Sqrt[4 a^2 - a b + b^2 - 7 a c - b c + 4 c^2],
           5 a + 2 b + 5 c + 2 Sqrt[4 a^2 - a b + b^2 - 7 a c - b c + 4 c^2]}

Let us diagram the eigenvalues of H (i.e., the energy eigenvalues) for the arbitrarily chosen parameters $a = 1, b = 2$, and $c = 5$:

```
(* Mathematica Version 2.2 Specifics *)

Show[
  Graphics[
    {{Line[{{1, #}, {2, #}}]& /@ one,
      Line[{{3, #}, {4, #}}]& /@ two,
      Line[{{5, #}, {6, #}}]& /@ three} /.
      {a -> 1, b -> 2, c -> 5},
    Text[FontForm["l = 1", {"Times", 9}], {1.5, 12}],
    Text[FontForm["l = 2", {"Times", 9}], {3.5, 28}],
    Text[FontForm["l = 3", {"Times", 9}], {5.5, 12}]}
      ],
```

```
      Frame -> True,
      FrameTicks -> {None, Automatic, None, None},
      FrameLabel -> {None, "Energy (h^2/(8 Pi^2 I1))"},
      PlotLabel -> "The Energy Spectrum"
         ];
```

In[20] := (* ⚜ 3.0 *)

```
          Show[
           Graphics[
             {{Line[{{1, #}, {2, #}}]& /@ one,
               Line[{{3, #}, {4, #}}]& /@ two,
               Line[{{5, #}, {6, #}}]& /@ three} /.
                     {a -> 1, b -> 2, c -> 5},
               Text[StyleForm["l = 1", ts], {1.5, 12}],
               Text[StyleForm["l = 2", ts], {3.5, 28}],
               Text[StyleForm["l = 3", ts], {5.5, 12}]}
                 ] /. ts ->
                     Sequence[FontFamily -> "Times", FontSize -> 9],
           Frame -> True,
           FrameTicks -> {None, Automatic, None, None},
           FrameLabel -> {None, "Energy (h^2/(8 Pi^2 I1))"},
           PlotLabel -> "The Energy Spectrum"
               ];
```

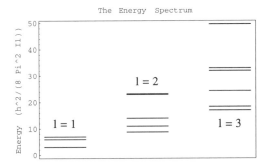

For $l = 2$, there are actually five energy levels; for $l = 3$, there are seven levels. In both cases, the lines for the two highest levels are too close together to be resolved.

With the function `eigenvaluesH`, we can also determine the eigenvalues of H for larger values of l. However, diagonalizing a $(2l+1) \times (2l+1)$ symbolic matrix can be time consuming unless l is small. For example, evaluating `eigenvaluesH[7]` involves the diagonalization of a 15×15 symbolic matrix and already takes 81 minutes on a Macintosh IIci and 11 minutes on a Gateway 2000 4DX2-66V. Of course, we would not even contemplate diagonalizing such a matrix with paper and pencil. (For possible *Mathematica* errors, see Problems 9 and 10 in Section 6.7.)

If we only wish to diagonalize numerically the Hamiltonian matrix with specified values for the principal moments of inertia I_1, I_2, and I_3 (i.e., a, b, and c), we can define the functions

```
In[21] := numericalmatrixH[l_, a_, b_, c_] :=
                            (a matrixLx[l].matrixLx[l] +
                             b matrixLy[l].matrixLy[l] +
                             c matrixLz[l].matrixLz[l])
             numericaleigenvaluesH[l_, a_, b_, c_] :=
                 Eigenvalues[numericalmatrixH[l, a, b, c]]
```

for which the arguments *must* be entered as approximate real numbers. As an example, let us determine the energy eigenvalues for $l = 12, a = 1, b = 5$, and $c = 2$:

```
In[23] := numericaleigenvaluesH[12., 1., 5., 2.]//Chop//Timing
          {140.65 Second, {738.45, 738.45, 658.862, 658.862, 586.315,
             586.315, 520.852, 520.852, 462.555, 462.554, 411.584, 411.555,
             368.453, 367.989, 335.258, 331.241, 313.359, 298.202, 292.985,
             263.765, 263.146, 224.49, 224.458, 179.724, 179.723}}
```

The evaluation takes only 141 seconds on a Macintosh IIci and 25 seconds on a Gateway 2000 4DX2-66V.

Molecules and nuclei often have rotational energy levels characteristic of a rigid rotator. For more information, see [McM94] and [Tho94].

```
In[24] := ClearAll[kdel, elementLz, elementLplus, elementLminus,
                   elementLx, elementLy, matrixLx, matrixLy,
                   matrixLz, matrixH, eigenvaluesH, one, two, three,
                   numericalmatrixH, numericaleigenvaluesH]
```

6.6   *The Kronig-Penney Model*

6.6.1   *The Problem*

Consider the Kronig-Penney model. Plot the energy of the electron as a function of the wave vector.

6.6.2   *Physics of the Problem*

Consider an electron in a one-dimensional crystal. The time-independent Schrödinger equation is

$$\left(-\frac{\hbar^2}{2m}\frac{d^2}{dx^2} + V(x)\right)\psi(x) = E\psi(x) \tag{6.6.1}$$

where the potential $V(x)$ has the periodicity of the crystal lattice, so that

$$V(x + na) = V(x) \tag{6.6.2}$$

for all integers n.

There is an important theorem, called Bloch's theorem, in solid state physics. For an electron in a one-dimensional periodic potential, it states that the solutions of the time-independent Schrödinger equation can be chosen so that associated with each ψ is a wave vector q such that

$$\psi(x + na) = e^{iqna}\psi(x) \tag{6.6.3}$$

for all integers n.

In the Kronig-Penney model, the potential is taken as a series of repulsive delta-function potentials:

$$V(x) = \frac{\hbar^2 \lambda}{2ma} \sum_{n=-\infty}^{\infty} \delta(x - na) \tag{6.6.4}$$

Away from the points $x = na$, the potential V is zero, and Equation 6.6.1 becomes a free-particle equation. In the region $(n - 1)a \le x \le na$, the solution can be written as

$$\psi(x) = A_n \sin k(x - na) + B_n \cos k(x - na) \tag{6.6.5}$$

and in the region $na \le x \le (n + 1)a$, it can take the form

$$\psi(x) = A_{n+1} \sin k[x - (n + 1)a] + B_{n+1} \cos k[x - (n + 1)a] \tag{6.6.6}$$

We have for the energy

$$E = \frac{\hbar^2 k^2}{2m} \tag{6.6.7}$$

It can be shown with the time-independent Schrödinger equation that if the potential contains a term like $\kappa\delta(x - na)$, the derivative of ψ is not continuous at $x = na$. Rather, it obeys the relation

$$\left[\frac{d\psi}{dx}\right]_{na+\varepsilon} - \left[\frac{d\psi}{dx}\right]_{na-\varepsilon} = \frac{2m}{\hbar^2}\kappa\psi(na) \tag{6.6.8}$$

with ε arbitrarily small and positive.

The wave functions in Equations 6.6.5 and 6.6.6 must satisfy the following conditions: the continuity of the wave function at $x = na$, Equation 6.6.8 for the derivative of the wave function, and Equation 6.6.3 with $n = 1$. Applying these conditions to the wave functions, we have, after some algebra, the eigenvalue condition

$$\cos qa = \cos ka + \frac{1}{2}\lambda \frac{\sin ka}{ka} \tag{6.6.9}$$

The energy E can now be determined as a function of the wave vector q. Given a wave vector q, we can use Equation 6.6.9 to find k, and Equation 6.6.7 gives us E. (For details of the Kronig-Penney model, see [Gas95], [Lib92], and [Par92].)

6.6.3   Solution with Mathematica

In[1] := `ClearAll["Global`*"]`

We would like to plot the energy E as a function of the wave vector q. To do so, we must solve Equation 6.6.9. However, this equation cannot be solved easily because, for every q, there are an infinite number of solutions for k. Fortunately, the programming capability of *Mathematica* can circumvent this difficulty. In what follows, we define a function for plotting the energy as a function of the wave vector. The function `kpmPlot` takes three arguments: the first is for the dimensionless constant λ that characterizes the strength of the potential in Equation 6.6.4, the second is for the number of energy bands to be plotted, and the third is for the number of points to be used for plotting the right branch of each energy band. Comments elucidating the construction of the function are included in its body in the form (* *comment* *).

In[2] :=
```
kpmPlot[lambda_?Positive, band_Integer?Positive,
         mf_Integer /; mf > 10] :=
  Module[{k, q, energy, r},

    (* value k of the right branches for each q
       [Eq.6.6.9] *)
    k[n_, 1] = n Pi;
    q[n_, m_] := n Pi - (m - 1) Pi/(mf - 1);
    k[n_, m_] := k[n, m] =
       FindRoot[
          Cos[q[n, m]] == Cos[x] + (lambda/2)(Sin[x]/x),
          {x, k[n, m - 1]}
             ][[1, 2]];

    (* value of the energy of the right branches for each k
       [Eq. 6.6.7] *)
    energy[n_, m_] := k[n, m]^2;

    (* coordinates of the right branches *)
    r[n_] := r[n] =
       Table[{q[n, m], energy[n, m]}, {m, 1, mf}];

    (* graphics *)
    Show[
       Graphics[
          Flatten[
             Table[{Line[r[n]],
                    Line[{-#[[1]], #[[2]]}& /@ r[n]]},
                   {n, 1, band}]
                ]
             ],
       Axes -> True,
       AxesLabel -> {" q (1/a)", "E [h^2/(8 Pi^2 m a^2)]"}
        ];
        ]
```

As an example, let us call the function with $\lambda = 3\pi$, *band* = 3, and *mf* = 25:

In[4] := **kpmPlot[3 Pi, 3, 25]**

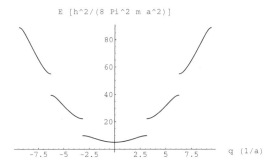

The graph shows an energy band structure. There are allowed energy bands separated by energy gaps, regions that are forbidden. The existence of energy gaps has important consequences in the transport properties of electrons. (For an in-depth discussion of energy bands, see, for example, [AM76]. For an earlier version of this notebook, see [Tam91].)

6.7 Problems

1. Consider a particle in a one-dimensional box with ends at $x = 0$ and $x = 2$. Suppose the probability density is given by

$$p(x) = \begin{cases} \dfrac{15}{16}(x^2 - \dfrac{1}{4} x^4) & 0 \le x \le 2 \\ \\ 0 & \text{elsewhere} \end{cases}$$

   (a) Plot the probability density $p(x)$.
   (b) Show that $p(x)$ is correctly normalized.
   (c) Calculate the expectation value of x.
   (d) Determine the root-mean-square deviation of x from the mean.

2. Consider a nonstationary state that is a superposition of the first two eigenstates of a one-dimensional harmonic oscillator:

$$\psi(x, t) = \frac{1}{\sqrt{2}}[\exp(-iE_0 t / \hbar)u_0(x) + \exp(-iE_1 t / \hbar)u_1(x)]$$

   where u_n and E_n are the normalized energy eigenfunctions and energy eigenvalues, respectively. Animate the time evolution of the probability density

$$P(x, t) = |\psi(x, t)|^2$$

   Indicate on the graphs the classically forbidden regions.

3. Consider a one-dimensional harmonic oscillator with mass m and spring constant k. Let the initial wave function $\psi(x, 0)$ be a Gaussian wave packet:

$$\psi(x, 0) = \left(\frac{1}{2\pi\sigma^2}\right)^{1/4} \exp\left[-\frac{(x - x_0)^2}{4\sigma^2}\right]$$

It can be shown (see [BD85]) that the probability density at time t is

$$P(x, t) = \frac{1}{\sqrt{2\pi}} \frac{2\sigma}{\sqrt{\sigma_0^4 s^2 + 4\sigma^4 c^2}} \exp\left[-\frac{2\sigma^2}{\sigma_0^4 s^2 + 4\sigma^4 c^2}(x - cx_0)^2\right]$$

where c and s are $\cos\omega t$ and $\sin\omega t$, respectively. Also,

$$\omega = \sqrt{k/m}$$

$$\sigma_0 = \sqrt{\hbar/m\omega}$$

Let $x_0 = 3\sigma_0$. On a three-dimensional graph and with $\sigma = \sigma_0/(2\sqrt{2})$, plot the probability density for times $t_n = (2\pi/\omega)(n/16)$, where $n = 0, 1, 2, \ldots, 31$. Repeat with $\sigma = \sigma_0/\sqrt{2}$ and $\sigma = 3\sigma_0/(2\sqrt{2})$.

4. A deuteron has spin 1. Consider two deuterons, with spin operators $\mathbf{S}_1$ and $\mathbf{S}_2$. Denote the one-particle spin eigenstates by $\xi_{+1}^{(i)}$, $\xi_0^{(i)}$, and $\xi_{-1}^{(i)}$, so that

$$\mathbf{S}_i^2 \xi_{m_s}^{(i)} = 1(1 + 1)\hbar^2 \xi_{m_s}^{(i)}$$

$$\mathbf{S}_{iz} \xi_{m_s}^{(i)} = m_s \hbar \xi_{m_s}^{(i)}$$

where $m_s = +1, 0, -1$ and $i = 1, 2$. The total spin $\mathbf{S}$ is defined as

$$\mathbf{S} = \mathbf{S}_1 + \mathbf{S}_2$$

If the orbital angular momentum of the two-deuteron system in their center-of-mass system is $\mathbf{L}$, the total angular momentum is defined as

$$\mathbf{J} = \mathbf{L} + \mathbf{S}$$

(a) Determine the eigenvalues and eigenstates of $\mathbf{S}^2$ and S_z. Express the eigenstates in terms of the single-particle spin eigenstates $\xi_{m_s}^{(i)}$. (b) Taking cognizance of the symmetrization requirement on the total state under a two-particle exchange, list the possible total angular momentum states in the spectroscopic notation

$$^{2S+1}L_J$$

for $L = 0, 1, 2$, and 3. (c) Express the $^5D_4$ states in terms of the eigenstates of $\mathbf{L}^2$ and L_z together with those of $\mathbf{S}^2$ and S_z.

5. Determine analytically and numerically the energy eigenvalue and the associated eigenfunction for the ground state of a particle of mass m moving in the potential

$$V(x) = \begin{cases} \infty & |x| > a \\ 0 & a/2 < |x| < a \\ V_0 & |x| < a/2 \end{cases}$$

with $V_0 > 0$. Let

$$V_0 = \frac{\pi^2 \hbar^2}{32ma^2}$$

6. Find the energy eigenvalues and the corresponding eigenfunctions for the ground state and the first two excited states of an anharmonic oscillator with potential energy of the form

$$V(x) = \frac{A}{2}x^2 + \frac{B}{2}x^4$$

First, convert the time-independent Schrödinger equation to the dimensionless form

$$\frac{d^2\psi}{du^2} = -(\epsilon - u^2 - \delta u^4)\psi$$

Then, consider the case $\delta = 0.25$.

7. In Section 6.3.3, the arguments L and k_0 of the function **particle** are restricted. Why? Explain.

8. In Section 6.4.3.2, we take $x_f = 4$. Determine the energy eigenvalues for the choices $x_f = 2, 3, 5, 6, 7$, and 8. What can be concluded from the results?

9. Consider the function **eigenvaluesH** in Section 6.5.3. (a) Evaluate **eigenvaluesH[4]**. What happens? (b) Now make the assignment **b = r a** where **r** is an unassigned variable. Evaluate **eigenvaluesH[4]** again and name the result **test**. What happens this time? (c) Make a new assignment **b = a + d** where **d** is also an unassigned variable. Evaluate **eigenvaluesH[4]** and name the result **newtest**. What happens here? (d) Evaluate **Im[test /. {a -> 1, r -> 3, c -> 4}]** and **N[test /. {a -> 1, r -> 3, c -> 4}]**. (e) Evaluate **Im[newtest /. {a -> 1, d -> 2, c -> 4}]** and **N[newtest /. {a -> 1, d -> 2, c -> 4}]**. (f) Now make the assignments **a = 1, b = 3**, and **c = 4**. Then evaluate **eigenvaluesH[4]** and name the result **moretest**. Evaluate **Im[moretest]** and **N[moretest]**. Compare the results with those in (d) and (e). (g) What can be concluded about *Mathematica* from the results in (a), (b), (c), (d), (e), and (f)?

10. Consider the function **eigenvaluesH** in Section 6.5.3. (a) Evaluate **eigenvaluesH[6]** and name the result **mytest**. (b) Evaluate **N[mytest /. {a -> 1, b -> 7, c -> 4}]**.

(c) Make the assignments `a = 1`, `b = 7`, and `c = 4`. Then evaluate `eigenvaluesH[6]` and name the result `ourtest`. (d) Evaluate `Im[ourtest]`. Any surprises? (e) Evaluate `N[ourtest]` and compare the result with that of (b). From the comparison, what can be concluded about *Mathematica*?

*11. In Section 6.6.3, the function `kpmPlot` plots the energy bands in the extended zone scheme. Modify the function to allow the inclusion of an option that specifies whether the energy bands are to be drawn in the extended, reduced, or periodic zone scheme. (For an introduction to energy bands, see [Kit86].)

The Last Ten Minutes

This book and the accompanying disk have been produced with prereleased versions of *Mathematica* 3.0. We can expect variations among the prereleased and official versions. For example, in converting the notebooks on the accompanying disk, the official versions undoubtedly will not preserve the vertical alignment of the lines in multiline inputs. For updates and corrections, check out the author's Web page at APNet (http://www.apnet.com/).

Here are last-minute amendments to two sections:

Section 1.1

Windows 95 versions of *Mathematica* do not have memory monitors, because *Mathematica* developers find the memory management of the operating system so efficient that monitors are unnecessary.

Section 2.5.4

After an in-place evaluation of a subexpression, % refers only to the result of the subexpression evaluation rather than the content of the entire cell. For example, type

$$\sqrt{2\,a + b - b\,Cos[\theta]^2 + b\,Sin[\theta]^2}$$

Highlight b - b Cos[θ]², choose AlgebraicManipulation in the Palettes submenu of the File menu, and click Simplify[■]. *Mathematica* returns in place

$$\sqrt{2\,a + b\,Sin[\theta]^2 + b\,Sin[\theta]^2}$$

Now, % refers to the result of the previous in-place evaluation:

```
In[2] := %

Out[2] = b Sin[θ]²
```

451

We have strived for perfection in the production of this book. Nevertheless, errors are always possible. If you have difficulties with any inputs or questions about any outputs, examine as well as execute, with *Mathematica*, the relevant code on the accompanying disk (which has been tested on Macintosh and Windows computers), and then check the result. Error reports will be appreciated. Please donate them to PTT1@axe.humboldt.edu.

Operator Input Forms

This appendix presents a table of some common operator input forms together with the corresponding full forms and examples. Those appropriate only for *Mathematica* version 3.0 are marked with the icon ⚹ **3.0**. To obtain more information on these input forms, invoke on-line help with the **?** operator, as described at the end of Section 1.3. For example, to obtain information on " << *filename*", enter "**?<<**".

Operator Input Form	Full Form Example
expr::*string*	`MessageName`[*expr*, "*string*"] `Sqrt::argx`
<<*filename*	`Get`["*filename*"] `<<Calculus`VectorAnalysis``
expr1?*expr2*	`PatternTest`[*expr1*, *expr2*] `n_Integer?NonNegative`
expr1[*expr2*, ...]	*expr1*[*expr2*, ...] `Exp[I k x]`
expr1[[*expr2*, ...]]	`Part`[*expr1*, *expr2*, ...] `mylist[[{3, 5}]]`
expr1⟦*expr2*, ...⟧	`Part`[*expr1*, *expr2*, ...] ⚹ **3.0** `ourlist⟦1, 2⟧`

Operator Input Form	Full Form Example
expr++	Increment[*expr*] i = 1; While[i <= 3, (Print[i^2]; i++)]
expr--	Decrement[*expr*] For[i = 7; t = x, i^2 > 10, i--, t = t^2 + i; Print[Expand[t]]]
++*expr*	PreIncrement[*expr*] For[i = 1, i <= 3, ++i, Print[i^2]]
--*expr*	PreDecrement[*expr*] --x
expr1 @ *expr2*	*expr1*[*expr2*] hamiltonian[V[x]] @ psi[x]
expr1 ~ *expr2* ~ *expr3*	*expr2*[*expr1*, *expr3*] {a, b} ~ Join ~ {c, d, e, f}

expr1 /@ *expr2*	Map[*expr1*, *expr2*]
	Line /@ {{{-13, 0}, {17, 0}},
	     {{0, -4}, {0, 4}},
	     {{-10, 0}, {-10, 2}}}

expr1 //@ *expr2*	MapAll[*expr1*, *expr2*] f //@ {{a, b}, {c, d}}
expr1 @@ *expr2*	Apply[*expr1*, *expr2*] Plus @@ {a, b, c, d}
expr!	Factorial[*expr*] n!
expr'	Derivative[1][*expr*] x'[t]
expr1 <> *expr2* <> *expr3*	StringJoin[*expr1*, *expr2*, *expr3*] "Happy" <> " New " <> "Year!"
expr1^*expr2*	Power[*expr1*, *expr2*] x^n
$expr1^{expr2}$	Power[*expr1*, *expr2*] �֎ **3.0** $e^{2\pi i}$
$\sqrt{expr}$	Sqrt[*expr*] ✷ **3.0** $\sqrt{a^2 + b^2}$

Operator Input Form	Full Form Example
$\int expr1\,\mathrm{d}expr2$	`Integrate[`*expr1*`, `*expr2*`]` ✿ **3.0** $\int \frac{3}{1+x^2}\,\mathrm{d}x$
$\int_{e1}^{e2} e3\,\mathrm{d}e4$	`Integrate[`*e3*`, {`*e4*`, `*e1*`, `*e2*`}]` ✿ **3.0** $\int_{1}^{4}\sqrt{1+x^2}\,\mathrm{d}x$
$\partial_{expr1}\,expr2$	`D[`*expr2*`, `*expr1*`]` ✿ **3.0** $\partial_x\,\mathtt{Cos[x]}$
expr1 `**` *expr2* `**` *expr3*	`NonCommutativeMultiply[`*expr1*`, `*expr2*`, `*expr3*`]` `p ** x ** x`
expr1 × *expr2* × *expr3*	`Cross[`*expr1*`, `*expr2*`, `*expr3*`]` ✿ **3.0** `{a1, b1, c1} × ({a2, b2, c2} × {a3, b3, c3})`
expr1 `.` *expr2* `.` *expr3*	`Dot[`*expr1*`, `*expr2*`, `*expr3*`]` `{x, y}.{{a, b}, {c, d}}.{x, y}`
-expr	`Times[-1, `*expr*`]` `-(x + y)^2`
+expr	*expr* `+3`
expr1/*expr2*	`Times[`*expr1*`, Power[`*expr2*`, -1]]` `x/y`
expr1 ÷ *expr2*	`Times[`*expr1*`, Power[`*expr2*`, -1]]` ✿ **3.0** $(a^2 + b^2) \div (a + b)$
expr1 *expr2* *expr3*	`Times[`*expr1*`, `*expr2*`, `*expr3*`]` `x y z`
expr1 `*` *expr2* `*` *expr3*	`Times[`*expr1*`, `*expr2*`, `*expr3*`]` `a * b * c`
expr1 × *expr2* × *expr3*	`Times[`*expr1*`, `*expr2*`, `*expr3*`]` ✿ **3.0** $2 \times \sigma \times \hbar$
expr1 + *expr2* + *expr3*	`Plus[`*expr1*`, `*expr2*`, `*expr3*`]` `1 + 2x + y`
expr1 - *expr2*	`Plus[`*expr1*`, Times[-1, `*expr2*`]]` `a^2 - b^2`
expr1 ∩ *expr2*	`Intersection[`*expr1*`, `*expr2*`]` ✿ **3.0** `{1, 2, 3} ∩ {2, 3, 4, 5}`

Operator Input Form	Full Form Example
$expr1 \cup expr2$	`Union[`*expr1*`, `*expr2*`]` ✤ **3.0** `{1, 2, 3} ∪ {2, 3, 4, 5}`
$expr1 == expr2$	`Equal[`*expr1*`, `*expr2*`]` `x^2 + 1 == 0`
$expr1 \mathrel{!}= expr2$	`Unequal[`*expr1*`, `*expr2*`]` `f[n_Integer /; n != 1] := 1/(n - 1)^2`
$expr1 \neq expr2$	`Unequal[`*expr1*`, `*expr2*`]` `f[n_Integer /; n ≠ 1] := ` $\frac{1}{(n-1)^2}$ ✤ **3.0**
$expr1 > expr2$	`Greater[`*expr1*`, `*expr2*`]` `Cases[%, x_Integer/;(x > 10 && x < 80)]`
$expr1 >= expr2$	`GreaterEqual[`*expr1*`, `*expr2*`]` `v[x_ /; Abs[x] >= 1] := 1`
$expr1 \geq expr2$	`GreaterEqual[`*expr1*`, `*expr2*`]` ✤ **3.0** `v[x_ /; Abs[x] ≥ 1] := 1`
$expr1 \geqslant expr2$	`GreaterEqual[`*expr1*`, `*expr2*`]` ✤ **3.0** `△x △p ⩾ ℏ/2`
$expr1 < expr2$	`Less[`*expr1*`, `*expr2*`]` `x < y`
$expr1 <= expr2$	`LessEqual[`*expr1*`, `*expr2*`]` `z <= 3`
$expr1 \leq expr2$	`LessEqual[`*expr1*`, `*expr2*`]` ✤ **3.0** $\alpha \leq \beta$
$expr1 \leqslant expr2$	`LessEqual[`*expr1*`, `*expr2*`]` ✤ **3.0** $\delta \leqslant 10^{-10}$
$expr1 === expr2$	`SameQ[`*expr1*`, `*expr2*`]` `Head[a + b] === Plus`
$expr1 =!= expr2$	`UnsameQ[`*expr1*`, `*expr2*`]` `While[1 + 1/x =!= x, x = 1 + 1/x]`
$!expr$	`Not[`*expr*`]` `!(NumberQ[x0] && Im[x0] == 0)`
$\neg expr$	`Not[`*expr*`]` ✤ **3.0** `¬(IntegerQ[η] && Positive[η])`

Operator Input Form	Full Form Example
expr1 **&&** *expr2* **&&** *expr3*	`And[`*expr1*`, `*expr2*`, `*expr3*`]` `x >= -1 && x < 1`
expr1 ∧ *expr2* ∧ *expr3*	`And[`*expr1*`, `*expr2*`, `*expr3*`]` ✤ **3.0** `θ ≥ 0 ∧ θ ≤ π`
expr1 **||** *expr2* **||** *expr3*	`Or[`*expr1*`, `*expr2*`, `*expr3*`]` `x < -3 || x >= 3`
expr1 ∨ *expr2* ∨ *expr3*	`Or[`*expr1*`, `*expr2*`, `*expr3*`]` ✤ **3.0** `ξ < -5 ∨ ξ ≥ 5`
expr`..`	`Repeated[`*expr*`]` `q:{{_, _, _}..}`
expr`...`	`RepeatedNull[`*expr*`]` `{{___}...}`
expr1 **|** *expr2*	`Alternatives[`*expr1*`, `*expr2*`]` `f[x_Rational | x_Integer]`
symb`:`*expr*	`Pattern[`*symb*`, `*expr*`]` `h:Sin[x_]`
expr1 `/;` *expr2*	`Condition[`*expr1*`, `*expr2*`]` `x_ /; Abs[x] <= 1`
expr1 `->` *expr2*	`Rule[`*expr1*`, `*expr2*`]` `PlotPoints -> 25`
expr1 → *expr2*	`Rule[`*expr1*`, `*expr2*`]` ✤ **3.0** `{m → 1, a → 1, ℏ → 1}`
expr1 `:>` *expr2*	`RuleDelayed[`*expr1*`, `*expr2*`]` `f[x_] :> x`
expr1 ⧴ *expr2*	`RuleDelayed[`*expr1*`, `*expr2*`]` ✤ **3.0** `χ[ω] ⧴ Random[]`
expr1 `/.` *expr2*	`ReplaceAll[`*expr1*`, `*expr2*`]` `2 x + 10 /. x -> 10`
expr1 `//.` *expr2*	`ReplaceRepeated[`*expr1*`, `*expr2*`]` `x^2 + y^2 //. {x -> b + 5, b -> 1}`
expr1 `+=` *expr2*	`AddTo[`*expr1*`, `*expr2*`]` `x += del`

Operator Input Form	Full Form Example
expr1 -= *expr2*	SubtractFrom[*expr1*, *expr2*] x -= 0.5
expr1 *= *expr2*	TimesBy[*expr1*, *expr2*] x *= 4 b^2
expr1 /= *expr2*	DivideBy[*expr1*, *expr2*] x /= 2
expr &	Function[*expr*] (1/(1 + #))&
expr1 // *expr2*	*expr2*[*expr1*] (x^2 + 1) // Fullform
expr1 = *expr2*	Set[*expr1*, *expr2*] a = 1
expr1 := *expr2*	SetDelayed[*expr1*, *expr2*] f[x_] := x^2 - 1
expr1 ^= *expr2*	UpSet[*expr1*, *expr2*] Im[a] ^= 0
expr1 ^:= *expr2*	UpSetDelayed[*expr1*, *expr2*] f[g[x_]] ^:= x^2
symb/: *expr1* = *expr2*	TagSet[*symb*, *expr1*, *expr2*] a/: Im[a] = 0
symb/: *expr1* := *expr2*	TagSetDelayed[*symb*, *expr1*, *expr2*] g/: f[g[x_]] := x^2
expr =.	Unset[*expr*] t =.
symb/: *expr* =.	TagUnset[*symb*, *expr*] g/: f[g[x_]] =.
expr1;*expr2*;*expr3*	CompoundExpression[*expr1*, *expr2*, *expr3*] r = 3; s = Sqrt[4] + r; r^2 + s^2
expr1;*expr2*;	CompoundExpression[*expr1*, *expr2*, Null] Plot[Sin[x], {x, 0, 2Pi}]; x = y;
#	Slot[1] 3#&

Operator Input Form	Full Form Example
#*n*	Slot[*n*] #1^2 + #2^2 &
##	SlotSequence[1] {##}&
##*n*	SlotSequence[*n*] Plus[##4]&
%	Out[] N[%, 2]
%%	Out[-2] Show[%, %%]
%%...% (*n* times)	Out[-*n*] %%%^2 + 1
%*n*	Out[*n*] x[t_] = x[t] /. %14
_	Blank[] _^_
_*expr*	Blank[*expr*] _Symbol
__	BlankSequence[] x:{_, __}
__*expr*	BlankSequence[*expr*] {__List}
___	BlankNullSequence[] {___, x_, x_, x_, ___}
___*expr*	BlankNullSequence[*expr*] ___Integer
_.	Optional[Blank[]] _^_.
*symb*_	Pattern[*symb*, Blank[]] x_
*symb*_*expr*	Pattern[*symb*, Blank[*expr*]] n_Integer

Operator Input Form	Full Form Example
*symb*__	`Pattern[`*symb*`, BlankSequence[]]` `f[x__]`
*symb*__*expr*	`Pattern[`*symb*`, BlankSequence[`*expr*`]]` `x__List`
*symb*___	`Pattern[`*symb*`, BlankNullSequence[]]` `f[a_, b_, c_, d___]`
*symb*___*expr*	`Pattern[`*symb*`, BlankNullSequence[`*expr*`]]` `opts___Rule`
*symb*_.	`Optional[Pattern[`*symb*`, Blank[]]]` `x_ + y_.`

APPENDIX C

The Infrastructure of *Mathematica*

Books

M. L. Abell and J. P. Braselton. **Differential Equations with *Mathematica*, Second Edition.** Academic Press, Boston, 1996.

M. L. Abell and J. P. Braselton. *Mathematica* **by Example, Second Edition.** Academic Press, Boston, 1996.

M. L. Abell and J. P. Braselton. **The *Mathematica* Handbook.** Academic Press, Boston, 1993.

A. O. Allen. **Introduction to Computer Performance Analysis with *Mathematica*.** Academic Press, Boston, 1994.

T. B. Bahder. *Mathematica* **for Scientists and Engineers.** Addison-Wesley, Reading, MA, 1995.

G. Baumann. *Mathematica* **in Theoretical Physics: Selected Examples from Classical Mechanics to Fractals.** Springer-Verlag, New York, 1995.

A. I. Beltzer. **Engineering Analysis with Maple/*Mathematica*.** Academic Press, Boston, 1996.

N. Blachman. *Mathematica*: **A Practical Approach.** Prentice-Hall, Englewood Cliffs, NJ, 1992.

N. Blachman. *Mathematica*: **Quick Reference Version 2.** Addison-Wesley, Reading, MA, 1992.

D. C. M. Burbulla and C. T. J. Dodson. **Self-Tutor for Computer Calculus Using *Mathematica*.** Prentice-Hall, Englewood Cliffs, NJ, 1992.

W. Burkhardt. **First Steps in *Mathematica*.** Springer-Verlag, New York, 1994.

K. R. Coombs, B. R. Hunt, R. L. Lipsman, J. E. Osborn, and G. J. Stuck. **Differential Equations with *Mathematica*.** John Wiley and Sons, New York, 1995.

461

R. E. Crandall. *Mathematica* **for the Sciences.** Addison-Wesley, Redwood City, CA, 1991.

R. E. Crandall. **Projects in Scientific Computation.** Springer-Verlag, New York, 1994.

R. E. Crandall. **Topics in Advanced Scientific Computation.** Springer-Verlag, New York, 1995.

P. Crooke and J. Ratcliffe. **A Guidebook to Calculus with** *Mathematica*. PWS, Boston, 1991.

B. Davis, H. Porta, and J. Uhl. **Calculus and** *Mathematica*. Addison-Wesley, Reading, MA, 1994.

W. Ellis and E. Lodi. **A Tutorial Introduction to** *Mathematica*. Brooks/Cole, Pacific Grove, CA, 1991.

J. M. Feagin. **Quantum Methods with** *Mathematica*. Springer-Verlag, New York, 1994.

J. K. Finch and M. Lehmann. **Exploring Calculus with** *Mathematica*. Addison-Wesley, Reading, MA, 1992.

J. A. Freeman. **Simulating Neural Networks with** *Mathematica*. Addison-Wesley, Reading, MA, 1994.

R. J. Gaylord and P. R. Wellin. **Computer Simulations with** *Mathematica*: **Explorations in Complex Physical and Biological Systems.** Springer-Verlag, New York, 1994.

R. J. Gaylord, S. N. Kamin, and P. R. Wellin. **Introduction to Programming with** *Mathematica*. Springer-Verlag, New York, 1993.

O. Gloor, B. Amrhein, and R. E. Maeder. **Illustrated Mathematics: Visualization of Mathematical Objects with** *Mathematica*. Springer-Verlag, New York, 1995.

A. Gray, M. Mezzino, and M. Pinsky. **Ordinary Differential Equations with** *Mathematica*: **A Media Approach.** Springer-Verlag, New York, 1996.

J. W. Gray. **Mastering** *Mathematica*: **Programming Methods and Applications.** Academic Press, Boston, 1994.

T. W. Gray and J. Glynn. **The Beginner's Guide to** *Mathematica* **Version 2.** Addison-Wesley, Reading, MA, 1992.

T. W. Gray and J. Glynn. **Exploring Mathematics with** *Mathematica*. Addison-Wesley, Reading, MA, 1991.

E. Green, B. Evans, and J. Johnson. **Exploring Calculus with** *Mathematica*. John Wiley and Sons, New York, 1994.

M. H. Höft. **Laboratories for Calculus I Using** *Mathematica*. Addison-Wesley, Reading, MA, 1992.

E. W. Johnson. **Linear Algebra with** *Mathematica*. Brooks/Cole, Pacific Grove, CA, 1995.

S. Kaufmann. *Mathematica* **as a Tool.** Birkhauser, Boston, 1994.

M. G. Kerckhove and V. C. Nall. **Calculus Laboratories with** *Mathematica*. 3 Volumes. McGraw-Hill, New York, 1992-1994.

C. Knoll, M. Shaw, J. Johnson, and B. Evans. **Discovering Calculus with** *Mathematica*, Second Edition. John Wiley and Sons, New York, 1995.

R. J. Korsan. **Decision Support Systems in** *Mathematica*. Springer-Verlag, New York, 1995.

E. Kreyszig and E. Normington. *Mathematica* **Manual to Accompany Advanced Engineering Mathematics.** John Wiley and Sons, New York, 1994.

R. E. Maeder. **The** *Mathematica* **Programmer.** Academic Press, Boston, 1994.

R. E. Maeder. **The** *Mathematica* **Programmer II.** Academic Press, Boston, 1996.

R. E. Maeder. **Programming in** *Mathematica*, Second Edition. Addison-Wesley, Redwood City, CA, 1991.

V. Ostovic. **Computer-aided Analysis of Electric Machines: A** *Mathematica* **Approach.** Prentice-Hall, Englewood Cliffs, NJ, 1994.

E. Packel and S. Wagon. **Animating Calculus:** *Mathematica* **Notebooks for the Laboratory.** WH Freeman, New York, 1994.

L. Parker and S. M. Christensen. **MathTensor: A System for Doing Tensor Analysis by Computer.** Addison-Wesley, Reading, MA, 1994.

A. Riddle and S. Dick. **Applied Electronic Engineering with** *Mathematica*. Addison-Wesley, Reading, MA, 1994.

J. S. Robertson. **Engineering Mathematics with** *Mathematica*. McGraw-Hill, New York, 1994.

C. C. Ross. **Differential Equations: An Introduction with** *Mathematica*. Springer-Verlag, New York, 1995.

W. T. Shaw and J. Tigg. **Applied** *Mathematica*: **Getting Started, Getting It Done.** Addison-Wesley, Reading, MA, 1994.

A. Shuchat and F. Schultz. **The Joy of** *Mathematica*: **A Point-and-Click Way to Use and Learn** *Mathematica*. Addison-Wesley, Reading, MA, 1994.

R. D. Skeel and J. B. Keiper. **Elementary Numerical Computing with** *Mathematica*. McGraw-Hill, New York, 1992.

S. Skiena. **Implementing Discrete Mathematics: Combinatorics and Graph Theory with** *Mathematica*. Addison-Wesley, Reading, MA, 1991.

C. Smith and N. Blachman. **The** *Mathematica* **Graphics Guidebook.** Addison-Wesley, Reading, MA, 1995.

A. G. Sparks, J. W. Davenport, and J. P. Braselton. **Calculus Labs Using** *Mathematica*. Harper Collins, New York, 1993.

K. D. Stroyan. **Calculus Using *Mathematica*.** Academic Press, Boston, 1993.

K. D. Stroyan. **Scientific Projects and Mathematical Background for Calculus Using *Mathematica*.** Academic Press, Boston, 1993.

M. Trott. **The *Mathematica* Guidebook: Concepts, Examples, and Applications.** Springer-Verlag, New York, 1995.

I. Vardi. **Computational Recreations in *Mathematica*.** Addison-Wesley, Reading, MA, 1991.

H. R. Varian, ed. **Economic and Financial Modeling with *Mathematica*.** Springer-Verlag, New York, 1993.

D. D. Vvedensky. **Partial Differential Equations with *Mathematica*.** Addison-Wesley, Wokingham, England, 1993.

S. Wagon. ***Mathematica* in Action.** WH Freeman, New York, 1991.

T. Wickham-Jones. ***Mathematica* Graphics: Techniques and Applications.** Springer-Verlag, New York, 1994.

R. S. Wolff and L. Yeager. **Visualization of Natural Phenomena.** Springer-Verlag, New York, 1993.

S. Wolfram. ***Mathematica*: A System for Doing Mathematics by Computer,** Second Edition. Addison-Wesley, Redwood City, CA, 1991.

S. Wolfram. **The *Mathematica* Book,** Third Edition. Wolfram Media, Champaign, IL, 1996.

S. Wolfram. ***Mathematica* Reference Guide (for *Mathematica*Version 2).** Addison-Wesley, Reading, MA, 1992.

R. L. Zimmerman and F. I. Olness. ***Mathematica* for Physics.** Addison-Wesley, Reading, MA, 1995.

Journals

The *Mathematica* **Journal,** Miller Freeman, Inc., 600 Harrison Street, San Francisco, CA 94107.

Mathematica **in Education and Research,** Springer-Verlag New York, Inc., 333 Meadowlands Parkway, Secaucus, NJ 07094.

Conferences

The *Mathematica* **Conference,** Wolfram Research, Inc., 100 Trade Center Drive, Champaign, IL   61820-7237. Email: conference@wolfram.com.

Electronic Archives and Forums

Usenet Newsgroup

Usenet is a part of the Internet, and newsgroups are electronic bulletin boards. Subscribers to the newsgroup **comp.soft-sys.math.mathematica** can read and post messages about *Mathematica*. Connection to the Internet and a newsreader (i.e., a newsreading program) are necessary for subscribing to the group.

MathSource

A collection of *Mathematica* packages, notebooks, examples, programs, courseware, and documentation is available free of charge via electronic mail, anonymous FTP, Gopher, World Wide Web, and direct dialup from Wolfram Research, Inc. For information, send the email message **Help Intro** to mathsource@wolfram.com.

APPENDIX D

Solutions to Exercises

This appendix provides solutions to selected exercises in Chapters 2 and 3.

Section 2.1.19

1.

```
(10 (10.8 10^3/300)^(1/2) + 4)^(1/3)
```

3.

```
N[400 Sin[35 Degree], 3]//ScientificForm
```

5.

```
({{1, 0, -1}, {2, 4, 7}, {5, 3, 0}}.
 {{6, 1}, {0, 4}, {-2, 3}})//MatrixForm
```

7.

```
data = {{0.608, 0.05}, {0.430, 0.10}, {0.304, 0.20},
        {0.248, 0.30}, {0.215, 0.40}, {0.192, 0.50}}
```

```
Fit[data, {1, 1/v, 1/v^2}, v]
```

```
Clear[data]
```

9.

```
Re[(4 + I)/(2 + 3 I)]
```

```
Im[(4 + I)/(2 + 3 I)]
```

11.

```
NSolve[x^7 + x^5 + 2 x^3 + x^2 + 1 == 0, x]
```

13.

```
NIntegrate[x^3/(Exp[x] - 1), {x, 0, 1}]
```

15.

```
NDSolve[{y''[x] == 2 x + y[x] + 3 y'[x], y[2] == 1,
         y'[2] == -1}, y, {x, 2.1, 2.3}]
```

```
y[2.2.] /. %
```

19.

```
Do[
   vdP = NDSolve[{x'[t] == v[t],
                  v'[t] == (1 - x[t]^2)v[t] - x[t],
                  x[0] == i,
                  v[0] == 0},
                 {x, v}, {t, 0, 7Pi}];
   ParametricPlot[Evaluate[{x[t], v[t]} /. vdP],
                  {t, 0, 7Pi},
                  AspectRatio -> Automatic,
                  AxesLabel -> {" x", "v"}],
   {i, 1, 3, 0.5}
   ]
```

Section 2.2.16

1.

```
Factor[x^4 + 2 x^3 - 3 x - 6]
```

3.

```
Factor[a x^2 + a y + b x^2 + b y]
```

5.

```
Together[1/(x^2 - 16) - (x + 4)/(x^2 - 3 x - 4)]
```

7.

```
Solve[{4 x + 5 y == 5, 6 x + 7 y == 7}, {x, y}]
```

9.

```
Solve[Sqrt[x + 2] + 4 == x, x]
```

```
Sqrt[x + 2] + 4 == x /. %
```

11.

```
D[Sin[Exp[x^2]], x]
```

13.

```
Integrate[5 Log[t] - Exp[-3 t], t]
```

We can also solve the problem with the function DSolve.

```
DSolve[r'[t] == 5 Log[t] - Exp[-3 t], r[t], t]
```

15.

```
Integrate[4/(1 + x^2), {x, 0, 1}]
```

17.

```
Series[(1 + x^4)^(1/3), {x, 0, 8}]
```

19.

```
Limit[Log[x - a]/((a - b)(a - c)) +
      Log[2 (x - b)]/((b - c)(b - a)) +
      Log[x - c]/((c - a)(c - b)),
      x -> Infinity]
```

21.

```
DSolve[{x'[t] == k(a - x[t])(a - x[t]), x[0] == 0},
       x[t], t]

Limit[%[[1,1,2]], t -> Infinity]
```

27.

```
Needs["Calculus`FourierTransform`"]

FourierTransform[
   (1 + x)(UnitStep[x + 1] - UnitStep[x]) +
   (1 - x)(UnitStep[x] - UnitStep[x - 1]),
   x, -p,
   FourierOverallConstant -> Sqrt[1/h],
   FourierFrequencyConstant -> ((2 Pi)/h)
               ] // Simplify // PowerExpand
```

Section 2.3.4

1.

```
data = {{0.608, 0.05}, {0.430, 0.10}, {0.304, 0.20},
        {0.248, 0.30}, {0.215, 0.40}, {0.192, 0.50}}

Fit[data, {1, 1/v, 1/v^2}, v]

plotfit = Plot[%, {v, 0.1, 0.7},
               Frame -> True,
               FrameLabel -> {"V (Martian L)",
                              "P (Martian atm)"}
              ];

plotdata = ListPlot[
               data,
               Frame -> True,
               FrameLabel -> {"V (Martian L)",
                              "P (Martian atm)"},
               PlotRange -> {{0.1, 0.65}, {0.001, 0.65}},
               PlotStyle -> PointSize[0.02]
                  ];

Show[plotfit, plotdata, PlotRange -> {0.001, 0.55}];

Clear[data, plotfit, plotdata]
```

3.

```
Plot3D[(x^2 - y^2), {x, -2, 2}, {y, -2, 2},
        BoxRatios -> {1, 1, 1},
        AxesLabel -> {"x", "y", "z"}];
```

5.

```
Plot[BesselJ[0, x], {x, 0, 16}];

mylist = {{2.44316, 0.003077}, {5.50913, 0.003077},
          {8.69301, 0.003077}, {11.8179, 0.003077},
          {14.9429, 0.003077}};

Table[FindRoot[BesselJ[0, x] == 0, {x, mylist[[i, 1]]}],
       {i, Length[mylist]}]

Clear[mylist]
```

7.

```
ParametricPlot[{2 Cos[theta], Sin[theta]}, {theta, 0, 2Pi},
                AspectRatio -> Automatic,
                AxesLabel -> {" x (b)", " y (b)"}];
```

9.

```
g[i_] := Plot[{Exp[-16 (x - i)^2] + 1.5 Exp[-(x + i)^2],
               1.5 Exp[-(x + i)^2] + 3,
               Exp[-16 (x - i)^2] + 5},
              {x, -3.0, 3.0},
              PlotStyle -> {{Thickness[0.008]},
                            {Thickness[0.005]},
                            {Thickness[0.005]}},
              PlotRange -> {0, 6.25}, Axes -> False,
              Frame -> True, FrameTicks -> None,
              DisplayFunction -> Identity
             ]

Show[
  GraphicsArray[
    Partition[Table[g[i], {i, -2.0, 3.5, 0.5}], 4],
    GraphicsSpacing -> {0.2, 0.2}
             ]
    ];

Clear[g]
```

Section 2.4.10

1.

```
Table[x^n - 1, {n, 1, 11, 2}]

Drop[%, {2, 3}]
```

3.

```
listA = {r, {s, t}, {w, {x, {y, z}}}};

listA[[1]]
listA[[2]]
listA[[3, 1]]
listA[[3, 2, 2]]
listA[[3, 2, 2, 2]]

Clear[listA]
```

5.

```
f[x_] := Sin[x]^2

{a, {b, c}, a, {d, {e, {a, c, g}}}}

MapAt[f, %, Position[%, a]]

Clear[f]
```

7.

```
Range[4]^3

Range[5, 8]^3

mylist = Join[%%, %]

Apply[Plus, mylist]/Length[mylist]

(mylist - %)^2
```

9.

```
Range[100]

Partition[%, Length[%]/2]
```

```
MapAt[Reverse, %, 2]

%[[1]] + %[[2]]

101 Length[%]

%/100
```

11.

```
Needs["Calculus`VectorAnalysis`"]

SetCoordinates[Cartesian[x, y, z]];

a = {ax[x, y, z], ay[x, y, z], az[x, y, z]};
f = func[x, y, z];

Simplify[Curl[f a] -
       f Curl[a] +
       CrossProduct[a, Grad[f]]
       ]

Clear[f, a]
```

Section 3.1.4

1.

a.

```
FullForm[Sqrt[x^2 + 8 x + 16]]
```

c.

```
FullForm[Sin[x + y]]
```

e.

```
FullForm[Hold[Integrate[Sin[x], {x, 0, Pi/2}]]]
```

3.

```
Level[x^2/(x - 1) == 1/(x - 1), 3]
```

5.

```
Position[Sqrt[(24 a^2 b)/(1 + 3 a^3)] Sqrt[a^3 b^2],
        a^3, {6}]
```

7.

```
Insert[h[e1, e2, e3, e4, e5, e6, e7], b,
      Table[{i}, {1, 2, 8}]]
```

9.

```
Position[a + b/c + (d + e)/(f + g), Times]

% /. {x_, 0} -> {x}

FlattenAt[a + b/c + (d + e)/(f + g), %]
```

11.

```
Position[a + b/c + (a + e/f)/(r + s/t), Times, {2}]

% /. 0 -> 1

MapAt[f, a + b/c + (a + e/f)/(r + s/t), %]
```

13.

```
complexConjugate[expr_] :=
    (expr /. Complex[x_, y_] -> Complex[x, -y])
```

Section 3.2.9

1.

```
mymean[{x__}] := (Plus[x])/Length[{x}]

yourmean[x:{__}] := (Plus @@ x)/Length[x]

ourmean[x_List /; Length[x] >= 1] := (Plus @@ x)/Length[x]

mylist = Table[Random[Integer, 1000],
              {Random[Integer, {1, 100}]}]

mymean[mylist]

yourmean[mylist]

ourmean[mylist]
```

3.

```
legdPoly[n_Integer?NonNegative, x_Symbol] :=
        ((1/(2^n n!)) D[(x^2 - 1)^n, {x, n}])//Simplify
legdPoly[n_Integer?NonNegative, x_Real /; Abs[x] <= 1] :=
        ((1/(2^n n!)) D[(y^2 - 1)^n, {y, n}] /. y -> x)//Simplify
assLPoly[n_Integer?NonNegative, 0, x_Symbol] := legdPoly[n, x]
assLPoly[n_Integer?NonNegative, 0,
        x_Real /; Abs[x] <= 1] := legdPoly[n, x]
assLPoly[n_Integer?NonNegative,
        m_Integer?Positive,
        x_Symbol] :=
                ((-1)^m (1 - x^2)^(m/2)
                D[legdPoly[n, x], {x, m}]//Simplify) /; m <= n
assLPoly[n_Integer?NonNegative,
        m_Integer?Positive,
        x_Real /; Abs[x] <= 1] :=
                ((-1)^m (1 - x^2)^(m/2)
                (D[legdPoly[n, z], {z, m}] /. z -> x)
                //Simplify) /; m <= n
assLPoly[n_Integer?NonNegative,
        m_Integer?Negative,
        x_Symbol] := (((-1)^(-m))((n + m)!/(n - m)!)
                assLPoly[n, -m, x])
assLPoly[n_Integer?NonNegative,
        m_Integer?Negative,
        x_Real /; Abs[x] <= 1] :=
                (((-1)^(-m))((n + m)!/(n - m)!)
                assLPoly[n, -m, x])

Plot[Evaluate[assLPoly[2, 1, x]], {x, -1, 1}];

Print["n      m           P\n"]
Do[Print[n, "      ", m, "           ", assLPoly[n, m, x]],
   {n, 0, 2}, {m, -n, n}]

Print["n     m          P\n"]
Do[Print[n, "     ", m, "          ", assLPoly[n, m, 0.5]],
   {n, 0, 2}, {m, -n, n}]

Table[assLPoly[n, m, 1.5], {n, 0, 2}, {m, -n, n}]

Table[Expand /@ (assLPoly[n, m, x] == LegendreP[n, m, x]),
     {n, 0, 2}, {m, -n, n}]

Table[Expand /@ (assLPoly[n, m, 0.5] ==
                LegendreP[n, m, 0.5]),
     {n, 0, 2}, {m, -n, n}]
```

5.

```
ClearAll[myfunc]

SetAttributes[myfunc, Orderless]

myfunc[x_Symbol, y_Integer, z_Complex] := {x, y, z}
```

Here is a function that preserves the order of the arguments:

```
ClearAll[func]

func[x_, y_, z_] := {x, y, z} /;
                        MatchQ[{x, y, z},
                                Alternatives @@
                                    Permutations[
                                        {_Symbol, _Integer, _Complex}
                                                ]
                                ]
                    ]
```

Here are alternative definitions for the function:

```
ClearAll[ourfunc]

ourfunc[x_, y_, z_] := ourfunc[{x, y, z}]

ourfunc[x:Alternatives @@
        Permutations[{_Symbol, _Integer, _Complex}]] := x
```

7.

```
x_. y_^n_?Negative
```

Section 3.3.7

1.

a.

```
f[x_] := 1 + x^3
```

c.

```
h[x_, y_, z_] := {x, y^z}
```

e.

```
Clear[f, g]
g[y_] := y^3
```

```
f[x_] := Map[g, x]
f[{3, 2, 7}]
```

g.

```
Clear[f]
f[x_] := Delete[x, Random[Integer, {1, Length[x]}]]
```

3.

```
Clear[a, x]

Cases[{1, a, 2.0, 5.0, 4, x^2, Sin[x]}, _Integer?(#>3&)]
```

5.

```
Clear[a, b, x, tabulate, mylist]

tabulate[x_List] := Transpose[{Union[x], Count[x, #]& /@ Union[x]}
                              ] // TableForm

mylist = {b, b, 1, a, a, a, Sin[x], Sin[x]}

tabulate[mylist]
```

7.

a.

```
Clear[hermite]
hermite[n_Integer /; n > 1, z_] :=
            Expand @ (2 z hermite[n - 1, z] -
                      2 (n - 1) hermite[n - 2, z])
hermite[0, z_] := 1
hermite[1, z_] := 2 z
totalCalls[n_Integer /; n > 1, z_] :=
            Trace[hermite[n, z], hermite[_Integer, z]] //
              Flatten // Length

totalCalls[20, z]
```

b.

```
partialCalls[n_Integer /; n > 1,
             k_Integer?NonNegative, z_] :=
                (Trace[hermite[n, z], hermite[k, z]] //
```

```
                            Flatten // Length) /; n >= k

    partialCalls[20, 3, z]
```

c.

```
    newTotalCalls[n_Integer /; n > 1, z_] :=
      (Clear[H];
       H[m_Integer /; m > 1, x_] := H[m, x] =
                    Expand @ (2 x H[m - 1, x] -
                              2 (m - 1) H[m - 2, x]);
       H[0, x_] := 1;
       H[1, x_] := 2 x;
       Trace[H[n, z], H[_Integer, z]] //
         Flatten // Length)

    newTotalCalls[20, z]

    newPartialCalls[n_Integer /; n > 1,
                    k_Integer?NonNegative, z_] :=
      (Clear[H];
       H[m_Integer /; m > 1, x_] := H[m, x] =
                    Expand @ (2 x H[m - 1, x] -
                              2 (m - 1) H[m - 2, x]);
       H[0, x_] := 1;
       H[1, x_] := 2 x;
       Trace[H[n, z], H[k, z]] //
         Flatten // Length) /; n >= k

    newPartialCalls[20, 3, z]
```

9.

```
    Clear[a, b, c, d]
```

a.

```
    FoldList[Plus, 0, {a, b, c, d}]
```

c.

```
    Fold[10 #1 + #2 &, 0, {1, 0, 3, 5, 7, 3}]
```

11.

```
    Clear[mu, f]

    f[x_] := mu (1 - 2 Abs[x - 1/2])
```

a.

```
mu = Random[Real, 1/2]

0 == f[0]

Table[Chop[
        FixedPoint[f, Random[],
                    SameTest -> (Abs[#1 - #2] < 10^-11 &)
                ], 10^-10
            ], {10}
        ]
```

b.

```
mu = Random[Real, {1/2, 1}]

0 == f[0]

ListPlot[
    Transpose[
      {Range[0, 100], NestList[f, 0.0001, 100]}
            ],
    PlotJoined -> True, PlotRange -> {0, 1}
        ];

(2 mu)/(1 + 2 mu) == f[(2 mu)/(1 + 2 mu)]

ListPlot[
    Transpose[
      {Range[0, 100],
        NestList[
          f, ((2 mu)/(1 + 2 mu)) + 0.0001, 100
                ]}
            ],
    PlotJoined -> True, PlotRange -> {0, 1}
        ];

Plot[{f[x], x}, {x, 0, 1}, PlotRange -> {0, 1}];
```

13.

```
Needs["Calculus`Vector Analysis`"]

CoordinateSystem

Coordinates[]
```

```
(* ✿ 3.0 *) SetCoordinates[Cartesian[x, y, z]]

a = {ax[x, y, z], ay[x, y, z], az[x, y, z]};
b = {bx[x, y, z], by[x, y, z], bz[x, y, z]};

Simplify[
  Curl[CrossProduct[a, b]] -
  (DotProduct[b, {D[#, x], D[#, y], D[#, z]}]&) /@ a +
  (DotProduct[a, {D[#, x], D[#, y], D[#, z]}]&) /@ b -
  a Div[b] +
  b Div[a]
      ]

ClearAll[a, b]
```

Section 3.4.6

1.

```
f[x_] := If[x > 0, x, -x]

g[x_ /; x > 0] := x
g[x_ /; x <= 0] := -x

{f'[0], g'[0]}
```

The answer -1 for f'[0] is wrong because *f* is not differentiable at 0.

```
Clear[f, g]
```

3.

```
v[x_] := Which[x < -2 || (x >= -1 && x < 1), 0,
               x >= -2 && x < -1, -2,
               x >= 1 && x < 3, -1,
               x >= 3, 1]

newv[x_ /; x < -2 || (x >= -1 && x < 1)] := 0
newv[x_ /; x >= -2 && x < -1] := -2
newv[x_ /; x >= 1 && x < 3] := -1
newv[x_ /; x >= 3] := 1
```

5.

```
Plot[x^2 - (1/12) x^3, {x, -5, 15}];

-D[x^2 - (1/12) x^3, x]
```

```
Plot[%, {x, -5, 13}];
```

Important: Enter the definition of motion1DPlot in Section 3.4.5 here.

```
motion1DPlot[-2 x[t] + (x[t]^2)/4, -3, 5, 4]
```

```
motion1DPlot[-2 x[t] + (x[t]^2)/4, 5, 0, 15.48]
```

```
ClearAll[motion1DPlot]
```

7.

```
ClearAll[binomialExpansion, caption, exp]
```

```
Options[binomialExpansion] = {caption -> False, exp -> 2};
```

```
binomialExpansion[x_Symbol, y_Symbol, opts___] :=
   Module[{caption, exp},
      Print[opts];
      Print[Options[binomialExpansion]];
      Print[caption];
      Print[exp];
      caption = caption/.{opts}/.Options[binomialExpansion];
      exp = exp/.{opts}/.Options[binomialExpansion];
      Print[caption];
      Print[exp];
      If[caption === True,
         Print["Expansion of ", (x + y)^exp]
         ];
      Expand[(x + y)^exp]
         ]
```

```
binomialExpansion[a, b]
```

```
binomialExpansion[a, b, exp -> 10, caption -> True]
```

Section 3.5.4

1.

```
g1 = {Circle[{0, 0}, {2, 1}]};
Needs["Graphics`Arrow`"]
g2 = {Arrow[{0.100, 1.00}, {-0.100, 1.00}],
      Arrow[{-0.100, -1.00}, {0.100, -1.00}]
      };
```

```
y[x_] := (1/2)Sqrt[4 - x^2]
g3 = Line /@ {
        {{1.3, 0}, {1.75, y[1.75]}},
        {{1.3, 0}, {1.75, -y[1.75]}},
        {{1.3, 0}, {-1.97, y[1.97]}},
        {{1.3, 0}, {-1.97, -y[1.97]}}
            };
g4 = {{RGBColor[1, 0.5, 0], Disk[{1.3, 0}, 0.16]}};
g5 = {{PointSize[0.04],
        Point /@ {{1.75, y[1.75]},
                  {1.75, -y[1.75]},
                  {-1.97, y[1.97]},
                  {-1.97, -y[1.97]}
                 }
      }
    };
g6 = {Text["A", {-2.15, y[1.97]}, {1, 0}],
      Text["B", {-2.15, -y[1.97]}, {1, 0}],
      Text["C", {1.95, -y[1.75]}, {-1, 0}],
      Text["D", {1.95, y[1.75]}, {-1, 0}],
      Text["Sun", {1.2, 0.25}, {0, -1}]
    };
Show[Graphics[Join[g1, g2, g3, g4, g5, g6]],
     PlotRange -> {{-2.5, 2.5}, {-1.2, 1.2}},
     DisplayFunction -> $DisplayFunction,
     AspectRatio -> Automatic];
ClearAll[g1, g2, g3, g4, g5, g6]
```

Section 3.6.4

3.

```
newbifurcation[x0_, mumin_, mumax_, del_,
               nmax_, opts___Rule] := (
    ListPlot[
        Flatten[
           Table[
            Thread[
             List[
              mu, Drop[
                     NestList[(# Exp[mu (1 - #)])&, x0, nmax],
                     100
                       ]
                 ]
                ],
```

```
                {mu, mumin, mumax, del}
                    ],
          1
                    ],
         opts
              ];
                                              )
```

```
  (* ❀ 3.0  *)

  newbifurcation[0.2, 2.0, 4.0, 0.01, 300,
                AxesOrigin -> {1.9, 0},
                PlotRange -> {{1.9, 4}, {0, 1}},
                AxesLabel -> {"λ", xₙ}]
```

```
  (* Mathematica Version 2.2 Specifics *)

  newbifurcation[0.2, 2.0, 4.0, 0.01, 300,
                AxesOrigin -> {1.9, 0},
                PlotRange -> {{1.9, 4}, {0, 1}},
                AxesLabel ->
                  {FontForm[" l", {"Symbol", 10}],
                   None},
                Epilog ->
                  {Text[FontForm["x", {"Times", 10}],
                        {1.96, 0.96}],
                   Text[FontForm["n", {"Times", 10}],
                        {2.00, 0.92}]}
                ]
```

```
ClearAll[newbifurcation]
```

4.

As an illustration, drop only five elements and use the values:

```
x0 = 0.2; mumin = 2.9; mumax = 3.0; del = 0.01; nmax = 10;

NestList[(# + del)&, mumin,
         Round[(mumax - mumin)/del]] (# (1 - #))

Map[Function, %]

Map[NestList[#, x0, nmax]&, %]

Map[Drop[#, 5]&, %]
```

```
Transpose[{NestList[(# + del)&, mumin,
                    Round[(mumax - mumin)/del]],
          %}]

Map[Thread, %]

Flatten[%, 1]

ListPlot[%]

ClearAll[x0, mumin, mumax, del, nmax]
```

5.

```
Unprotect[NonCommutativeMultiply];

A_ ** U := A
U ** A_ := A

A_ ** (B_ + C_) := A ** B + A ** C
(A_ + B_) ** C_ := A ** C + B ** C

number3Q[x_, y_, n_] := NumberQ[x]&&NumberQ[y]&&NumberQ[n]

A_ ** (B_ (x_. y_^n_. /; number3Q[x, y, n])) :=
                                   ((x y^n) A ** B)
(A_ (x_. y_^n_. /; number3Q[x, y, n])) ** B_ :=
                                   ((x y^n) A ** B)

Protect[NonCommutativeMultiply];

commutator[A_, B_] := A ** B - B ** A

NumberQ[hb] ^= True;

xp3DCommutator[expr_] :=
  ExpandAll[expr //.
            {px ** x :> x ** px - I hb U,
             py ** y :> y ** py - I hb U,
             pz ** z :> z ** pz - I hb U,
             x ** y :> y ** x,
             x ** z :> z ** x,
             y ** z :> z ** y,
             py ** x :> x ** py,
             pz ** x :> x ** pz,
             px ** y :> y ** px,
```

```
                   pz ** y :> y ** pz,
                   px ** z :> z ** px,
                   py ** z :> z ** py,
                   px ** py :> py ** px,
                   px ** pz :> pz ** px,
                   py ** pz :> pz ** py}]

lx = y ** pz - z ** py;
ly = z ** px - x ** pz;
lz = x ** py - y ** px;

(commutator[lx, ly] - I hb lz == 0) // xp3DCommutator

(commutator[ly, lz] - I hb lx  == 0) // xp3DCommutator

(commutator[lz, lx] - I hb ly == 0) // xp3DCommutator

(commutator[lz, lx ** lx + ly ** ly +
          lz ** lz] == 0) // xp3DCommutator

ClearAll[number3Q, commutator, hb,
        xp3DCommutator, lx, ly, lz]
```

6.

```
ClearAll[f]

f[{x_, y_, z___}] := {x}~Join~f[{z}]

f[{x_}] := {x}

f[{}] = {};

Table[Random[Integer, {-100, 100}],
      {Random[Integer, 20]}]

f[%]
```

Section 3.7.5

5.

```
Needs["Miscellaneous`PhysicalConstants`"]

<<Miscellaneous`PhysicalConstants`

Needs["Miscellaneous`PhysicalConstants`"]
```

The "`<<`" command requires fewer key strokes to enter than the **Needs** command does. On the other hand, **Needs** has the advantage over "`<<`" in that it reads in a package only if the package is not already loaded, whereas "`<<`" reads in a package even if it has been loaded. The "`<<`" command may trigger an error message because some packages are not set up to be loaded twice.

APPENDIX E

Solutions to Problems

This appendix provides solutions to selected problems in Chapters 2, 4, 5, and 6. When working with limited computer memory, consider starting a new notebook for each problem.

Section 2.6

1.

```
Sum[1/(2 n - 1)^4, {n, Infinity}]//N
```

3.

```
Solve[{2 x + y == 3 z,
       6 x + 24 + 4 y == 0,
       20 - 5 x + 2 z == 0},
      {x, y, z}]

{2 x + y == 3 z, 6 x + 24 + 4 y == 0,
 20 - 5 x + 2 z == 0} /. %
```

5.

```
Plot3D[x^2 + y^2, {x, -2, 2}, {y, -2, 2},
       BoxRatios -> {1, 1, 1},
```

```
             AxesLabel -> {"x", "y", "z"}];
Show[%, ViewPoint -> {-0.079, -3.353, 0.447}];
Show[%, ViewPoint -> {-0.000, -0.000, 3.384}];
Show[%, ViewPoint -> {-2.171, 0.515, 2.544}];
```

7.

```
Show[Plot3D[0.25x + 0.25y + 2.5,
            {x, -3, 3}, {y, -3, 3},
            DisplayFunction -> Identity],
     Plot3D[x^2 + y^2,
            {x, -2, 2}, {y, -2, 2},
            DisplayFunction -> Identity],
     BoxRatios -> {1, 1, 1}, Boxed -> False,
     Axes -> False,
     DisplayFunction -> $DisplayFunction];
```

9.

```
Do[Plot[{1, -1, Cos[x] + (i Pi/2)(Sin[x]/x)},
        {x, -5 Pi, 5 Pi}, PlotRange -> {-4, 15},
        AxesLabel -> {"x", "y"}
       ];,
   {i, 3, 9}
  ]
```

11.

```
Needs["Graphics`Shapes`"]
Needs["Graphics`Animation`"]
SpinShow[Graphics3D[Sphere[]], Boxed -> False,
         Frames -> 6,
         SpinRange -> {0 Degree, 90 Degree}]
```

13.

```
NDSolve[{i'[t] + 2i[t] == Sin[t], i[0] == 0},
        i, {t, 0, 1.1}]


numericali = (i /. %)[[1]]


DSolve[{i'[t] + 2i[t] == Sin[t], i[0] == 0},
       i[t], t]
```

```
analyticali[t_] = (i[t] /. %)[[1]]

Print["t          numerical    analytical"]
Table[{t, numericali[t], analyticali[t]},
      {t, 0, 1, 0.1}]//Chop//MatrixForm

Clear[numericali, analyticali]
```

15.

```
Series[Sin[x] Cos[x], {x, 0, 7}]

approx = Normal[%]

Plot[{approx, Sin[x] Cos[x]}, {x, -Pi, Pi},
     AxesLabel -> {"x", "sin(x)cos(x)"}];

Clear[approx]
```

19.

```
Needs["Calculus`FourierTransform`"]

FourierTransform[
          Sin[w0 t](UnitStep[t + T] - UnitStep[t - T]),
          t, -w,
          FourierOverallConstant -> Sqrt[1/(2 Pi)]
              ]
```

Section 4.5

9.

We cannot. The full form of v_0 is Subscript[v, 0], where v is an element of the expression.

Section 5.4

9.

The keyboard letters x and y cannot be the subscripts of the Greek letter E. The full form, for example, of E_x[x, y] is Subscript[E,x][x,y], where the two x's are identical; consequently, the definitions for f and ε are no longer valid.

Section 6.7

4.

a.

```
Off[ClebschGordan::phy]
Print["\n\nAddition of Two Spins\n"]
Do[Print["s = ", s, "    m = ", m, "\n"];
    Print[Sum[
            ClebschGordan[{1, m1}, {1, m - m1}, {s, m}]
            ket1[m1] ket2[m - m1],
            {m1, 1, -1, -1}
            ],
          "\n\n\n\n"
        ],
    {s, 2, 0, -1}, {m, s, -s, -1}
    ]
On[ClebschGordan::phy]
```

b.

$^1S_0, {}^1D_2, {}^3P_{2,1,0}, {}^3F_{4,3,2}, {}^5S_2, {}^5D_{4,3,2,1,0}$

c.

```
Off[ClebschGordan::phy, General::spell1]
Print["\n\nAddition of a Spin with\n
an Orbital Angular Momentum\n"]
Do[Print["J = 4", "    M = ", m, "\n"];
    Print[Sum[
            ClebschGordan[{2, m1}, {2, m - m1}, {4, m}]
            ketS[m1] ketL[m - m1],
            {m1, 2, -2, -1}
            ],
          "\n\n\n\n"
        ],
    {m, 4, -4, -1}
    ]
On[ClebschGordan::phy, General::spell1]
```

9.

```
kdel[n_, m_] := 1 /; n == m; kdel[n_, m_] := 0 /; n != m
elementLz[n_, m_] := m kdel[n, m]
elementLplus[l_, n_, m_] := (Sqrt[l(l + 1) - n m]
                            kdel[n, m + 1])
```

```
elementLminus[l_, n_, m_] := elementLplus[l, m, n]
elementLx[l_, n_, m_] := (elementLplus[l, n, m] +
                          elementLminus[l, n, m])/2

elementLy[l_, n_, m_] := (elementLplus[l, n, m] -
                          elementLminus[l, n, m])/(2I)
matrixLx[l_] := Table[elementLx[l, n, m], {n, 1, -1, -1},
                      {m, 1, -1, -1}]

matrixLy[l_] := Table[elementLy[l, n, m], {n, 1, -1, -1},
                      {m, 1, -1, -1}]

matrixLz[l_] := Table[elementLz[n, m], {n, 1, -1, -1},
                      {m, 1, -1, -1}]

matrixH[l_] := (a matrixLx[l].matrixLx[l] +
                b matrixLy[l].matrixLy[l] +
                c matrixLz[l].matrixLz[l])
eigenvaluesH[l_] := Eigenvalues[matrixH[l]]
```

a.

```
eigenvaluesH[4]
```

b.

```
b = r a;

(test = eigenvaluesH[4])//Timing
```

c.

```
b = a + d;

(newtest = eigenvaluesH[4])//Timing
```

d.

```
Im[test /. {a -> 1, r -> 3, c -> 4}]

N[test /. {a -> 1, r -> 3, c -> 4}]
```

e.

```
Im[newtest /. {a -> 1, d -> 2, c -> 4}]

N[newtest /. {a -> 1, d -> 2, c -> 4}]
```

f.

```
a = 1; b = 3; c = 4;

(moretest = eigenvaluesH[4]) // Timing

Im[moretest]

N[moretest]

Sort[N[test /. {a -> 1, r -> 3, c -> 4}]] ==
Sort[N[newtest /. {a -> 1, d -> 2, c -> 4}]] ==
Sort[N[moretest]]

Clear[a, b, c]
```

References

[Ada95] R. K. Adair. The Physics of Baseball. **Physics Today,** 48(5):26–31, 1995.

[AAB93] V. Adamchik, J. Adams, and A. Bocharov. **Guide to Standard** *Mathematica* **Packages,** Third Edition. Wolfram Research, Champaign, IL, 1993.

[AHR94] J. M. Aguirregabiria, A. Hernández, and M. Rivas. Are We Careful Enough When Using Computer Algebra? **Computers in Physics,** 8(1):56–61, 1994.

[AM76] N. W. Ashcroft and N. D. Mermin. **Solid State Physics.** Holt, Rinehart and Winston, New York, 1976.

[BG90] G. L. Baker and J. P. Gollub. **Chaotic Dynamics: An Introduction.** Cambridge University Press, Cambridge, England, 1990.

[Bla92] N. Blachman. *Mathematica:* **Quick Reference Version 2.** Addison-Wesley, Reading, MA, 1992.

[Bra85] P. J. Brancazio. Trajectory of a Fly Ball. **The Physics Teacher,** 20–23, January 1985.

[BD85] S. Brandt and H. D. Dahmen. **The Picture Book of Quantum Mechanics.** John Wiley and Sons, New York, 1985.

[BG96] V. Bush and J. Grohens. **Getting Started with** *Mathematica* **on Macintosh Systems.** Wolfram Research, Champaign, IL, 1996.

[BWG96] V. Bush, J. Walsh, and J. Grohens. **Getting Started with** *Mathematica* **3.0 under Microsoft Windows.** Wolfram Research, Champaign, IL, 1996.

[CDSTD92a] D. Cook, R. Dubisch, G. Sowell, P. Tam, and D. Donnelly. A Comparison of Several Symbol-Manipulating Programs: Part I. **Computers in Physics,** 6(4):411–419, 1992.

[CDSTD92b] D. Cook, R. Dubisch, G. Sowell, P. Tam, and D. Donnelly. A Comparison of Several Symbol-Manipulating Programs: Part II. **Computers in Physics,** 6(5):530–540, 1992.

[Cop91] V. T. Coppola. Working Directly with Subexpressions. **The** *Mathematica* **Journal,** 1(4):41–44, 1991.

[CS60] H. C. Corben and P. Stehle. **Classical Mechanics,** Second Edition. John Wiley and Sons, New York, 1960.

[Cra91] R. E. Crandall. *Mathematica* **for the Sciences.** Addison-Wesley, Redwood City, CA, 1991.

[DG94] D. W. DeLand and J. Grohens. *Mathematica* **User's Guide for the Macintosh,** Fourth Edition. Wolfram Research, Champaign, IL, 1994.

[DK67] P. Dennery and A. Krzywicki. **Mathematics for Physicists.** Harper and Row, New York, 1967.

[DeV94] P. L. DeVries. **A First Course in Computational Physics.** John Wiley and Sons, New York, 1994.

[Eis61] R. M. Eisberg. **Fundamentals of Modern Physics.** John Wiley and Sons, New York, 1961.

[ER85] R. Eisberg and R. Resnick. **Quantum Physics of Atoms, Molecules, Solids, Nuclei, and Particles,** Second Edition. John Wiley and Sons, New York, 1985.

[Fea94] J. M. Feagin. **Quantum Methods with** *Mathematica.* Springer-Verlag, New York, 1994.

[FGT93] P. M. Fishbane, S. Gasiorowicz, and S. T. Thornton. **Physics for Scientists and Engineers.** Prentice-Hall, Englewood Cliffs, NJ, 1993.

[Gar94] A. L. Garcia. **Numerical Methods for Physics.** Prentice-Hall, Englewood Cliffs, NJ, 1994.

[Gas74] S. Gasiorowicz. **Quantum Physics.** John Wiley and Sons, New York, 1974.

[Gas95] S. Gasiorowicz. **Quantum Physics,** Second Edition. John Wiley and Sons, New York, 1995.

[GKW93] R. J. Gaylord, S. N. Kamin, and P. R. Wellin. **Introduction to Programming with** *Mathematica.* Springer-Verlag, New York, 1993.

[Gos92] A. Goswami. **Quantum Mechanics.** Wm. C. Brown, Dubuque, IA, 1992.

[GT88] H. Gould and J. Tobochnik. **An Introduction to Computer Simulation Methods, Part 1.** Addison-Wesley, Reading, MA, 1988.

[Gra94] J. W. Gray. **Mastering** *Mathematica***: Programming Methods and Applications.** Academic Press, Boston, 1994.

[Gri89] D. J. Griffiths. **Introduction to Electrodynamics,** Second Edition. Prentice-Hall, Englewood Cliffs, NJ, 1989.

[Hal58] P. R. Halmos. **Finite-Dimensional Vector Spaces,** Second Edition. D. Van Nostrand Company, Princeton, NJ, 1958.

[Has91] S. Hassani. **Foundations of Mathematical Physics.** Allyn and Bacon, Boston, 1991.

[Jea66] J. H. Jeans. **The Mathematical Theory of Electricity and Magnetism,** Fifth Edition. Cambridge University Press, Cambridge, England, 1966.

[Kit86] C. Kittel. **Introduction to Solid State Physics,** Sixth Edition. John Wiley and Sons, New York, 1986.

[KM90] S. E. Koonin and D. C. Meredith. **Computational Physics,** Fortran Version. Addison-Wesley, Redwood City, CA, 1990.

[Lib92] R. L. Liboff. **Introductory Quantum Mechanics,** Second Edition. Addison-Wesley, Reading, MA, 1992.

[LC70] P. Lorrain and D. R. Corson. **Electromagnetic Fields and Waves,** Second Edition. W. H. Freeman, San Francisco, 1970.

[Mae91] R. E. Maeder. **Programming in** *Mathematica,* Second Edition. Addison-Wesley, Redwood City, CA, 1991.

[MT95] J. B. Marion and S. T. Thornton. **Classical Dynamics of Particles and Systems,** Fourth Edition. Saunders College Publishing, Fort Worth, TX, 1995.

[McM94] S. M. McMurry. **Quantum Mechanics.** Addison-Wesley, Wokingham, England, 1994.

[Mes62] A. Messiah. **Quantum Mechanics.** Translated from the French by G. M. Temmer. 2 Volumes. John Wiley and Sons, New York, 1961–1962.

[Moo87] F. C. Moon. **Chaotic Vibrations.** John Wiley and Sons, New York, 1987.

[Mor90] M. A. Morrison. **Understanding Quantum Physics.** Prentice-Hall, Englewood Cliffs, NJ, 1990.

[MF53] P. M. Morse and H. Feshbach. **Methods of Theoretical Physics.** McGraw-Hill, New York, 1953.

[Oha90] H. C. Ohanian. **Principles of Quantum Mechanics.** Prentice-Hall, Englewood Cliffs, NJ, 1990.

[Par92] D. Park. **Introduction to the Quantum Theory,** Third Edition. McGraw-Hill, New York, 1992.

[Pat94] V. A. Patel. **Numerical Analysis.** Saunders College Publishing, Fort Worth, TX, 1994.

[PJS92] H. Peitgen, H. Jürgens, and D. Saupe. **Chaos and Fractals: New Frontiers of Science.** Springer-Verlag, New York, 1992.

[Ras90] S. N. Rasband. **Chaotic Dynamics of Nonlinear Systems.** John Wiley and Sons, New York, 1990.

[RHK92] R. Resnick, D. Halliday, and K. S. Krane. **Physics,** Fourth Edition. John Wiley and Sons, New York, 1992.

[Sch55] L. I. Schiff. **Quantum Mechanics,** Second Edition. McGraw-Hill, New York, 1955.

[Ser87] R. A. Serway. **Physics for Scientists and Engineers with Modern Physics,** Second Edition. Saunders College, Philadelphia, 1987.

[Sym71] K. R. Symon. **Mechanics,** Third Edition. Addison-Wesley, Reading, MA, 1971.

[Tam91] P. Tam. Physics and *Mathematica.* **Computers in Physics,** 5(3):342-348, 1991.

[Tho92] W. J. Thompson. **Computing for Scientists and Engineers.** John Wiley and Sons, New York, 1992.

[Tho94] W. J. Thompson. **Angular Momentum.** John Wiley and Sons, New York, 1994.

[Tip91] P. A. Tipler. **Physics for Scientists and Engineers,** Third Edition. Worth Publishers, New York, 1991.

[Tow92] J. S. Townsend. **A Modern Approach to Quantum Mechanics.** McGraw-Hill, New York, 1992.

[Vve93] D. D. Vvedensky. **Partial Differential Equations with** *Mathematica.* Addison-Wesley, Wokingham, England, 1993.

[WMS94] Waterloo Maple Software. **Theorist®,** Version 2.0. Waterloo Maple Software, Waterloo, Ontario, 1994.

[WW94] T. Williams and J. Walsh. *Mathematica* **User's Guide for Microsoft Windows,** Third Edition. Wolfram Research, Champaign, IL, 1994.

[Wit91] D. Withoff. *Mathematica* **Warning Messages.** Wolfram Research, Champaign, IL, 1991.

[Wol91] S. Wolfram. *Mathematica*: **A System for Doing Mathematics by Computer,** Second Edition. Addison-Wesley, Redwood City, CA, 1991.

[Wol96] S. Wolfram. **The** *Mathematica* **Book,** Third Edition. Wolfram Media, Champaign, IL, 1996.

[WR96] Wolfram Research. *Mathematica* **3.0 Standard Add-on Packages.** Wolfram Research, Champaign, IL, 1996.

[Won92] S. S. M. Wong. **Computational Methods in Physics and Engineering.** Prentice-Hall, Englewood Cliffs, NJ, 1992.

Index